Macroevolution

Macroevolution

Pattern and Process

Steven M. Stanley

THE JOHNS HOPKINS UNIVERSITY PRESS

Baltimore and London

Johns Hopkins Paperbacks edition, 1998
9 8 7 6 5 4 3 2 1

The Johns Hopkins University Press
2715 North Charles Street
Baltimore, Maryland 21218-4363
The Johns Hopkins Press Ltd., London

Library of Congress Cataloging-in-Publication Data

Stanley, Steven M.
   Macroevolution, pattern and process / Steven M. Stanley.
      p.   cm.
   Originally published : San Francisco : W. H. Freeman, c1979. With new introd.
   Includes bibliographical references (p.   ) and index.
   ISBN 0-8018-5735-X (pbk.   :   alk. paper)
   1. Macroevolution. 2. Paleontology. I. Title.
QH371.5.S835   1998
576.8—dc21   97-40163
                                                                CIP

A catalog record for this book is available from the British Library.

482   2611

Contents

Preface

It has long been fashionable in paleontology to pay homage to the actualistic idea that the present is the key to the past—that the fossil record can best be brought to life by injecting it with large doses of conceptual serum from the field of biology. Like many other young paleontologists, I embarked on a career with this idea in mind and, though trained primarily in geology, put the belief into practice by undertaking a dissertation on living animals. My efforts left no cause for regret, yet somehow, during the past few years, I have found myself aiming more and more in the direction of bringing fossil data to bear on biologic questions—in particular, questions relating to the process of evolution. Certainly, I am not alone in this posture. (Some of the reasons for the timeliness of the approach are given in the first chapter of this book.)

Quite simply, I have come to believe that paleontologic data tell us things about evolution that have not gained general acceptance through the collective biologic effort known as the Modern Synthesis. The fossil record, in certain places and for certain purposes, is more instructive in the field of evolution than many have believed. Here, perhaps, is the fundamental message of this book.

Macroevolution is designed to be used in advanced courses and seminars. More generally, it is meant to stimulate interaction between neontology and paleontology. The ten chapters to some extent represent an amalgamation of techniques and ideas published during the past few years. Part of my original motive for uniting published and unpublished material under a single cover was a belief that the whole would exceed the sum of the parts: Certain arguments could be conveyed most effectively in association with others. Furthermore, a great deal of expository material of a sort not normally permissible in journal articles could be mixed in, for clarity. The book is meant to be intelligible not only to paleontologists but also to readers with biologic training and no knowledge of geology.

A personal belief that flavors the following chapters is that concepts of evolution should be studied with an appreciation of the natural history of organisms. Much is lost in the treatment of taxa as identical chips or marbles that come and go in a vast numbers game. Thus, while I will argue that the species is the natural (if imperfect) unit of macroevolution, I by no means wish to obscure the fact that each unit has unique biologic properties, without knowledge of which our analytic efforts are greatly impaired. A pervasive influence in the development of ideas here, culminating in the final two chapters, is a conviction that a comparative approach is essential. Contrasts among taxa in the nature of large-scale evolution (including rates of change) often point to fundamental principles that would otherwise elude us. Cosmopolitan versus endemic distribution; marine versus terrestrial occurrence; simple versus complex behavior—contrasting qualities of these kinds relate to differences in rates and patterns of evolution. The bivalve mollusks remain my touchstone here. As the animals I know best, they serve as a standard for comparison.

A central theme of this book is that fossil information can be used fruitfully to investigate macroevolution at the level of the species. While data for higher taxa are valuable and historically have received much more attention, it is both possible and desirable to seek a finer degree of resolution. Certainly, neontologists of a reductionist bent can move to much closer levels of observation, but I will argue that the discontinuities separating many species and the operation of random factors in the origin of these discontinuities decouple macroevolution from microevolution, forcing us to focus certain kinds of analysis at the level of the species.

While much that is included here might be called "theoretical paleontology," it should be appreciated that the general approach followed is heavily empirical. Even where appropriate data are now lacking or sparsely available, there is reason to believe that the situation will improve. In other words, I have sought to erect hypotheses that are or will become testable and to utilize techniques that employ meaningful data.

To some, the book's title may seem unduly general or comprehensive. It is difficult to provide anything with a name that is both terse and accurate, however, and the rationale for my choice is as follows. The word *Macroevolution* directs attention to the book's point of departure from the Modern Synthesis: the previously mentioned notion that evolution above the species level is decoupled from microevolution within established lineages. *Pattern* refers to the book's fundamental claim that macroevolution has a more strongly rectangular fabric than the diagonally branching configuration traditionally envisioned. Finally, *Process* refers to the mechanisms by which macroevolutionary trends develop, the most important of which I believe to be selection among species rather than persistent microevolution within established lineages.

At my request or that of the publisher, seven colleagues evaluated the original draft of this book, giving this task far more time and effort than could possibly have been expected. To these workers—Robert T. Bakker, Léo F. Laporte, Ernst Mayr, Everett C. Olson, David M. Raup, G. Ledyard Stebbins, and James W. Valentine—I am more thankful than I can say. Of course, not all of their advice has been heeded, and since their assessment, errors may have crept in. Thus, I bear responsibility for any remaining flaws of fact or logic.

To give the reader a sense of direction, I have begun each chapter with a précis or abstract. A more detailed summary is appended to each chapter but the first. This pattern of organization comes dangerously close to the legendary outline for a military briefing: "First ya tell em what yer gonna tell 'em. Then ya tell 'em. Then ya tell 'em what ya told 'em." I maintain more respect for my audience than does the stereotyped officer of the cliché, but also believe my message to be sufficiently heterodox that the repetition will not be found excessive.

In order to maintain brevity and thereby make the book more readable, I have kept examples to a minimum. Often the choice of a case study has been arbitrary, and I regret not having been able to cite the work of a larger number of fellow workers. I owe a particular debt of gratitude to my colleague Robert T. Bakker, who has unselfishly placed at my disposal his extraordinary knowledge of vertebrate life and directed me to many useful examples. The comparative aspects of this writing have also put me in contact with many other experts who have kindly shared taxonomic and stratigraphic information. I express warm thanks to all of these people, including any whose names may inadvertently have been omitted from the following list: Thomas G. Gibson (benthic forams); William A. Berggren and Richard Cifelli (planktonic forams); William W. Hay (nannoplankton); Jean P. Chevalier, Stanley H. Frost, Judith C. Lang, Carden Wallace, and John W. Wells (reef corals); Alan H. Cheetham, Jeremy B. C. Jackson, and Erik Thomsen (bryozoans); Norman F. Sohl (gastropods); Bernhard Kummel (ammonites); Porter M. Kier (echinoids); Daniel C. Fisher (horseshoe crabs); James H. Stitt (trilobites); William L. Brown and Frank M. Carpenter (insects); Ted M. Cavender and Gerald R. Smith (freshwater fishes); Storrs L. Olson (birds); Frank C. Whitmore (whales); Björn Kurtén and Richard H. Tedford (terrestrial mammals). I am also grateful to three assiduous students for directing me to pertinent literature: Debbie L. Lakin (coevolution), Robert Hershler (freshwater mollusks), and Phillip Signor (marine gastropods). Mary Carrington typed the manuscript with great skill.

A number of the concepts of this book—many of which have been published elsewhere—were developed and documented during the tenure of National Science Foundation Grant GB 76-09774, for which I served as Principal Investigator.

May 1979   *Steven M. Stanley*

Introduction 1998

MACROEVOLUTION THEN AND NOW

I preface this paperback edition of *Macroevolution: Pattern and Process* with comments on what has happened in the book's primary subject areas since its original publication. Happily, various new developments have reinforced my arguments for such things as the widespread stability of established species, the importance of rapid evolutionary branching, and the production of large-scale evolutionary trends by selection at the species level.

Though best known for advocating punctuational evolution and exploring its implications, this book makes a more fundamental contribution. It represents the first general attempt to employ fossil data at the level of the species to examine rates, trends, and patterns of large-scale evolution. At the time of its genesis, paleontologists, exemplified especially by George Gaylord Simpson, had customarily by-passed species when studying evolution that transcended species boundaries (macroevolution). The genus and family were deemed more appropriate units for analysis because the geologic record of these higher taxa was, on average, more complete. Furthermore, in the era dominated by gradualism, the so-called species problem loomed large: most paleontological species were assumed to be arbitrarily delimited segments of evolutionary lineages. The punctuational model, however, assigns species a different evolutionary role. In recognizing remarkable stability for most species, it opens the door for their use as units of macroevolution. To take advantage of this opportunity, it is necessary to devise techniques for circumventing the incompleteness of the fossil record of species, and the most basic purpose in writing this book was to introduce techniques of this kind. My further purpose was to explore patterns of macroevolution using species as units.

With regard to the punctuational model, initially articulated by Eldredge (1971) and Eldredge and Gould (1972), the writing of this book had two purposes. My first goal was to employ fossil data to evaluate the relative merits of the punctuational and gradualistic models of evolution. I further developed tests that I had previously introduced in 1975 in a small paper in response to the statement by Eldredge and Gould (1972, p. 99) that the data of the fossil record were too incomplete to favor either gradualism or punctuational equilibria, as these authors had termed the two alternative patterns for large-scale evolution. When portrayed graphically, the first pattern—the traditional one—showed most lineages evolving at significant rates, with no general tendency for evolution to accelerate during speciation (branching) events. The second, punctuational alternative was more rectangular, with evolution concentrated in speciation events that were so brief on a geological scale of time that the trends they depicted were virtually horizontal on a graph of morphologic change against time.

Early in the controversy engendered by the writings of Eldredge and Gould, much confusion resulted from the emergence of widely disparate views about exactly what alternatives were being debated. Some skeptics rejected the notion of punctuational equilibria, citing examples in which the fossil record seemed to document phyletic evolution (meaning evolution within established species, also known as anagenesis). The implication was that punctuational equilibria represented an extreme scenario that attributed virtually all nontrivial evolutionary change to speciation events while all but denying the existence of phyletic evolution. It seemed necessary to explicitly define two alternative models in such a way as to give each a piece of the middle ground. Thus, first in the 1975 paper and again in this book, I defined the traditional, gradualistic model as asserting that phyletic evolution has accounted for most evolutionary change in the history of life, whereas the punctuational model assigns the larger overall role to speciation. These two models are not overlapping, and yet their joint assessment is nonpolarizing.

The tests that I devised favored the punctuational model, and this led to the second purpose of my book: to explore the important implications of this model, which I review briefly here in light of recent developments. If, as I conclude, the phenotypic change associated with speciation greatly exceeds the amount contributed by phyletic evolution, we must rethink many aspects of evolution and ecology, such as the structure and origins of large-scale evolutionary trends, the prevalence of sexual reproduction among eukaryotic organisms, and the ability of established species to respond to environmental change by evolving.

THE GREAT PARADOX OF THE MODERN SYNTHESIS

Ingrained in the Modern Synthesis of evolution, but never widely addressed or even acknowledged, was a remarkable paradox. In the laboratory, geneticists were orchestrating the evolution of fruit flies, an investigation that entailed selection coefficients so large that they would have transformed any animal into a radically different taxon in 10^4 to 10^5 generations. Paleontologists, on the other hand, had data in hand showing that fossils similar enough to be assigned to single species encompassed very little change over millions of years; in fact, George Gaylord Simpson himself estimated that an average species of animals had survived for about 5 million years! Strangely, neither geneticists nor paleontologists blinked an eye when they encountered the other group's data. No one attempted to reconcile the incompatible rates. The emergence of the punctuational model awakened researchers to the reality of evolutionary stability and led to research showing that this phenomenon is far more common than almost all modern evolutionists—paleontological or neontological—had previously envisioned.

The punctuational model can be viewed as an attempt to resolve the great paradox of the Modern Synthesis, but this was not the motivation behind its formulation. In fact, Eldredge and Gould (1972, p. 99) favored it "because it is more in accord with the process of speciation as understood by modern evolutionists." Of course, had it actually seemed in accord with what biologists generally understood, there would not have been such a great outcry by skeptics and downright opponents. Powerful evidence that detractors of the punctuational model were off-target was the long-standing failure of evolutionists of all stripes to acknowledge the widespread stability of species on geological scales of time. A large portion of the present book is devoted to documenting this stability.

DARWIN'S POSITION

Here and there, the claim has been made that Darwin himself was a punctuationalist. Any fair reading of *On the Origin of Species,* however, reveals Darwin to have been a thoroughgoing gradualist. I have quoted one small statement that departs from the overtly gradualistic message of the rest of the *Origin* and offers a punctuational sentiment.

> It is a more important consideration, clearly leading to the same result, as lately insisted on by Dr. Falconer, namely, that the periods during which species have been undergoing modification, though very long as measured by years, have probably been short in comparison with the periods during which these same species remained without undergoing any change.

The sentence appears in the third edition and all subsequent editions and was added in response to a critical letter from Hugh Falconer. I have more recently uncovered the relevant correspondence in which Falconer noted that the European mammoth had persisted through the recent ice age without noticeable evolutionary change (Stanley, 1981, p. 102). Falconer argued that because of its climatic vicissitudes, the ice age was a time when, according to Darwin's theory, well-established species should have been evolving significantly—unless species typically undergo substantial evolutionary change only when they originate but then evolve rather little for the remainder of their stay on earth. Darwin paid only lip service to Falconer's punctuational proposition with his single sentence, leaving dozens of strongly gradualistic statements in place throughout the *Origin* (three are quoted below, on pp. 4–7). Only because of his gradualistic orientation was Darwin so perplexed about the sudden appearance of numerous phyla in the Cambrian period and of diverse angiosperms in mid-Cretaceous time.

In all editions, Darwin retained two chapters that denigrated the fossil record for being too incomplete to confirm his gradualistic portrayal of evolution (pp. 7–8). A punctuational explanation for the persistence of species through great thicknesses of strata was an alternative he never considered. For him the only evident alternative to gradual evolution, creationism, was anathema.

Elsewhere I have more thoroughly examined Darwinism (and neo-Darwinism) in light of the punctuational model (Stanley, 1981). There are several cogent explanations for Darwin's staunch gradualism; in fact, he had no choice but to adopt this posture (Stanley, 1981, pp. 47–53). For one thing, he would have been hard pressed to differentiate a punctuational portrayal of evolution empirically from Creationism, the pervasive concept he sought to displace. For another, the reigning taxonomic dogma, based on the Platonic ideal, dictated that species in nature display no significant variation, except in the form of freakish sports. Variation was the raw material of evolution, and yet naturalists believed it to be almost nonexistent. Darwin was at pains to argue otherwise in the *Origin,* and he could only hope to gain followers by granting his process long stretches of time to achieve its results. Finally, throughout the *Origin,* when Darwin assured the reader that natural evolution operated with extreme slowness, he was implicitly underscoring why his colleagues should not argue against his theory simply because, while they could witness artificial selection in the barnyard, they could not observe selection in nature.

DOCUMENTING EVOLUTIONARY STASIS

In making a case for the punctuational model in this book, I relied heavily on evidence of great geologic longevities of species. When first considering how one might pit the gradualistic and punctuational models against one another in tests that employed fossil data, I had focused on evolutionary radiations because they are where so much of the action is—where most adaptive innovations arise (Stanley, 1975a). The test of adaptive radiation compares these longevities to the spans of time during which evolutionary radiations have created numerous, markedly divergent higher taxa. A small number of species stacked end-to-end do not encompass nearly enough evolution to produce a new family, for example— yet, as it turns out, this is all that the great longevity of species allows. We must therefore look to rapid-branching events to explain the origins of distinctive new adaptations.

Some authors have alleged that the designation of species in the fossil record is too shaky for recognized species to offer evidence about rates of evolution— that a species is only a species in the eye of the beholder. The key point, however, is that when a competent taxonomist assigns two or more well-preserved populations to a single species, we can be sure that these populations are so similar as to be separated by only a very small amount of evolution. The suggestion that we cannot use species to document evolutionary stability in the fossil record because sibling species are unrecognizable there (Schopf, 1982) has the logic backwards. If we measure little or no evolution in an alleged fossil lineage and that lineage actually consists of ten species instead of one, then we have ten examples of approximate stasis instead of just one (Stanley, 1985). The actual formation of sibling species is no problem: Some speciation events produce almost no evolutionary divergence, while others produce a great deal.

Certainly, we must take seriously as evidence for the punctuational model the high frequency of recognized fossil species that survived for ten million years or more in such groups as the foraminifera, bivalves, and gastropods and one or two million years in the more rapidly evolving mammals. (This book provides evidence of similar longevity for numerous additional taxa.) Such remarkable stability points to minuscule net selection coefficients for almost all measurable features.

COMPREHENSIVE STUDIES REVEALING A PREVALENCE OF STASIS

To circumvent skepticism of the taxonomically based approach to the documentation of stasis, I designed a project with a student, Xiangning Yang, to assess rates of evolution in strictly morphological terms (Stanley and Yang, 1987). This project had a second goal as well. We were concerned that many scientists had weighed in on one side of the controversy with examples that were not objective samples of phylogeny. Traditionally, nearly all studies purporting to document phyletic evolution were stimulated by the discovery of what appeared in advance to be examples of phyletic trends. Furthermore, in the era of gradualism, it seemed legitimate to assume that phyletic evolution connected populations that might in fact have been separated by one or more speciation events. On the other hand, innumerable examples of evident stability for species had gone undocumented because they seemed uninteresting.

This neglect of approximate stasis called for new, more objective analyses that treated large, unbiased samples of evolutionary lineages. The study Yang and I undertook attempted to do just this, and it provided evidence of remarkable evolutionary stability. Our starting point was a group of species of early Pliocene age (about 4 million years old). We included all species that met an arbitrary set of criteria, including availability of about twenty well-preserved specimens; the specimens used were quite pristine, resembling modern sea shells. Our criteria yielded nineteen fossil species, each of which was compared to the most similar living species. We characterized shapes with twenty-four morphologic variables normalized for size. These variables provided information not only on shell form but, indirectly, on a wide array of traits of soft anatomy and behavior. There is no reason to assume that the collective rate of evolutionary change for these traits should have differed from that of other adaptive features.

We compared each of the fossil species to its most similar living species in order to assess the maximum amount of evolution that might have occurred in 4 million years. The choice of the living form was a matter of parsimony; we assumed that the most similar living species was the most likely living descendant. In twelve cases, the most similar living species had conventionally been assigned to the same species as the fossil form. In four other cases, the living species had traditionally gone by a different name but had been widely assumed to be the phyletic descendant of the fossil form. And finally, in three cases, the living species was a Pacific form that almost certainly was not the direct phyletic descendant of the fossil species, meaning that it was separated from it by at least one speciation event, which had possibly occurred more that 4 million years ago. Our approach was intentionally conservative; we almost certainly erred in attributing more change to phyletic evolution than it had produced in 4 million years.

We compared the fossil and living populations using the Mahalanobis multivariate statistic. As a yardstick for our comparisons, we made similar comparisons of geographically separated populations belonging to each living ("descendant") species in the study. The result was striking evidence for approximate evolutionary stasis. Nearly all comparisons of the 4-million-year-old and recent populations yielded multivariate distances within the range of intraspecific variability (Figure P-1). A comparison based on individual morphologic parameters produced a comparable result.

Alan Cheetham's (1986) study of bryozoans resembled ours in testing the gradualistic model against the punctuational model by undertaking an objective morphometric analysis of a large, arbitrarily chosen group of lineages. The difference was that Cheetham's method entailed constructing a phylogeny—complete with evolutionary rates—from all available fossil populations that belonged to a single radiating clade (the marine bryozoan genus *Metrarabdotos*) in one region (the Dominican Republic). Cheetham took advantage of the great abundance of fragments of *Metrarabdotos* in a remarkably continuous stratigraphic section representing the interval from about 8 to 3.5 million years. The mean sampling interval was 0.16 million years, although no species was found in every sample within its range. Cheetham's morphometric analysis was based on a set of forty-six characters previously shown to depict soft-part morphology faithfully. He used cluster analysis to group populations into species and produced a phylogeny with a stratophenetic analysis that was borne out by a cladistic analysis. Minimum rates of evolution required to form descendant species from ancestral species were enormously larger than rates within established species. In fact, for no established species was the mean rate of evolution

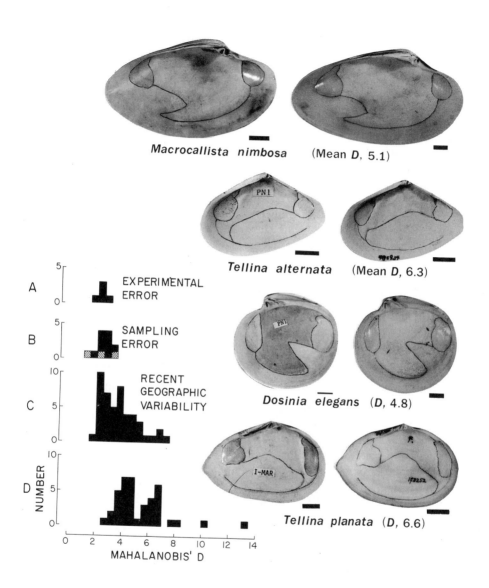

FIGURE P-1

Multivariate morphometric differences between 4-million-year-old and recent populations of marine bivalves (*D*) compared to geographic variability for living species included
in the comparison (*C*) and to experimental and sampling error (*A* and *B*). The left-hand
member of each pair of photographs shows a fossil, and the right-hand member shows a
modern specimen. The large majority of fossil-recent comparisons fall within the range of
modern geographic variation. (From Stanley and Yang, 1987.)

significantly different from zero. In short, species appeared abruptly and then hardly changed at all before their extinction. Not only did phyletic evolution go virtually nowhere, but very few new species could possibly have arisen through the phyletic transformation of others because the radiating pattern provided few potential ancestors. Subsequent research showed Cheetham's morphologic traits, when used to group living bryozoan populations into species, to be in accord with genetic data (Jackson and Cheetham, 1990).

The two morphometric studies I have just described illustrate what I view as the only valid approach for assessing the relative frequencies of stability and gradual change for species. We must evaluate a large, unbiased sample of lineages, and we must avoid a simple "connect-the-dots" exercise that presupposes that two morphologically distinct fossil populations that are separated by a substantial interval of geologic time form parts of a single, gradually evolving lineage. I also avoided this problem with a third example, using published records for Eocene mammals of the Big Horn Basin. Here numerous lineages were represented by closely spaced samples encompassing 2 to 3 million years (Stanley, 1982), and virtually no phyletic evolution was evident. Most of the fossil remains in this example are teeth and jaws, but as Prothero and Heaton (1996) have documented, Oligocene mammals of the White River Group of the High Plains, which are represented by extensive cranial and postcranial material, also exhibit a pervasive pattern of stasis. Most of the exceptionally well preserved White River skeletons come from the Big Badlands of South Dakota, where they represent a 6-million-year interval spanning the Chadronian, Orellan, and Whitneyan ages. Figure P-2 depicts stratigraphic ranges for the artiodactyl species of this fauna, but Prothero and Heaton have provided comparable range charts for all known taxa, including numerous rodents and carnivores. Mean duration for the 177 recognized species of White River mammals is 2.4 million years. New species appear abruptly throughout the White River interval. They do not diverge slowly from others; in fact, all but three of the 177 species experienced virtually no evident morphologic change throughout their documented existence. Furthermore, evolution within the three nonstatic species was largely a matter of change in body size. Prothero and Heaton have emphasized that the overall stability is especially striking because sixty-two of the species survived a great climatic change about 33 million years ago, during which the regional mean annual temperature dropped by about 13° C, and a shift to drier and more seasonal climates caused open forested grasslands to replace dense forests. In other words, mammalian morphology was stable even in a highly perturbed environment.

ABORTED SPECIES AND THE LONGEVITY BREAK

Observations in the modern world indicate that many species must arise and die out almost instantaneously on a geologic scale of time. Coupling this point with

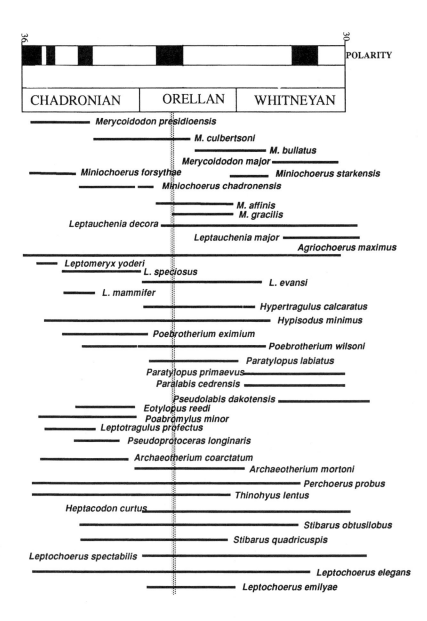

FIGURE P-2

Ranges of artiodactyl ungulate species that form part of the White River fauna of the High Plains. Individual species, most of which are known from extensive skeletal remains, experienced remarkable evolutionary stability, even across an interval of major climatic and vegetational change (vertical dotted band). Numbers along the top of the magnetic reversal scale give approximate ages in millions of years. (Reprinted by permission from Prothero and Heaton, 1996.)

paleontological evidence that modal duration for the well-established species of a higher taxon does not lie along the Y-axis of a histogram, one can conclude that a typical species, once widespread and populous, is likely to survive for 10^6 to 10^7 years (pp. 174–176). The valley in a histogram of species longevies between the very slender mode for what I have termed *aborted species* and the broader mode for established species that might be termed the *longevity break*.

PHYLETIC CHANGES IN BODY SIZE

The morphometric study that Yang and I conducted showed that, although phyletic changes in shape were minor, size was quite labile in bivalve evolution. Actually, the issue of size change is difficult to treat rigorously because, for many invertebrates, growth is indeterminate and size at death varies with growth rate and external mortality factors. Even so, our findings supported the assertion of Hallam (1975) that body size is much more labile than shape in animal evolution. The morphogenetic simplicity of moderate size changes makes this generalization easy to explain.

The relative evolutionary lability of body size is of special significance in the history of macroevolutionary research. When this book was written, the large majority of alleged rates of morphological evolution represented nothing more than apparent changes in body size. A summary of published evolutionary rates by Van Valen (1974) included seventy-four ostensible examples, sixty-four of which represented only body size. Although size change often has ecological consequences, it does not produce adaptive innovations or distinctive new higher taxa. In short, it is not where the real action is.

PUNCTUATED GRADUALISM (STEPWISE EVOLUTION)

The pattern that I labeled *stepwise evolution* in this book (p. 27) was later dubbed *punctuated gradualism* by other workers (Malmgren et al., 1983). In this pattern evolution is phyletic (occurs without branching) but is concentrated in brief intervals. As in the study by Malmgren and colleagues, which focused on the foraminiferan genus *Globorotalia*, most documented examples of this kind of pattern entail unicellular planktonic marine lineages, which, for reasons I will describe later, may make them atypical of evolution in general.

The radiating phylogenetic pattern in Cheetham's study rules out punctuated gradualism for nearly all substantial evolutionary shifts because overlap between ancestral and descendant species requires branching. One might still ask whether staircase patterns are more common in other segments of phylogeny. Though descriptively punctuational, such patterns fall outside the original definition of punctuational equilibrium (Eldredge and Gould, 1972). The test of living fossils—which, like the test of adaptive radiation, opposes gradualism—also testifies against a pervasive presence of staircase patterns. This test reveals that persistently narrow clades—ones that experience little or no branching—tend to undergo little phyletic evolution (gradual or stepwise) even if they survive for many millions of years (pp. 122–132).

NEW EVIDENCE OF MAJOR PUNCTUATIONAL STEPS

In my 1975 paper and in this book, my strategy was to identify two potential loci for evolution and then show that one of these (generation-by-generation descent within established species) was generally characterized by such slow evolution that we were compelled to look to the other site (speciation) to account for the large majority of evolutionary innovations and higher taxonomic transitions— that is, most total evolution. In weighing the gradualistic and punctuational models against one another in tests, I employed the classic hypothetico-deductive method. This approach yielded support for one descriptive component of the punctuational model: widespread stability for established species. It therefore implied a dominant role for the other descriptive component: rapidly divergent speciation events. Though lacking any *direct* method of assessing the total contribution of these punctuational events, I offered examples of several, including the origins of new species of pupfishes in the American West within the past 20,000–30,000 years, and of new species of banana-feeding moths of the genus *Hedylepta* in Hawaii since Polynesians introduced the banana plant there about a thousand years ago (pp. 43–44). Recently, new evidence has painted a compelling picture of dramatically divergent speciation events in several taxa.

For many years, all that could be said of the age of the famous flock of cichlid species in Lake Victoria, based on the maximum age estimated for the lake, was that the evolutionary radiation that produced it began no more than about 750,000 years ago. Even this maximum time span for the remarkable cichlid radiation was quite brief. After all, an average Neogene species of freshwater fishes has survived for about 3 million years after its evolutionary birth (p. 253), and the Lake Victoria radiation produced about 170 species, many of them markedly divergent and deserving of unique generic status, even though they live in close proximity to their imputed ancestral species (pp. 45–47). New evidence reveals that the rate of radiation here was even more rapid than previously understood—and that it must have been powerfully punctuational. Coring of Lake Victoria's floor has penetrated a layer of fossil grass beneath the present lake's sediments that, by the radiocarbon method, dates to only about 12,000 years ago (Johnson, *et al.,* 1996). This establishes the earliest possible time for the start of the cichlid radiation. There is no evidence that lakes in general harbor more dramatic evolutionary radiations than those of less sharply bounded regions. Like islands, lakes simply confine the products of radiation conveniently for our analysis.

Refined stratigraphic analysis has recently revealed that dramatically divergent speciation events have occurred even in the broad planktonic realm of the ocean. It turns out that the terminal Cretaceous extinction event triggered an explosive evolutionary radiation of planktonic foraminifera (D'Hondt et al., 1996). Only three or four of about seventy species in this group survived the extinction event. Then, within the first 100,000 years of the Cenozoic era, seventeen new species assigned to eight new genera evolved from *Guembelitria cretacea* and two species of *Hedbergella* (S. L. D'Hondt, personal communication, 1997). *Guembelitria cretacea* had survived for more than 20 million years before the terminal Cretaceous event, illustrating the remarkable stability of a well-established species relative to the rapid shifts that occurred when it budded off new species. The longevity of this ancestral form is typical for planktonic foraminifera. I estimated that an average species in the group has survived for at least 20 million years (p. 244). (For refined estimates of species durations for Neogene globigerinids and globorotaliids, see Stanley, et al., 1988.)

RAPID ORIGIN OF A FAMILY OF ANIMALS

Seldom can paleontologists trace a family of animals to a small ancestral group or a narrow geographic area of origin. An exception, brought to light by my student Steven Beadle (1989, 1991) since this book was first published, is the geologically sudden emergence of the Dendrasteridae, a distinctive family of sand dollars in which *Dendraster* is the only living genus. Members of this family are notable for their unusual, eccentric morphology. On the animal's aboral side, where it gathers food, the anterior portion of the flattened test is much larger than the posterior portion. *Dendraster*'s mode of life reveals the function of this shape. Unlike other sand dollars, which are flat-lying deposit feeders, *Dendraster* stands on edge, only partly buried in the sand, and feeds on suspended organic matter with its enlarged anterior aboral region. Early in its ontogeny, the genus *Echinarachnius* develops morphologic eccentricity resembling that of the dendrasterids. Subsequently, a change in anterior and posterior growth rates produces a more-or-less symmetrical adult. There is good evidence that the Dendrasteridae evolved from the Echinarachniidae neotenically by retaining their eccentric juvenile morphology into the adult stage. The presence of aberrant, eccentric sand dollars within fossil and living groups of Echinarachniidae reveal the ease with which this family could give rise to a new, fully eccentric family (Figure P-3).

Today the Dendrasteridae dwell along the Pacific coast from southern Alaska to Baja California. An excellent fossil record, however, shows the group's geographic range constricting progressively backward in time to a first occurrence in a relatively small region of central California. Evidently the new family arose there during a brief drop in sea level. It appears abruptly in the uppermost

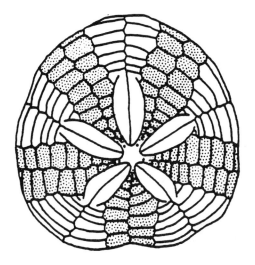

FIGURE P-3
Reconstruction of the aboral surface of the late Miocene sand dollar species *Echinarachnius alaskense*. This aberrant form retains an enlarged anterior region, normally a juvenile trait, into the adult stage, illustrating the ease with which this neotenic adult trait can evolve. (From Beadle, 1989.)

Miocene above an unconformity, beneath which potential echinarachniid ancestors left an abundant fossil record.

Behavioral evolution of the upright life position of *Dendraster* is easily explained. Overturned sand dollars with a flat-lying habit right themselves by rotating 180°. In the transition to the Dendrasteridae, animals simply terminated the normal righting behavior when they reached a vertical position. Food was plentiful for feeding in the new position: The family arose in a region characterized by strong currents, which carry abundant suspended organic matter.

In summary, Beadle constructed a powerful case for the rapid origin of a distinctive new family of animals in a narrow geographic area about 6 million years ago. Members of the obvious ancestral group were abundant in the region, and the evolutionary transition is readily explained as a simple neotenic and behavioral shift that permitted the emerging group to stabilize itself in a new life position and avail itself of a locally plentiful food supply. Radiation from this abrupt beginning has produced four living species, and the family has progressively expanded its geographic range. Parenthetically, it is worth noting that the average duration for Neogene echinoid species is roughly 5 to 7 million years (p. 240) in both eastern and western North America.

In another study capitalizing on the excellent fossil record of echinoids, Kier (1982) concluded that the order that includes all sand dollars (Clypeasteroida) also evolved rapidly from cassiduloid ancestors in a small region of West Africa during the Paleocene. A rich fossil record in the region has yielded no intermediate forms in what was apparently an abrupt evolutionary transition.

STASIS AND PUNCTUATIONAL EVENTS IN HUMAN EVOLUTION

When first contemplating the punctuational model, I noted that the traditional gradualistic posture of evolutionary biology has nowhere been more pronounced than in the study of human origins (p. 30*)*. I called attention to the evidence of great geologic duration for fossil hominid species—evidence that had been accumulating with little effect on gradualistic portrayals of human origins (pp. 80–82). Subsequent discoveries, which I have reviewed in a recent book on the origin of the human genus (Stanley, 1996), must force any portrayal of human evolution even further in a punctuational direction. I will now provide a brief summary of my views on the pattern of human evolution, referring the interested reader to my recent book and an earlier paper (Stanley, 1992) for details and literature citations.

One can contrast *Australopithecus afarensis* and *A. africanus* in minor ways, but from the perspective of macroevolution the two species are remarkably similar. Together, they form a segment of hominid phylogeny in which very little evolution occurred between about 3.9 and 2.6 to 2.4 million years ago: the net evolutionary trajectory moved morphology hardly at all toward the condition of *Homo*. Throughout its long existence, *Australopithecus* retained a brain only slightly larger than that of a chimpanzee, as well as numerous postcranial adaptations for climbing.

The evolution of *Homo* from *Austalopithecus* about 2.4 million years ago was abrupt, and I have attributed it to a climatic shift toward cooler, and especially dryer, conditions during the onset of the modern ice age. This shift caused forests to shrink and grasslands to expand in Africa. *Australopithecus* was presumably forced to abandon its habit of climbing trees in order to find food and, more importantly, to avoid swift predators on the ground. The sudden inability of populations to the rely on climbing must have devastated most populations—in fact, *Australopithecus* became extinct—but it did make possible the evolutionary origin of *Homo*. Here is the key point: Only after human ancestors had abandoned habitual climbing was it possible for females to carry and tend to helpless infants, which are a by-product of the delayed development by which *Homo* grows its large brain. One population of *Australopithecus* survived long enough to undergo this evolutionary change, but this was an event that might well never have happened.

Homo erectus differed little from earliest *Homo* (for which *Homo rudolfensis* may be the appropriate name). Once on the scene, *Homo erectus*, like *Australopithecus*, experienced relative evolutionary stability. Skull OH9 of *Homo erectus*, from Olduvai Gorge, has an estimated cranial capacity of 1,067 cubic centimeters. This is higher than the mean for the conspecific Zhoukoudian specimens of China, which are three-quarters of a million to a million years younger!

Interestingly, the cultural evolution of hominids, evidenced by stone tools, also displays a punctuational pattern. In general, the distinctive cultures coincided temporally with species but, as might be expected, appeared slightly later than their fabricators.

The sluggish nature of phyletic evolution within hominid species should not be surprising, in that an average species of Cenozoic mammals has persisted for 1 to 2 million years (see below, pp. 113–116). Notable, too, is the morphological similarity of the isolated West African population of chimpanzees and the larger eastern population. Molecular genetic evidence indicates isolation of the western population about 1.6 million years ago (Morin *et al.*, 1994), which implies approximate evolutionary stasis for both populations since that time.

Recently, an important vestige of physical anthropology's traditional gradualism suffered a mortal blow. This was the idea, fervently defended by a few researchers, that modern humans and Neanderthals constituted components of a single, anastomosing lineage that emerged gradually from *Homo erectus* and eventually turned into modern *Homo sapiens*. The remarkable discovery of well-preserved mitochondrial DNA in a bone of Neanderthal (Krings *et al.*, 1997) has demonstrated that Neanderthal was a genetically isolated, discrete species, as its distinctive morphology led some of us to believe long ago (Stanley, 1981, pp. 152–154). According to the molecular clock, the most recent common ancestor of *Homo neanderthalensis* and *Homo sapiens* lived 590,000 to 660,000 years ago. Not unexpectedly, this is somewhat earlier than the date for the first morphologically identified members of the Neanderthal lineage. This new evidence leaves intact the genetic and morphologic conclusion that *Homo sapiens* arose in Africa between 120,000 and 150,000 years ago. Fossil evidence is meager for the early history of the modern human species, but the morphology of early Cro-Magnon populations indicates remarkable stability for our species during the most recent 25 percent of its existence.

WHY ARE SPECIES SO STABLE?

Why have so many populous, well-established species changed so little in the course of millions of years, and why has so much change occurred rapidly (presumably in small, localized populations)? These are complementary questions. The answer to one will give the answer to the other. At present, we have no certain answer, but part of the explanation must lie in the complexity of living organisms. A species is an incredibly intricate, self-regulating, self-replicating entity. Only very rare genetic accidents confer substantial phenotypic changes that are in some way useful to such an entity without disrupting another aspect of the development of its coadapted system. It seems likely that in small populations that occupy unusual habitats wherein competition and predation are relaxed, natural selection sometimes has the opportunity to eliminate potentially deleterious side effects that arise when an otherwise beneficial new trait evolves. Such settings must be the sites of fixation of distinctive traits that would have next to no chance of spreading throughout a very large population.

There has been some debate as to whether stabilizing selection or developmental canalization is primarily responsible for conserving the morphology of so many species over millions of years. It seems often to have been overlooked that both factors must actually be operating together. Canalization, though probably severely confining, is imperfect. Freakish organisms—what were once called "sports of nature"—appear in all species. Their failure to survive is a matter of stabilizing selection. What seems evident is that there is not generally an adequate supply of *continuous* variation to allow unidirectional natural selection to modify traits persistently. If such lability were normally present (if pleiotropy and morphogenetic entanglements created no barriers), then changing biotic environments, as well as opportunities for improved adaptation without environmental change, would have produced much more conspicious phyletic evolution than we observe in the fossil record.

PROTOZOANS AS RINGERS: FLEXIBILITY THROUGH UNICELLULARITY

Earlier I mentioned an example of so-called punctuated gradualism in planktonic foraminifera. Several additional examples of phyletic evolution in unicellular planktonic organisms have come to light (e.g., Prothero and Lazarus, 1980; Malmgren and Kennett, 1981). These examples by no means demonstrate a dominant role for phyletic evolution in organisms of this type. I have previously noted the explosive diversification of planktonic foraminifera that occurred immediately after the terminal Cretaceous extinction and also the great mean longevity of foraminiferan species.

Even so, it stands to reason that unicellular organisms should display greater evolutionary plasticity than do multicellular organisms. The ontogenies of single-celled forms are generally much simpler than the complicated sequences of development that multicellular organisms undergo as their cells multiply and differentiate to form a coadapted complex of tissues and organs. Not only is multicellular development complicated, but the result is generally an intricate organism that functions and interacts with its environment in many different ways. A change in one element of the system is likely to disrupt others.

The point is that, although some unicellular planktonic organisms have provided a splendidly complete record of their morphologies through time, we cannot take patterns of evolution for these organisms to be representative of life in general. Adult morphology results from morphogenesis, and if morphogenesis entails canalization, then simple morphogenesis must produce weaker canalization than complex morphogenesis.

DEFINING SPECIES SELECTION

Much of *Macroevolution* is devoted to species selection, a topic that has attracted much interest because it focuses on the nature of large-scale evolutionary trends. I will address the philosophical basis for defining species selection shortly, but, simply put, the process entails variation among species in rates of speciation and extinction. I have also explored macroevolutionary analogs of genetic drift (phylogenetic drift) and of mutation pressure (directed speciation, one source of which is isolate selection—selection among small populations that determines which kinds blossom into species [pp. 195–197]). Such comparisons serve as stimulating intellectual exercises for assessing the full range of causation for large-scale trends. Nearly all trends, for example, entail some degree of phylogenetic drift, just as natural selection in small populations is likely to be leavened with some genetic drift.

Some negative reactions to the concept of species selection have been based on the misconception that, in advocating it, one by-passes selection at the level of the individual. Perhaps part of the difficulty is my assertion that a random component of speciation separates macroevolution from microevolution (Stanley, 1975a). "Random" here does not refer to genetic drift or accidental mutation but to unpredictability. The point is that if one could stand back and view a phylogeny unfolding, at any time it would be impossible to predict the location or direction of the next speciation event, even if all possible events were not equally likely. I italicized a clause in this book (p. 189) stating that this notion is quite compatible with a situation in which all evolution taking place during speciation results entirely from natural selection among individuals.

I also emphasized that species selection would operate even if the gradualistic model were valid—even if phyletic trends prevailed in phylogeny. In fact, a close reading of Darwin shows that, in discussing selection, he moved back and forth between the level of the individual and the level of the species. What he failed to note explicitly is that two distinct, analogous processes operate, with speciation and extinction at the higher level being equivalent to reproduction and death at the lower level; actually, the precise analogy is between species selection and selection among asexual organisms, because species, by definition, do not interbreed (Stanley, 1975a).

It is untrue that I consider only external, ecological factors as determinants of species selection, ignoring the influence of intrinsic traits on rates of speciation (Vrba, 1980, 1984). Note, for example, that I relate propensity to speciate to such variables as levels of behavioral complexity in animals and modes of reproduction in both plants and animals (pp. 260–268). Also, my macroevolutionary analysis of the value of sexual reproduction (Chapter 8) is based on the idea that sexual species win out over asexual species in species selection simply because asexual species lack the ability to speciate in the normal sense of the word.

In a series of papers, Elisabeth Vrba and colleagues have asserted that species selection must be defined in a narrow way that restricts its application to a subset of the many cases of what they term *sorting* among species in macroevolution (Vrba, 1980, 1984; Vrba and Eldredge, 1984; Vrba and Gould, 1986). Following a reductionist principle, these workers have focused on the biological level at which traits emerge: A heritable trait that is emergent at a given level is subject to selection at that level, but not at higher levels. They conclude that group traits, such as population size and distribution, often vary in ways that influence rates of speciation differentially within higher taxa, producing what can legitimately be termed species selection. On the other hand, they argue, many variable traits that influence rates of extinction are traits of individuals, so differential rates of extinction resulting from such traits can only be regarded as species sorting. Selection here is among individuals, even though this process ends up preferentially eliminating certain kinds of species.

The problem with the formulation of Vrba and colleagues is that it is only applicable if selection can be explained in terms of a single property or a set of properties that are emergent at a single level. Actually, the world is too messy to be analyzed so pristinely. I would argue that almost every example of a trend that results from differential rates of speciation and/or extinction—and does not represent phylogenetic drift—entails the influence of one or more traits that are emergent at the species level. The most important of these group traits are population size, spatial distribution of individuals, patterns of behavior that bear on reproductive isolation, and gene pool composition.

First, let me explain why traits such as the size and spatial distribution of a population are, to a significant degree, heritable. An appropriate analogy here is with the genotype and phenotype at the level of the individual. At the species level, we can view the gene pool as a *collective genotype* and all the external traits of species as a *collective phenotype*. The latter includes not only morphology and behavior but also such things as the size and distribution pattern of the population. The numerical and spatial traits of a population reflect traits of the gene pool that influence the way in which a species relates to its environment— where it can live, for example, and how readily it disperses. Not only do limiting factors of a physical nature enter in here, so do such biological factors, such as relationships to food resources, competitors, and predators. Of course, environmental variables also exert a strong influence over numerical and spatial traits of populations, but the same is true for phenotypes of individual organisms. The point is that in most species the nature of the collective genotype contributes significantly to group features, such as the size and distribution of populations, which in turn exert a strong influence over rates of speciation and extinction.

Let us consider the role that speciation plays in the generation of macroevolutionary trends. Speciation entails the evolutionary separation of one population from another. This means that differential rates of speciation that do not result from chance factors (such as fragmentation of a habitat) necessarily entail group traits. Traits that can powerfully influence rates of speciation, such as complex reproductive behavior or weak dispersal ability (pp. 260–268), have generally evolved through natural selection at the individual level because they confer adaptive benefits unrelated to whatever influence they have on rate of speciation. Any phylogeny embodying a trend that develops from differential rates of speciation (and is not the result of phylogenetic drift or arbitrary habitat fragmentation) includes some kinds of species that are winners and other kinds that are losers, and these fates are largely the result of heritable traits. Selection at the level of the individual endowed the losers with any trait that led to their decline, and selection at this level failed to reverse what amounted to a flaw at the higher level. In other words, a negative consequence of the trait emerged at the level of the species. The trait influenced rate of speciation in ways that were unrelated to and thus unreduceable to the reproductive success and survival of individuals. Such a trend can only be explained in terms of selection at the species level.

Next, let us consider differential extinction. Extinction might be viewed as amounting to the evolutionary failure of the individual members of a species because they were the units of its declining population. Thus, if one or more heritable traits of individual organisms led to this decline, to some degree the extinction represented selection against individual organisms. However, heritable group properties will inevitably have entered in. An impoverished gene pool is the most obvious of these, contributing to nearly all extinctions as the population shrinks and fragments. The small population size, fragmented distribution, and genetic deficiency may have arisen because of selection at the level of the individual, but these are group traits, emergent at the species level. Because such group properties play a role in nearly all extinction events, we cannot fully analyze these events by assessing selection at the level of the individual.

We must analyze differential rates of branching and termination (not resulting from chance factors) at the level of the species in order to take into account traits, such as genetic impoverishment, that *virtually always* emerge at this level, in addition to traits whose effects flow upward to the species level, having emerged at lower levels. In other words, species selection subsumes relevant selection at lower levels.

More generally, a complete explanation of differential rates of branching and termination at any level (not resulting from drift or other chance factors) is possible only at the highest level at which any of the causal traits emerges. Selection at lower levels lacks full explanatory power.

Only in the extreme case of instantaneous, catastrophic extinction that strikes some species and not others might strictly group properties fail to come into play. Even here, however, one might make a case for species selection on the basis of parsimony: If some species suddenly disappear because all of their members share one or more traits, whereas other species of the same higher taxon survive because all of their members have alternative traits, for simplicity, why not make the species the unit of selection?

While this in itself is a cogent argument for a broad definition of species selection, a more fundamental point should also be made. No reductionist principle has dissuaded physicists, chemists, or engineers from recognizing innumerable processes at levels of organization far above those at which the motion of matter can ultimately be resolved. In the era of nuclear physics, scientists still recognize messy kinds of change, such as chemical equilibration and fluid convection, as *processes*. The reason, of course, is that causation in nature is subject to rigorous analysis at many different levels, and even if we recognize future possibilities for analysis at a lower level than is currently feasible, to make progress we must operate at this feasible level. In fact, we will never be able to analyze large-scale trends in the fossil record at the level of individual selection. Quite feasible, however, is analysis at the level of species selection. If convection is a process, species selection is a process. Why should organismal biology, any more than physics or chemistry, be straitjacketed by extreme reductionism?

Enough philosophizing. Let me offer a new example of the explanatory power of species selection.

HYPSODONTY IN HORSES: A CLASSIC TREND REANALYZED AS SPECIES SELECTION

I have long been on the lookout for an excellent example of species selection entailing a continuously varying morphologic feature with obvious adaptive significance. The chief obstacle has been the difficulty of finding a phylogeny with a sufficiently complete and well-studied fossil record to permit detailed evaluation. Finally, I can claim success. The example I have uncovered is especially appealing for two reasons. First, it entails a classic subject for study of large-scale evolutionary trends: horse phylogeny. Second, the trend was driven by an identifiable environmental shift: the drying of climates, which, in two stages, altered the fodder that nature offered North American horses.

Paleontologists have long recognized that horse molars, on average, became longer in the course of the Neogene period. This trend has been attributed to the spread of harsh grasses. Grasses wear down teeth of herbivores more rapidly than leafy vegetation does. Old grazers with heavily worn teeth die of malnutrition, which is why it is desirable, if tactless, to look a gift horse in the mouth and also why paleontologists use tooth wear to estimate the life spans of fossil horses. It is well documented that North American grasslands expanded and animals adapted to them diversified as climates became drier during the Neogene (for a review, see Webb and Opdyke, 1995).

Figure P-3, displaying data from a paper by Richard Hulbert (1993), shows a persistent net increase in average molar length for the horse fauna of North America between 18 and 1 million years ago. The composition of the fauna

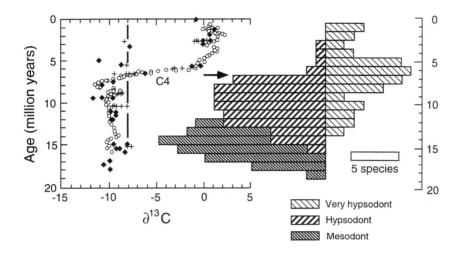

FIGURE P-4
Macroevolutionary change in mean molar length for North American horses. The trend between 18 and 7 million years ago resulted from species with relatively long teeth experiencing relatively high rates of speciation (Data from Hulbert, 1993.) The plot of carbon isotope ratios shows an abrupt shift toward heavier values between 7 and 6 million years ago in the central and southwestern United States. (Data from Cerling *et al.*, 1993 [circles]; Wang *et al.*, 1994 [diamonds]; and Latorre, 1996 [plus symbols]). This shift, which reflects the spread of highly siliceous C4 grasses into North America from the south, coincides with the extinction of most merely hypsodont species; the two surviving species survived in a moist Gulf Coast refuge.

changed from purely mesodont species to a mixture of mesodont and hypsodont (long-toothed) species and then to a mixture of hypsodont and very hypsodont species. All but two of the species that lived after 6 million years ago were very hypsodont.

An assessment of the trend for tooth length that is apparent in Hulbert's data will be presented in detail elsewhere, but in brief, the trend resulted primarily from relatively high rates of speciation for species with relatively long teeth. Statistically, it is highly unlikely that such a trend could represent phylogenetic drift, but there is also evidence from functional morphology and paleoenvironmental analysis that it represents species selection: It was clearly driven by well-documented, large-scale changes in vegetation that resulted from global climatic change.

A different kind of macroevolutionary shift took place between 7 and 6 million years ago. An extinction event selectively eliminated species of North American horses that were merely hypsodont (Figure P-4). This was part of the biggest mammalian extinction event of the past 30 million years in North America—an event that took place as savannahs abruptly turned into grasslands.

The most conspicuous pattern was the preferential elimination of a variety of herbivorous species—among them horses, rhinos, and camels—that were not extremely hypsodont (Webb *et al.*, 1995).

The puzzle is why hypsodont species, which were already adapted to grazing, died out in such large numbers. (Note that Figure P-4 shows the species richness of this group remaining almost unchanged for about 6 million years before the sudden extinction; during this earlier interval, when grasses were already abundant, species selection mainly entailed the net addition of very hypsodont species.) The answer, I believe, is that the grasses themselves underwent an important change between about 7 and 6 million years ago. At this time, a dramatic carbon isotope shift occurred for soils and the teeth of large herbivores, in both southern Asia and North America (Figure P-4). This shift resulted from the spread of so-called C4 grasses northward from subtropical and tropical areas (Cerling *et al.*, 1993). These grasses are more tolerant of dry conditions than are C3 grasses, which they partly displaced. The isotope shift resulted from C3 and C4 grasses partitioning carbon isotopes differently in photosynthesis.

The primary agents of wear for herbivore teeth in grasses are silica bodies (phytoliths). It turns out that C4 grasses, on average, contain about three times as many silica bodies as C3 grasses (Kaufman *et al.*, 1985). What apparently happened was that, with the arrival of C4 grasses at mid-latitudes, the demography of many species of grazer was drastically altered. Presumably, individuals had their lives shortened to the degree that they failed to produce enough offspring to perpetuate their species. Only two horse species with merely hypsodont dentition survived the crisis, and they occupied a relatively moist refuge along the Gulf Coast (Webb *et al.*, 1995).

The sudden extinction event for horses near the end of the Miocene amounted to what I have called catastrophic species selection (p. 208). It contrasted with the earlier trend toward longer molars, which entailed species selection in the form of differential rates of speciation throughout some 10 million years as grassy terrain progressively replaced woodlands.

RATES OF SPECIATION AND EXTINCTION: THE GENERAL CORRELATION

Having estimated mean species durations and typical rates of adaptive radiation for many animal taxa, I found a general correlation between the two (pp. 231–237). Groups that experience high rates of nearly exponential radiation also tend to suffer high rates of extinction. Species selection, by tending to increase rates of speciation and decrease rates of extinction, constantly strains this correlation. The correlation evidently persists because, fortuitously, rates of speciation and extinction are governed by common variables, including level of behavioral complexity, dispersal ability, and size and stability of populations. I have explored this subject more thoroughly elsewhere (Stanley, 1986, 1990; Stanley *et al.*,

1988). Certainly, as implied by the previously cited example of the explosive evolution of a group of planktonic foraminifera immediately after the terminal Cretaceous extinction, under highly unusual circumstances a taxon can radiate more rapidly than its usual rate.

It has been satisfying to see a similar correlation between rates of evolutionary radiation and rates of extinction documented for terrestrial plants (Niklas, *et al.*, 1983).

WHY SEX?

Early on, in contemplating the ramifications of the punctuational model, I recognized that one of its most important implications is related to the prevalence of sexual modes of reproduction among eukaryotic organisms (Stanley, 1975b). The punctuational model gives recombination its primary role in divergent speciation. It follows that asexual clones cannot diversify rapidly, whereas sexual clades can, because in the absence of sexual reproduction there is no true speciation. This means that, all else being equal, asexual species are less likely than sexual species to be perpetuated in phylogeny. Adaptively constricted clones are more likely to be wiped out by major environmental changes than are rapidly diversifying clades. In other words, few asexual groups diversify rapidly enough to survive environmental vicissitudes. These arguments remain fully laid out in Chapter 8, and I still believe that they hold water, in part because they provide explanations for otherwise problematical phenomena, including the remarkable success of the asexual bdelloid rotifers.

IMPLICATIONS FOR ECOLOGY

Species innertness has implications for our ecological world view that are among the most profound and revolutionary intellectual contributions of the punctuational model.

As a student, I was led to believe that species commonly adjust to major environmental changes by evolving. The implication was that evolution stabilizes ecosystems through the *in*stability of their component species. This was part of the central dogma of ecology. The punctuational model offers a different view of ecosystem. In the face of major environmental change, established species have more limited options; they may migrate to more favorable areas, survive in place with a bit of modification, or suffer extinction. They do not readily survive by undergoing substantial evolution.

A corollary of this punctuational view is that modern communities are not coadapted entities whose component species have evolved in concert over long stretches of time. In fact, those of us who study species that have lived through the modern Ice Age can see evidence in the fossil record for a continual reshuffling of species that assembles and disassembles transitory communities without evident evolution of the species themselves.

WHERE THINGS STAND

Since this book first appeared, its most important conclusions have been bolstered by a variety of developments. It is gratifying to have been able to turn a well-known evolutionary trend—elongation of horses' teeth during the Neogene—into a two-tiered example of species selection. More fundamentally, the two basic elements of the punctuational pattern have received support from various new kinds of evidence. Comprehensive multivariate morphometric studies have documented pervasive patterns of species stability. Survival for millions of years with little evolutionary change is clearly the norm for species within a wide variety of higher taxa. In complementary fashion, geologically rapid origins of species and higher taxa, ranging from planktonic foraminifera to cichlid fishes, have come to light.

New evidence of remarkable morphogenetic effects for single genetic changes suggests that studies that combine developmental biology and genetics will soon revolutionize the study of macroevolution. Certainly, recent morphogenetic discoveries have already opened biologists' eyes to the possibility of rapid morphological restructuring in small populations. In the meantime, I think it is fair to say that Charles Darwin would have been surprised by the abundant evidence now in hand both for dramatically punctuational speciation events and for widespread stability of species through millions of years. Equally surprised would have been most of the forgers of evolution's Modern Synthesis. After all, Julian Huxley himself wrote (1942, p. 32) that "it is further obvious that only abundant and widespread species will be of any service in tracing the detailed course of past evolution. . . . In the first place, abundant species will have a larger reservoir of inheritable variation, both actual and potential." While he attributed this major role to phyletic evolution, he envisioned only a minor role for speciation, stating (p. 389) that "a large fraction of it is in a sense an accident, a biological luxury, without bearing on the major and continuing trends of the evolutionary process." Thirty years later, Theodosius Dobzhansky still clung to this view (1972, p. 667) portraying the gradual divergence of a new species from its ancestor as the "usual, and by now, orthodox view."

References Not Cited in the Bibliography

Beadle, S. C. (1989) Ontogenetic regulatory mechanisms, heterochrony, and eccentricity in dendrasterid sand dollars. *Paleobiology, 15*: 205–222.

Beadle, S. C. (1991) The biogeography of origin and radiation: Dendrasterid sand dollars in the northeastern Pacific. *Paleobiology, 17*: 325–339.

Cerling, T. C., Yang, W., Quade, J. (1993) Expansion of C4 ecosystems as an indicator of global ecological change in the late Miocene. *Nature, 361*: 344–345.

Cheetham, A. H. (1986) Tempo of evolution in a Neogene bryozoan: Rates of morphologic change within and across species boundaries. *Paleobiology, 12*: 190–202.

Dobzhansky, T. (1972) Species of *Drosophila*. *Science, 177*: 664–669.

D'Hondt, S. L., Herbert, T. D., King, J., and Gibson, C. (1996) Planktonic foraminifera, asteroids, and marine production: Death and recovery at the Cretaceous-Tertiary Boundary. *Geol. Soc. Amer. Spec. Paper, 307*: 303–317.

Hulbert, R. C. (1993) Taxonomic evolution in North American Neogene horses (subfamily Equinae): The rise and fall of an adaptive radiation. *Paleobiology, 19*: 216–234.

Jackson, J. B. C., and Cheetham, A. H. (1990) Evolutionary significance of morphospecies: Test with cheilostome Bryozoa. *Science, 248*: 579–583.

Johnson, T. C., Scholz, C. A., Talbot, M. R., Kelts, K., Ricketts, R. D., Ngobi, G., Beuning, I. S., and McGill, J. W. (1996) Late pleistocene desiccation of Lake Victoria and rapid evolution of cichlid fishes. *Science, 273*: 1091–1093.

Kaufman, P. B., Dayanandan, P., and Franklin, C. I. (1985) Structure and function of silica bodies in the epidermal system of grass shoots. *Annals of Botany, 55*: 487–507.

Kier, P. M. (1982) Rapid evolution in echinoids. *Palaeont., 25*: 1–9.

Krings, M., Stone, A., Schmitz, R. F., Krainitzki, H., Stoneking, M., and Paabo, S. (1997) Neanderthal DNA sequences and the origin of modern humans. *Cell, 90*: 19–30.

Latorre, C., Quade, J., and McIntosh, W. C. (1996) The expansion of C4 grasses and global change in the late Miocene: Stable isotope analysis from the Americas. *Earth and Plan. Sci. Letters, 146*: 83–96.

Malmgren, B. A., and Kennett, J. B. (1981) Phyletic gradualism in a late Cenozoic planktonic Foraminferal lineage: DS DP Site 284, Southwest Pacific. *Paleobiology, 7*: 230–240.

Malmgren, B. A., Berggren, W. A., and Lohmann, G. P. (1983) Evidence for punctuated gradualism in the late Neogene Globoratalia tumida lineage of planktonic foraminifera. *Paleobiology, 9*: 377–389.

Morin, P. A., Moore, J. J., Ranajit, C., Jin, L., Goodall, J., and Woodruff, D. G. (1994) Kin selection, social structure, gene flow, and the evolution of chimpanzees. *Science, 265*: 1193–1201.

Niklas, K. J., Tiffney, B. H., and Knoll, A. H. (1983) Patterns in vascular and plant diversification. *Nature, 303*: 614–616.

Prothero, D. R. and Heaton, T. H. (1996) Faunal stability during the Early Oligocene climatic crash. *Palaeogeography, Palaeoclimatology, Palaeoecology, 127*: 257–283.

Prothero, D. R., and Lazarus, D. B. (1980) Planktonic microfossils and the recognition of ancestors. *System. Zool., 29*: 119–129.

Schopf, T. J. M. (1982) A critical assessment of punctuated equilibria. I. Duration of taxa. *Evolution, 36*: 1144–1157.

Stanley, S. M. (1981) *The New Evolutionary Timetable: Fossils, Genes, and the Origin of Species*. New York, Basic Books, 222 pp.

Stanley, S. M. (1982) Macroevolution and the fossil record. *Evolution, 36*: 460–473.

Stanley, S. M. (1985) Rates of evolution. *Paleobiology, 11*: 13–26.

Stanley, S. M. (1986) Population size, extinction, and speciation: The fission effect in Neogene Bivalvia. *Paleobiology, 12*: 89–110.

Stanley, S. M. (1990) The general correlation between rate of speciation and rate of extinction: Fortuitous causal linkages. *In* Ross, R. M., and Allmon, W. D., eds., *Causes of Evolution: A Paleontological Perspective*. Chicago, Univ. of Chicago Press, pp. 103–127.

Stanley, S. M. (1992) An ecological theory for the origin of *Homo*. *Paleobiology 18*: 237–257.

Stanley, S. M. (1996) *Children of the Ice Age: How a Global Catastrophe Allowed Humans to Evolve*, New York, Harmony Books.

Stanley, S. M., and Yang, X. (1987) Approximate evolutionary stasis for bivalve morphology over millions of years. *Paleobiology, 13*: 113–139.

Stanley, S. M., Wetmore, K. L., and Kennett, J. P. (1988) Macroevolutionary differences between the two major clades of Neogene planktonic foraminifera. *Paleobiology, 14*: 235–249.

Van Valen, L. (1974) Two modes of evolution. *Nature, 252*: 298–300.

Vrba, E. S. (1980) Evolution, species, and fossils: How does life evolve? *South African Jour. Sci., 76*: 61–84.

Vrba, E. S. (1984) What is species selection? *System. Zool., 33*: 318–328.

Vrba, E. S., and Eldredge, N. (1984) Individuals, hierarchies, and processes: Towards a more complete evolutionary theory. *Paleobiology, 10*: 146–171.

Vrba, E. S., and Gould, S. J. (1986) The hierarchical expansion of sorting and selection: Sorting and selection cannot be equated. *Paleobiology, 12*: 217–228.

Wang, Y., Cerling, T. E., and MacFadden, B. J. (1994) Fossil horses and carbon isotopes: New evidence for Cenozoic dietary, habitat, and ecosystem changes in North America. *Palaeogeography, Palaeoclimatology, Palaeoecology, 107*: 269–279.

Webb, S. D., and Opdyke, N. D. (1995) Global climatic influence on Cenozoic land mammal faunas. *In Effects of Past Global Change on Life*. Washington, D. C., National Academy Press, pp. 184–208.

Webb, S. D., Hulbert, R. C., and Lambert, W. D. (1995) Climatic implications of large-herbivore distributions in the Miocene of North America. In Vrba, E. S., Denton, G. H., Partridge, T. C., and Burckle, L. C., eds., *Paleoclimate and Evolution, with Emphasis on Human Origins*. New Haven, Yale Univ. Press, pp. 91–108.

1

Introduction

In the study of evolution, paleontology has traditionally been valued primarily for its general documentation of large-scale rates, trends, and patterns of change. Its contribution has been quite distinct from that of biology, which has focused on small-scale evolution within populations. What has been lacking in both fields is an intensive study of the species—particularly its mode of origin and subsequent evolutionary fate. The species is, in fact, the natural unit of large-scale evolution and should become a common focus for paleontology and neontology. In part, the role of paleontology in evolutionary research has been defined narrowly because of a false belief, tracing back to Darwin and his early followers, that the fossil record is woefully incomplete. Actually, the record is of sufficiently high quality to allow us to undertake certain kinds of analysis meaningfully at the level of the species. Such analysis shows that many ideas now enjoying widespread support among biologists are in need of re-examination.

Paleontology, or paleobiology, differs fundamentally from geophysics and geochemistry in the manner in which it relates to its cognate discipline outside the earth sciences. Geophysics and geochemistry draw sustenance from physics and chemistry but provide little in return, while paleontology feeds vital material back into biology. In fact, paleontology is sometimes classified within biology as the sister science to neontology, the study of living organisms. The mutualistic relationship between paleontology and neontology exists because evolution is the great unifying theme of biology, endowing the field with an immense historical dimension that cannot be ignored. While many inferences about evolution are derived from living organisms, we must look to the fossil record for the ultimate documentation of large-scale change. In the absence of a fossil record, the credibility of evolutionists would be severely weakened. We might wonder whether the doctrine of evolution would qualify as anything more than an outrageous hypothesis.

There has, however, been some criticism of paleontology for its limited contribution to evolutionary theory (Kitts, 1974; Hecht, 1974). The basic role of the fossil record in Darwin's general paradigm was simply to provide evidence of large-scale biotic turnover and long-term increase in the complexity and variety of life. More than a century transpired following Darwin's contribution with little expansion of this role. While the broad outlines of the history of life fell into place, paleontologists did little to elucidate the underlying mechanisms and processes of evolutionary change. Even so, there have been highlights in the progress of paleontology within its traditional bounds, perhaps the brightest of these being the publication of G. G. Simpson's *Tempo and Mode in Evolution* (1944) and, to a lesser extent because of overlap, its sequel, *The Major Features of Evolution* (1953). Using taxa above the species level as units of analysis, Simpson fleshed out the skeleton of classical paleontology with facts and interpretations relating to rates, trends, and patterns of large-scale evolution. In the first book, but not in the second, Simpson adopted the idea of Goldschmidt (1940) that evolutionary research could be divided into the study of microevolution, or changes within species, and the study of macroevolution, or evolution above the species level. Goldschmidt believed that a natural discontinuity actually exists within the evolution of life—that species and higher taxa arise only through sudden chromosomal changes, while the conventional process of natural selection acts upon genes to produce only lesser modifications within species. Simpson's temporary acceptance of the division into micro- and macroevolution did not stem from acceptance of Goldschmidt's beliefs, but seems instead to have manifested a practical recognition that evolution was being studied in two separate ways. Biologists, limited by their own generation times, examined infinitestimal changes within living populations. Paleontologists, saddled with an incomplete fossil record, adopted a coarse-grained approach, concerning themselves mainly with genera, families, and orders. Unfortunately, a substantial hiatus had remained between the two scales of observation and analysis. The journal *Evolution* was founded in 1947 with the aim of mending the division. While its early issues contained some paleontologic articles, many paleontologists tended to publish their evolutionary contributions elsewhere, and soon the journal drifted largely into the domain of neontology.

The species stands astride the gulf between the higher taxa of most paleontologic studies and the small populations upon which neontologists tend to focus. Clearly, if we are to find a common ground, it must be at the level of the species. A partial foundation for a bridge across the gulf has been laid from the neontologic side by workers like Ernst Mayr (1942; 1954; 1963; 1970), Harlan Lewis (1962; 1966), Verne Grant (1963; 1971), Hampton Carson (1968; 1975), and Guy Bush (1975). To varying degrees, these workers have viewed the species as the basic unit of large-scale evolution and have concerned themselves with the ways in which new species arise. For Mayr, and to some extent for the others, this focus upon species has followed logically from two ideas. One of them is that species are rather stable in an evolutionary sense: once formed, they tend to evolve slowly. If transpecific evolution is assumed to occur, the second idea is a necessary consequence of the first; it is that most evolution occurs as species originate. The point here is not that all species differ considerably from their parent species, but that those that do differ markedly usually develop their distinctive features rapidly, in the process of budding off from ancestral species.

In other words, the validity of the species as the fundamental unit of large-scale evolution depends upon the presence of discontinuities between many species in the tree of life. Some workers have denied the importance of discontinuities, contending that species tend to intergrade in space and time to such a degree that they cannot be considered natural units. The following chapter will review the controversy regarding these aspects of the formation of species. In subsequent chapters, paleontologic evidence will be presented favoring the idea that species are indeed the units of large-scale evolution. The most important of this evidence relates to the longevity of fossil species. It will be shown that species have tended to last for such long intervals of geologic time that, once formed, they must have evolved very slowly—far more slowly than most biologists have realized. This condition, when compared to the rapid pace of large-scale evolution, implies that most sizable evolutionary steps in the history of life must have occurred cryptically from a paleontologic vantage point, during the rapid origination of certain species from small, localized populations of pre-existing species. This idea, asserted by Ernst Mayr since 1954, has been popularized within the paleontologic community by Eldredge (1971) and Eldredge and Gould (1972) as what can be termed the **punctuational model** of evolution. **Phyletic gradualism,** or the **gradualistic model,** is the opposing scheme.

In this book, I will attempt to trace briefly the origins and historical progress of some of these ideas. The failure of the punctuational view to gain substantial support until so late in the present century will be touched upon in the chapters that follow. Why this view did not emerge earlier, during the last century, merits immediate consideration, because it seems related to the early rift between paleontology and evolutionary biology.

Rudwick (1972) recently lamented a decline in the intellectual momentum of paleontology before the turn of the century:

> It would seem that paleontology—and indeed geology too—failed to gain a proportionate number of recruits of first-class intellectual calibre. The output of specialist papers and monographs continued to rise exponentially, but their character became routine and their intellectual level stagnant. Only recently have there been hopeful signs that paleontology may be recovering, in its younger generation, the broad interests and outlook that it possessed so markedly earlier in its history. (Rudwick, 1972, pp. 265–266.)

In large part, Rudwick attributed the stagnation he described to the failure of paleontology to shed light on human origins, which had become a matter of particular interest, but he cited as another factor a growing belief in the imperfection of the fossil record. It is this second factor that I wish to consider briefly.

The disenchantment with the fossil record that followed publication of *On the Origin of Species* is epitomized by the fate of Heinrich-Georg Bronn, whose contribution to science Rudwick reviewed. At the time of publication of the *Origin*, Bronn stood pre-eminent among paleontologists. Though a German, in 1857 he was awarded the Grand Prix for Physical Sciences in a contest of the Academy of Sciences in Paris for essays that treated the appearance and disappearance of fossil organisms through time and the relationship between fossil and living forms. Despite Bronn's great prominence in life, few extant paleontologists recognize his name. The reason is that although in some ways he laid groundwork for the acceptance of Darwin's scheme, Bronn argued against the gradual transmutation of species. The motto of his prize-winning essay was "To be taught by nature," and what nature seemed to teach him was that there was little evidence for more than minor temporal variation within fossil species. Quite simply, he believed species to be discrete entities. Bronn's death in 1862 left him little time to react to the *Origin*, and his adversary position caused his reputation to fade rapidly in the flood tide of scientific support for Darwinism.

Darwin, in fact, found the fossil record, as elucidated by Bronn and others, a great disappointment. In the *Origin* he wrote:

> Geological research, though it has added numerous species to existing and extinct genera, and has made the intervals between some few groups less wide than they otherwise would have been, yet has done scarcely anything in breaking down the distinction between species, by connecting them together by numerous, fine, intermediate varieties; and this not having been affected, is probably the gravest and most obvious of all the many objections which may be urged against my views. (Darwin, 1859, p. 299.)

Darwin analyzed at length the allegedly incomplete nature of the fossil record, and his views were underscored to an extreme degree by Lyell, the most prominent geologist of the century. Not only had Lyell failed to propose natural selection, he had refused to accept the very idea of evolution until Darwin made his contribution. It was from this awkward position that Lyell denigrated the fossil record, the particular source of evidence about evolution that fell within his domain. Thus, Darwin and his followers came to view the fossil record as being their enemy as much as their friend. The positive contribution of fossil data—the documentation of large-scale biotic transformation—remained, but was augmented by little new evidence of gradual transformation within specific lineages.

Here then we seem to witness the beginning of the schism between paleontology and neontology. The most productive areas of early evolutionary research turned out to be purely biologic: areas like embryology, comparative anatomy, geographic ecology, and, later, genetics. Paleontology found a home primarily within geology, where it contributed especially to the temporal correlation of rocks. Certainly, some paleontologists remained lodged in departments of biology and a few contributions were made to evolutionary theory, but the biologic luster of the science was lost.

In the decade after publication of the *Origin*, a few students of the fossil record, such as A. Gaudry, H. F. Osborn, W. B. Scott, and J. L. Wortman, documented crude lines of descent at the genus and family level. Their work refined the contribution of paleontology, culminating in the twentieth century works of G. G. Simpson and others, which gave the field greater stature than it had been allotted by Lyell. Still, as just described, the gulf between paleontology and evolutionary biology persisted: Evolutionary paleontologists primarily employed higher taxa rather than species as building blocks for constructing phylogenies and measuring rates of evolution.

If, as will be argued here, Mayr and his followers are correct that most evolutionary change is related to speciation, a sobering conclusion follows: The exciting young science of paleontology, personified by Heinrich-Georg Bronn, was badly mistreated in the last century. Bronn had a valid objection to Darwin's scheme, which demanded continuous change within established species at what we now recognize to be rather high rates—rates sufficient to account for the sequential origins of major new higher taxa documented by the fossil record. Of course, absolute dating techniques were lacking in 1859, but based on rates of erosional denudation, Darwin estimated the age of the latest Cretaceous to be about 300 My (million years), which was excessive by a factor of nearly five. Thus, it was in part an exaggerated view of the immensity of geologic time that fostered Darwin's gradualism.

More importantly, it would seem that the punctuational view was simply too subtle a refinement of Darwin's theory to have taken root in the decades following 1859. Lyellian geology, which called upon everyday processes to reshape the earth gradually and continuously, had gained so much momentum that gradual evolution at rates that were significant in terms of large-scale transformation seemed a natural outgrowth (Eldredge and Gould, 1972). Certainly, also, the claim of a deficient fossil record represented a comfortable refuge for the fragile new theory if the alternative was a contention that extinct species commonly appeared rapidly and thereafter evolved so slowly that it was difficult to detect substantial modification of their form before extinction. Any claim that natural selection operated with greatest effect exactly where it was least likely to be documented—in small, localized, transitory populations—would have seemed to render Darwin's new theory untestable against special creation, and perhaps almost preposterous as a scientific proposition. One of Darwin's greatest assets, in an age when fact and speculation were often hopelessly intermingled, was that he characteristically searched for hypotheses that were testable.

Gould and Eldredge (1977) have noted that in personal correspondence (November 23, 1859) T. H. Huxley chided Darwin for excluding saltation (evolutionary jumps) from his scheme. A fact that has not been appreciated by most punctuationalists, however, is that Darwin actually expressed their basic view, if only briefly and in speculation. Present-day evolutionists have a penchant for poring over *On the Origin of Species* in search of concepts that are popular today. Often the germ of a modern idea will indeed be found there, yet, as with examination of the *Bible*, careful search will also yield contradictory or alternative assertions. Darwin (1859, pp. 286–298) wrote extensively about the incompleteness of the stratigraphic record as a reason for poor fossil documentation of evolutionary links between taxa, but he also contributed the following brief suggestion, of which the opening clause, placed in brackets here, was deleted from the fourth edition (1866) and all subsequent editions.

> [One other consideration is worth notice:] with animals and plants that can propagate rapidly and are not highly locomotive, there is reason to suspect, as we have formerly seen, that their varieties are generally at first local; and that such local varieties do not spread widely and supplant their parent-forms until they have been modified and perfected in some considerable degree. According to this view, the chance of discovering in a formation in any one country all the early stages of transition between any two forms, is small, for the successive changes are supposed to have been local or confined to some one spot.

Below this paragraph in all editions after the third, Darwin inserted the following idea, which was reiterated in the summary of the first eleven chapters:

> It is a more important consideration, clearly leading to the same result, as lately insisted on by Dr. Falconer, namely, that the periods during which species have been undergoing modification, though very long as measured by years, have probably been short in comparison with the periods during which these same species remained without undergoing any change.

Thus, after publication of the first edition, Darwin strengthened his punctuational hypothesis. He nonetheless continued to place much greater emphasis on the poor quality of the fossil record, and the remainder of his evolutionary writing was strongly gradualistic in tone, perhaps for the reasons discussed above. He wrote, for example, in all editions (making only minor modifications after the first, including the insertion of the word "metaphorically," here placed in parentheses):

> It may be said (metaphorically) that natural selection is daily and hourly scrutinising, throughout the world, every variation, even in the slightest; rejecting that which is bad, preserving and adding up all that is good; silently and insensibly working, whenever and wherever opportunity offers, at the improvement of each organic being in relation to its organic and inorganic conditions of life. We see nothing of these slow changes in progress, until the hand of time has marked the long lapse of ages.... (Darwin, 1859, p. 84.)

> That natural selection will always act with extreme slowness, I fully admit.... (Darwin, 1859, p. 108.)

It was by stressing the imperfection of the fossil record, the immensity of geologic time, and the slowness of natural selection that Darwin left a gradualistic heritage.

The rapid divergence of paleontology from biology that followed publication of the *Origin* has now ceased. The unfortunate nineteenth century trend has been reversed in part because paleontologists have increasingly come to study fossils as biologic entities, but in part also because attempts have been made to infuse data from the fossil record into biologic theory. Until quite recently, however, few paleontologic efforts have been made to study evolution at the level of the species. This kind of study will be the focal point of the chapters that follow.

Mayr (1963) and Grant (1971) have lamented the absence of any formally recognized subdiscipline of biologic science concerned with the study of species. Mayr (1963, p. 429) noted the irony of Darwin's naming his great book *On the Origin of Species* and then failing to come to grips with multiplicative speciation. In their own focus upon the origin of species, Mayr and Grant and others like Lewis (1962, 1966), Carson (1968, 1970, 1975), Wilson *et al.* (1974; 1975), and Bush (1975) have drawn data and concepts from numerous subdisciplines, including taxonomy, genetics, biogeography, and ecology. The need for an eclectic approach would seem to have represented a major barrier to this sort of endeavor. Many scientists, having initially specialized in one established field, have resisted professional expansion into unfamiliar conceptual domains of the sort required to develop a science of species. Given the failure of such a science to take root, it is hardly surprising that few biologists, which is to say few specialists, have assimilated the arguments that many species arise rapidly and that the species is the basic unit of macroevolution.

The paleontologic approach summarized in the present volume is meant to complement the efforts of biologists. The aim is to present material for construction of a bridge from the paleontologic side of the traditional academic gulf between macroevolution and microevolution. The basic message is that fossil data can be used to study rates, trends, and patterns of evolution *at the level of the species.* Harper (1976) has challenged this approach, claiming that the answers sought can come only from neontology. He compares the attempt to convert fossil data into information about the formation of species to an attempt to fashion a science of sociology from scraps of archaeological material. The analogy seems unfair, however, because sociologists can fully observe in their lifetimes many of the processes they seek to analyze. A rich store of information also has been supplied by people of preceding generations who lived during the growth of modern social institutions. Biologists have occupied such a thin slice of evolutionary time that they are in a far less advantageous position: They need the fossil record. As for the ability of paleontology to meet this need, the remainder of this book is designed to build a favorable case.

Some of the methods presented here are analogous to simple techniques used in population biology. In the general analogy, species take the place of individuals, and the processes of speciation and extinction substitute for birth and death. The techniques are applicable primarily to Cenozoic taxa. This limitation might be viewed as potentially restricting or biasing their utility. There does exist, however, a good fossil record for a variety of Cenozoic taxa, which, taken together, display a spectrum of biologic traits. Furthermore, not all living groups are ancient or highly diversified: the Recent time plane slices through taxa in varying stages of diversification and decline, providing biologic data for a varied sample of late Cenozoic life. Finally, it should be appreciated that the general insights developed in the classic works of Simpson (1944; 1953) were derived largely from study of Cenozoic mammals, with only limited contributions from the fossil records of invertebrates, plants, and pre-Cenozoic vertebrate taxa.

There is no question that fossil entities recognized as species are not strictly comparable to species of the living world. The Cenozoic orientation of the present study avoids this problem to a considerable degree. Many of the techniques employed here actually rest primarily on the diagnosis of living species. For example, rates of phyletic evolution are evaluated by tracing extant mammalian lineages backward from the present; rates of extinction are estimated by calculating for Cenozoic faunas of particular ages the percentage of species that are recognized in the Recent; and net rate of adaptive radiation for a higher taxon is typically determined by calculations based on the number of *living* species that have been produced since the time of origin of the higher taxon—a time that the fossil record documents to a good approximation.

A number of biologists, implicitly or explicitly, have opposed the puctuational view of evolution with actualistic arguments—arguments projecting biologic observations and conclusions into the past. The problem here is that brief intervals of observation in the modern world are not easily extrapolated to geologic scales of time. An important lesson of history is that we should not easily abandon strong circumstantial evidence of the geologic record in favor of theoretical or actualistic assertions that may seem weightier simply because of their rigor. Wegener's scheme of continental drift, though based on geologic evidence too strong to be readily dismissed, nonetheless was generally rejected early in this century in part because of the discovery of a rigid crust of dense rock beneath the oceans. Physicists could conceive of no mechanism by which continents might plough through such a barrier. Today it is believed that the oceanic crust simply slides along with the continents, posing no problem. Belatedly, Wegener is a hero. Similarly, Lord Kelvin's calculation of the earth's age, based on estimation of its rate of cooling from a molten state, frustrated Darwinists unneccessarily for many years, until radioactive decay was shown to represent an additional heat source. As already discussed, it will be argued here that the interval of time from the late Precambrian to the present during which modern life evolved (perhaps 700 My) was much too brief for Darwin's gradualistic scheme to apply, especially when we consider the brevity of intervals during which many distinctive new taxa have formed. Even so, early evolutionists should have been allotted far more geologic time for the unfolding of modern life than the 20 to 40 My that Kelvin and other physicists were claiming: It seems incredible that the physicists' attack on Darwinism actually cast doubt, in some scientific circles, on the fundamental geologic evidence that had led James Hutton (1795, vol. I, p. 200) a century earlier to envision "no vestige of a beginning,—no prospect of an end." Hutton's views had, after all, been strengthened by the work of Lyell and numerous other geologists over a period of several decades.

If we accept these lessons of the past, it hardly seems unreasonable to call for a reassertion of fossil data in an era when neontologists are overturning a number of their own traditional beliefs. The recent recognition of regulatory genes, for example, which will be discussed in Chapter 6, has led Mayr (1970, p. 183) to write: "The day will come when much of population genetics will have to be rewritten in terms of the interaction between regulatory and structural genes."

It is true that the fossil record of species is much poorer than the record of higher taxa. Raup (1972), in particular, has elegantly examined the relationship between taxonomic level and completeness of the record. It does not follow, however, that species are useless as units for analyzing large-scale evolution. The techniques that will be described in the following chapters are aimed at circumventing problems traditionally associated with the enumeration and biostratigraphic evaluation of fossil species. These techniques provide a highly imperfect quantitative picture of species and speciation, yet their application yields provocative inferences. To be sure, the techniques developed to date represent but a beginning. The bridge between neontologic and paleontologic approaches to the study of evolution is not yet passable and will never be a thoroughfare. It would be nice, however, to believe that a few pylons are in place and that the future will bring an increase in the flow of research from both directions.

2

The Fabric of Evolution

A central theme of Darwinism and the Modern Synthesis has been the idea that most evolutionary transition takes place within established species—that while the branching off of new species adds new vectors to evolution, the process involved (speciation) accounts for a relatively small amount of large-scale change. The prevalence of this belief is indicated, among other ways, by (1) conventional views on the function of sexuality, (2) the focus of modern population genetics, (3) the traditional interpretation of human evolution as having taken place within a single lineage, and (4) the commonly held notion that coevolution represents a race of gradual, correlated change within interacting species. Such gradualistic views, which have been challenged only sporadically in the past, now seem untenable in light of fossil evidence.

INTRODUCTION

It is in documenting rates of evolution that the fossil record provides evidence about the discrete nature of many species. To understand why this is true we must examine the structure of the tree of life and the vocabulary used to describe this structure.

All evolutionists agree that rates of evolution vary in space and time, but exactly how they vary has been the subject of much debate. A basic question about any segment of phylogeny is how the tempo of evolution has changed through time. Another question is what factors have controlled this tempo. Some kinds of organisms, like coelacanths and ginkgo trees, have survived for long spans of geologic time with little visible change. These we call **living fossils.** Other living organisms bear little resemblance to apparent ancestors that lived only a few tens of millions of years ago. If we turn our attention from rate of morphologic change to rate of increase in number of species within taxa like families or orders, we find similar contrasts. Some higher taxa have persisted for long intervals at low diversity. Others have proliferated so rapidly as to contain dozens of species within a few million years after originating. The study of stagnant and "explosive" episodes in the history of life is currently one of the most active areas of evolutionary research (Mayr, 1970, p. 6).

ALTERNATIVE SHAPES
FOR THE TREE OF LIFE

Rates of phenotypic evolution can be measured using either morphologic or taxonomic criteria. In general, morphologic change and taxonomic change are strongly correlated because taxa are distinguished on the basis of morphologic features. Any measured rate of evolution takes on meaning only in relation to other rates of evolution. Thus, a central topic of this book will be the manner in which rates of evolution have been distributed in phylogeny, as normally portrayed graphically, with morphologic change plotted against geologic time. On this kind of graph, the slope of a line represents rate of evolutionary change within a single line of descent, or **lineage.** In other words, an established species that persists, generation after generation, forms a lineage. Many lineages documented by fossil data display little or no discernible morphologic change.

Living representatives of species represent the terminal portions of extant lineages. Evolution within a lineage is commonly called **phyletic** evolution. Lineages may be terminated by extinction, but they also may branch, to increase the number of species. Branching of lineages is termed **speciation** and is accomplished when a population that belonged to an established species has become reproductively isolated from the remaining populations of the established species. The prevailing view is that most species arise from populations of pre-existing species to which gene flow has been cut off or at least slowed. Some aspects of the process of speciation will be considered in Chapter 6.

Branching from a single lineage will ultimately produce a cluster of lineages known as a **clade.** Thus, branching is sometimes termed **cladogenesis.** When referring to a clade or to a group of related clades, it is common to use the adjective **phylogenetic** (not to be confused with the similar word *phyletic* defined above). Phylogenetic pertains to a portion of phylogeny consisting of more than one lineage, whereas phyletic refers to a single lineage.

A lineage reconstructed from fossil data may exhibit sufficient evolutionary change that a taxonomist deems it appropriate to divide it into two intergrading species. Such species are known as **chronospecies**—or successional species, paleospecies, or evolutionary species. The taxonomic division of a continuum into chronospecies is necessarily subjective and arbitrary (Figure 2-1). An actual example of such a taxonomic division is illustrated in Figure 2-2. We will consider the nature of chronospecies more fully in Chapter 4. Paleontologists sometimes refer to taxonomic transition from one chronospecies to another as "phyletic speciation," but I will avoid this use in order to conform to the biologic convention of restricting the term **speciation** to the *splitting of lineages*. A line drawn subjectively across a lineage to demarcate the origin of one chronospecies automatically establishes the arbitrary end of another chronospecies. This kind of "extinction" is sometimes called "phyletic extinction," but for parallelism (having rejected the phrase "phyletic speciation") I will apply the word **pseudoextinction** to this phenomenon. **Extinction** as used in this book will refer only to the termination of a lineage.

As introduced in the previous chapter, the most fundamental question I will consider concerns the relationship between speciation and rate of evolution. Do species commonly evolve most rapidly as they emerge from ancestral species, or is the pace of evolution unrelated or weakly related to the process of branching? At one extreme is the possibility that all evolution occurs in association with the branching process (Figure 2-3,A). At the other is the possibility that speciation is only incidental to rate of change. If the second alternative represented reality, then every new species, once reproductively isolated, would depart from its ancestral species along a new evolutionary pathway, yet the initial rate of evolution of the descendant species would be no more rapid, on the average, than its own subsequent rate of change or than the ongoing rate for the ancestral species (Figure 2-3,B). Speciation would do nothing but add a new direction of evolution. A full spectrum of intermediate possibilities exists. The question is, where in this spectrum does the condition typical of phylogeny lie? In other words, is most evolution in the history of life associated with speciation or is more evolution of the phyletic variety? Additionally, how much variation is there about the mean condition, and what factors account for the variation?

In effect, the preceding paragraph suggests that evolution can be separated into two components. One is the **phyletic component,** representing evolution within established species. The other is the component associated with specia-

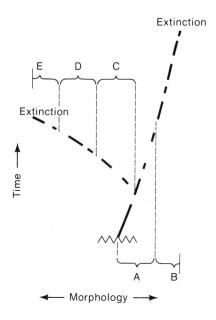

FIGURE 2-1
Diagram illustrating the subjective division of hypothetical lineages into chronospecies. The zigzag line at the base marks the appearance of the original lineage in the fossil record. This lineage branches before going extinct. Chronospecies A is arbitrarily recognized as originating at the zigzag line. It turns into chronospecies B at a point at which a taxonomist judges the range of morphologies produced by phyletic evolution to exceed that permissible within a species of the kind of organism forming the lineage. Chronospecies B is of shorter duration because of termination of the lineage. Chronospecies C, D, and E are of shorter temporal duration than A or B because the lineage to which they belong is evolving more rapidly.

tion. On a geologic scale of time this **speciational component** must sometimes be contributed almost instantaneously, as a new species becomes reproductively isolated from its parent species. As we shall see, durations of most species are measured in millions of years, whereas the period of emergence of some species may be measured in tens of thousands of years, or possibly even hundreds of

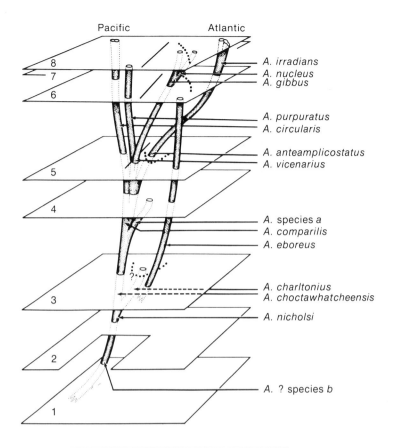

	Age	Years ago
1	Early Middle Miocene	19,000,000
2	Middle Middle Miocene	16,000,000
3	Late Middle Miocene	13,000,000
4	Late Miocene	8,000,000
5	Early Pliocene	6,000,000
6	Early Pleistocene	1,800,000
7	Late Pleistocene	250,000
8	Present	0

FIGURE 2-2

Cenozoic phylogeny of scallops of the genus *Argopecten* in North America. Numbers 1 through 8 identify chronologic planes spaced according to absolute time; each plane represents a stratigraphic unit that has contributed fossils and other data used to construct the phylogeny. The straight solid line on planes 5, 6, and 8 symbolizes the geographic barrier separating the Atlantic Ocean, Gulf of Mexico, and Caribbean from the Pacific Ocean after the Miocene. The curved dotted line on planes 3, 5, 6, and 8 symbolizes the ecologic barrier separating enclosed bay environments (upper right) from open marine environments (lower left) on the Atlantic side. (From Waller, 1969).

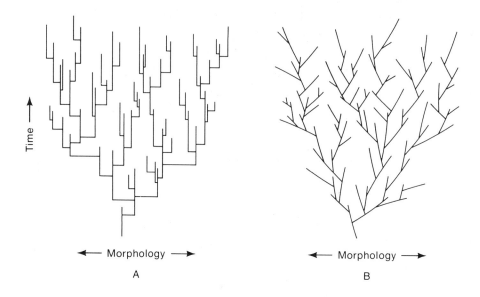

FIGURE 2-3
Hypothetical phylogenies representing extreme views. In A, all evolution is concentrated in speciation. In B, all evolution is phyletic.

years or less. Thus, in the context of geologic time, speciation is often an instantaneous event. Another important point is that there is no *a priori* reason to believe that the time required for reproductive isolation should correlate inversely with the rate of morphologic divergence. Even a population that quickly becomes reproductively isolated might diverge very slowly in form and habit from its parent species.

To restate a fundamental point, a variety of evidence will be advanced in the pages that follow in support of the idea that the speciational component of evolution is much more important than the phyletic component—that most evolution in the history of life has been associated with the multiplication of species. A different way of stating this idea is to assert that most major evolutionary transitions have been temporally associated with speciation. Eldredge and Gould (1972) referred to the implied pattern of phylogeny as punctuated equilibria. As noted in Chapter 1, the scheme of evolution representing this pattern will be referred to here as the **punctuational model.** In Figure 2-4, this model is contrasted with **phyletic gradualism,** or the **gradualistic model,** in which most evolution is considered to be of the phyletic variety, with rapidly divergent speciation playing a lesser role.

It is important here to clarify a misunderstanding about terminology that has hindered communication during the past few years. The phrase "phyletic

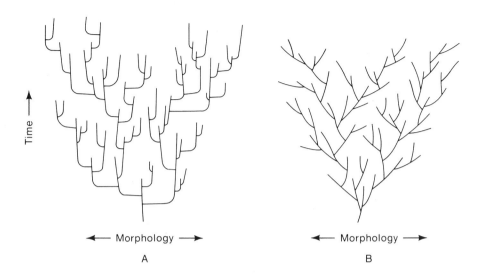

FIGURE 2-4
Hypothetical phylogenies representing the punctuational model (A) and the gradualistic model (B). Note that some phyletic evolution is represented in A and that some speciation events in B display accelerated evolution.

gradualism" has sometimes been employed as if it were synonymous with phyletic evolution. Had its originators (Eldredge and Gould, 1972) meant it to be synonymous, they would have had no reason to introduce it in the first place because the phrase "phyletic evolution" has long been part of the language of evolutionary paleontology. The term "gradualism" does not denote a single example of phyletic change but a general doctrine holding that phyletic evolution dominates in phylogeny. In fact, no matter how slowly phyletic evolution may be operating, it must be ubiquitous, because for any species the maintenance of perfect stasis from generation to generation (the so-called Hardy-Weinberg equilibrium) is a virtual impossibility, as is fluctuation about an unchanging mean condition. The punctuational model does not deny the existence of phyletic evolution, but relegates this mode of change to a subordinate role. Similarly, the gradualistic model, as defined here, is compatible with the idea that evolution is accelerated in association with some speciation events.

   Another source of confusion has been the incorrect inference that the punctuational model implies that most entities recognized as species arise through the multiplication of lineages, or through speciation, as I have defined the term (Harper, 1975). Figure 2-5 shows that this need not be the case. Even if most recognized species have formed by phyletic transition, the smaller number of rapidly divergent speciation events can be responsible for most of the net change in a segment of phylogeny. The phrase "net change" is also important. It is

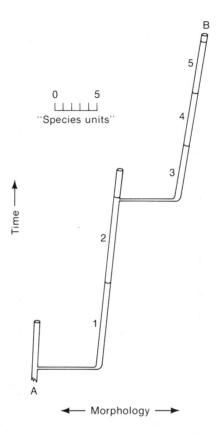

B

5

0   5
⌞⌴⌴⌴⌴⌴⌋
"Species units"

4

3

Time →

2

1

A

← Morphology →

FIGURE 2-5
An idealized segment of phylogeny illustrating how the speciational component of evolution can dominate even
if most species originate by phyletic
transition. A "species unit" is the
range of morphologic variation tolerated within a single species by the
taxonomist assigning populations to
species. Of the five hypothetical
species recognized in the evolutionary
pathway between A and B, three
originate by phyletic transition and
only two (1 and 3) originate by speciation (branching), yet speciation accounts for three-fourths of the evolution that occurs (15 of 20 "species
units" of change).

possible that much phyletic change is canceled out by reversals of the direction of change. Minor oscillations are to be expected, so that the sum of all phyletic evolution occurring between the origin and termination of a lineage may be substantially greater than the net amount.

The thesis will be developed here that phyletic evolution represents the fine tuning of adaptation. It may accomplish modest adjustments in various aspects of life history, such as breeding time and number of eggs or offspring produced, in addition to producing slow modifications in physical form (especially body size). Quite often such changes accumulate to the point where, with sufficient data, a competent paleontologist will demarcate a new chronospecies. It will be concluded, however, that the speciational component of evolution so outweighs the phyletic component, in general, that it accounts for the large majority of transitions between genera. The role of phyletic evolution in the origin of families and orders is even smaller.

The punctuational model has other profound implications for the interpretation of transpecific evolution. One is that large-scale evolutionary trends are not produced primarily by phyletic evolution, as has commonly been envisioned, but rather by selection among lineages. This topic is considered in Chapter 7. Another is that most higher organisms reproduce sexually for a different reason than has generally been favored. This corollary is the subject of Chapter 8.

As mentioned in Chapter 1, the punctuational model emerged in its modern form primarily through the writings of Ernst Mayr (1942; 1954; 1963). We will begin our evaluation of the idea with a glimpse of the historical framework in which it surfaced.

THE MODERN SYNTHESIS

By the early 1950's, the biologic profession had settled into a much more comfortable posture with respect to evolutionary thinking than it had enjoyed in the early 1900's when debates raged between supporters of Darwin and their opponents, some of whom favored the idea of dramatic transformation by instantaneous genetic mutation. The consensus in the early 1950's represented what Julian Huxley had earlier labeled the Modern Synthesis. As presented in the famous book by Huxley (1942), the Modern Synthesis, or Synthetic Theory, was an amalgamation of the data and concepts of genetics, taxonomy, embryology, biogeography, and other disciplines. Of these, genetics was the newest and had caused the greatest difficulty for neo-Darwinians very early in the twentieth century. By 1942, however, genetics had been alloyed with the other subjects in the forging of a widely accepted modern evolutionary theory grounded in Darwinian natural selection. Genetic mutation was seen as a largely random process, providing small bits of raw material for selection.

For our purposes, two particular attitudes of the Modern Synthesis, deserve consideration. The first, found in Huxley's book, is the belief that rate of evolution by natural selection is not generally accelerated in association with speciation events. Huxley believed that distinctive new species are likely to appear abruptly only if they arise by processes like hybridization, which is regarded as producing new species very rarely in the animal world and uncommonly in the plant world. By the very definition of the word "species," hybridization is deemed abnormal. Mayr (1940) defined species as "groups of actually or potentially interbreeding natural populations, which are reproductively isolated from other such groups." Most new species, Huxley asserted, differ little, at first, from their parent species. Speciation, according to this view, does little more than change the direction of evolution by isolating a portion of the gene pool, which, according to geographic circumstances and the chance appearance of new mutations and gene combinations, will inevitably move along a different evolutionary path from the remaining gene pool of the original species. Huxley wrote:

> Species-formation constitutes one aspect of evolution; but a large fraction of it is in a sense an accident, a biological luxury, without bearing upon the major and continuing trends of the evolutionary process. (Huxley, 1942, p. 389.)

Thus, branching of lineages was seen as being largely incidental, while persistent evolution within lineages accounted for most large-scale change. As discussed in the previous chapter, this view was essentially the same as Darwin's. Part of the diagram representing his gradualistic conception of phylogeny in *On the Origin of Species* is reproduced here as Figure 2-6.

The second idea of the Modern Synthesis that is important to our discussion complements the first and was originally expressed by Sewall Wright (e.g., 1931, 1940). This is the idea that rates of evolution tend to be especially high in large species that are divided into numerous, partly isolated subpopulations. Wright reasoned that in such populations there would be many opportunities for local production of adaptations that could be of general value to the species. In effect, the situation could represent an ideal balance between the frequent "testing" of new genotypes within local subpopulations and the spread of useful ones, by gene flow, throughout the total population. The opportunity for continual incorporation into the general gene pool of new features was expected to quicken the pace of evolution. The implication was that large, well-established species have commonly undergone rapid phyletic evolution.

That the Modern Synthesis placed great emphasis upon phyletic evolution is hardly surprising, because this historic confluence of biologic approaches represented a reassertion of natural selection as the dominant source of evolution. Earlier in the twentieth century several prominent biologists and paleontologists had expressed the radical view that species arise suddenly, fully formed, by dramatic genetic accidents. The prominent geneticist Richard B. Goldschmidt (1940), in his book *The Material Basis of Evolution*, was among the most recent to take this position. As mentioned in Chapter 1, Goldschmidt believed that muta-

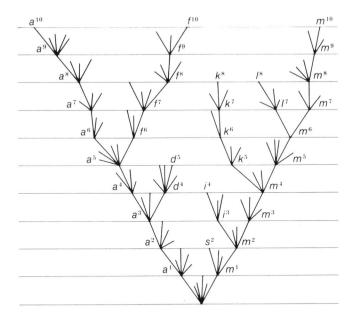

FIGURE 2-6
A segment of the idealized gradualistic phylogeny that Darwin published. The original species gives rise to a fan of lineages that represent diverging populations: "When a dotted line reaches one of the horizontal lines, and is there marked by a small numbered letter, a sufficient amount of variation is supposed to have been accumulated to have formed a fairly well-marked variety, such as would be thought worthy of record in a systematic work" (Darwin, 1859, p. 117). The pattern of gradual divergence continues to the top of the diagram.

tions of individual genes accomplish only minor transformations within species. He claimed that new species arise in single births, always by way of chromosomal mutations (changes in the arrangement of genes within chromosomes). He also believed that some new species arise full-blown as bizarre creatures that are quite distinct from their progenitors. While admitting that most of these "hopeful monsters" must suffer extinction, Goldschmidt claimed that others succeed in new modes of life, becoming initial members of major new higher taxa. Today it is almost universally agreed that Goldschmidt's views contained exaggerations and oversimplifications. Unfortunately, however, contributors to the Modern Synthesis, zealous in their defense of natural selection, cast aside not only Goldschmidt's genetic mechanisms but also his idea that most evolution occurs in association with speciation. Almost forgotten was the possibility that established species might be rather stable, in an evolutionary sense, and that natural selection or other agents of change might operate most intensively during the emergence of certain species.

DIVERGENT SPECIATION

The fact that the fossil record seemed often to document sudden origins of higher taxa troubled Darwin but was largely ignored by the architects of the Modern Synthesis, few of whom were paleontologists. Much of the momentum of neo-Darwinism in the early 1900's, and of the Modern Synthesis that followed, came from the apparent removal of two impediments that had greatly troubled Darwin and his supporters late in the last century.

One development was the general recognition of particulate inheritance, which led to the growth of modern genetics. This recognition eliminated the anti-Darwinian argument that sexual reproduction would automatically blend into oblivion the germ plasm underlying a useful new adaptation. Darwin's second major obstacle had been related to the scope of geologic time. Assuming the earth to be cooling from its time of origin, Lord Kelvin and other physicists calculated that the planet's present temperature permitted its age to be no greater than perhaps 20 to 40 My. In proposing that natural selection could not form new species, De Vries (1905) cited Kelvin's calculations. He concluded that speciation must instead occur instantaneously, by macromutation. As noted in Chapter 1, however, the discovery of radioactive heating soon showed the earth to be much older than the physicists had estimated. The vast expanse of geologic time that Lyellian geologists and followers of Darwin had envisioned was no longer threatened. Buoyed by this development and the rise of genetics, gradualistic Darwinism entered a renaissance.

In the chapters that follow, it will be shown that the temporal strictures that had frustrated early Darwinists have by no means been fully removed. As already noted, careful study of rates of evolution and the modern scale of geologic time indicate that phyletic transformation is painfully sluggish. This conclusion implies that the resurgence of gradualistic thinking after the turn of the century represented an overly optimistic reaction.

The Age of Mammals (Cenozoic Era) is now considered to have lasted about 65 My. The entire Phanerozoic, extending from the Cambrian (when metazoans first appear in abundance) to the present, spans about 570 My. Such stretches of time seem long until compared to the intervals within which many large-scale taxonomic transitions seem to have taken place. For example, discrete new orders of mammals have arisen in intervals of less than 10 My. It was this kind of observation that led Simpson (1944) to postulate **quantum evolution.** He suggested that this mode of evolution accounted for the origins of most higher taxa by the rapid crossing of adaptive thresholds. The population or taxon making the transition was seen as passing from one **adaptive zone,** or set of related ecologic niches, to another by way of an unstable, inadaptive phase of evolution. In 1944, Simpson suggested that "most radical types of quantum evolution probably begin by random fixation of inadaptive mutations in very small, isolated populations" (p. 211). This disequilibrium phase, he suggested, was followed by a phase in which natural selection carried the group into a new adaptive zone. What seems clearly to have been described is a form of geographic speciation. Simpson also expressed the idea that quantum evolution "may be involved in either speciation or phyletic evolution, and ... certain patterns within those modes intergrade with quantum evolution" (p. 206). Thus, he did not restrict quantum evolution to speciation. In his sequel to the 1944 book, Simpson (1953) abandoned the notion that speciation is a primary site of quantum evolution. He did remark that "species ... often arise in this way [by quantum evolution]" (p. 389). On the other hand, he focused upon phyletic evolution as the source of major adaptive breakthroughs, viewing quantum evolution, in effect, as greatly accelerated phyletic evolution. In discussing the origin of higher taxa occupying distinct adaptive zones, Simpson stated that "it is, normally, phylogenetic [phyletic, as defined in the present book] progression rather than speciation that leads into the new zone" (p. 350). He then added that "dichotomy or multiple speciation may also occur but it is not the essential feature of the pattern" (p. 354).

In a paper, entitled "Change of Genetic Environment and Evolution," Mayr (1954) established the modern punctuational view with a number of ideas that he later elaborated upon in larger works (Mayr, 1963; 1970). His primary conclusion was that the origin of species from small, locally isolated populations of pre-existing species is often associated with rapid evolutionary divergence. He extrapolated this idea to paleontology, pointing out that this condition may account for many evolutionary gaps in the fossil record:

> Rapidly evolving peripherally isolated populations may be the place of origin of many evolutionary novelties. Their isolation and relatively small size may explain phenomena of rapid evolution and lack of documentation in the fossil record hitherto puzzling to the paleontologist. (Mayr, 1954, p. 179.)

Mayr opposed the previously cited idea of Sewall Wright that large populations deployed as partly isolated subpopulations should evolve rapidly; instead he suggested that gene flow among populations of a species tends to prevent local populations from adjusting by natural selection to local conditions. Basically, Mayr's idea is as follows: A species is typically deployed as an array of subpopulations that are partly isolated by geographic barriers. Each subpopulation must face a unique set of environmental conditions, but because of some gene flow among populations, local selection pressures will tend to cancel out. It will be very unlikely that the entire species will evolve rapidly in any particular direction, In support of these ideas, Mayr cited studies by Dice and Blossom (1937) and Hooper (1941) of small mammals living on lava flows in the western United

States. Endemic blackish races of these creatures have developed only on lava flows surrounded by inhospitable sandy desert. Where light colored rocks are in contact with the lava flows, gene flow with light-colored populations inhabitating the rocks prevents localized blackish races from appearing on the lava flows.

In recent years, the notion that gene flow plays a primary role in stabilizing species has fallen into disfavor (Eldredge and Gould, 1972). The reason is that gene flow has been judged to be ineffective. Spread of new genetic material throughout a large species appears to be quite slow (Ehrlich and Raven, 1969; Mayr, 1970, p. 300). In the next chapter, however, it will be argued that gene flow may indeed be sufficient to stabilize many species, even though spread of useful mutations may be quite sluggish.

Mayr has placed special emphasis on another condition (1954; 1963; 1970; 1975) that he believes confers stability upon established species—"the unity of the genotype." Obviously, any configuration of genetic material within an established species is adaptively successful: The species exists. Organisms are exceedingly complex entities, and most changes in a genotype produced by random mutations of genes or chromosomes are likely to be deleterious. Given these conditions, Mayr has argued that major changes are not likely to be brought about by piecemeal restructuring of the genotype through the periodic fixation of mutations and new gene combinations. Only sudden restructuring is likely. The **adaptive peak** concept of Wright (1932) is instructive here. Wright has viewed species as entities that occupy peaks in an adaptive landscape (Figure 2-7). Peaks represent genotypes associated with suites of morphologic characters that have high fitness. The genotypes that underlie "valley morphologies," on the other hand, are unsuccessful. In essence, Mayr's argument, like that of Simpson in 1944, is that passage from peak to peak is perilous. It is accomplished only by a rapid genetic revolution occurring within a small, localized population. This isolated population is likely to be occupying an atypical habitat that will impose strong, uniform selection pressures upon individuals of the small population. Mayr sees chromosomal restructuring as a common feature of speciation. Heterozygous chromosomal combinations are usually deleterious, so that there is a premium on rapid passage through the heterozygous bottleneck. In part, the rapid divergence may also stem from the fact that the population is a nonrepresentative sample of the gene pool of the parent species. The idea that a biased genetic sample may contribute to divergent speciation is known as the **founder principle** (Mayr, 1942).

The view that evolution can proceed most rapidly in small populations had been championed earlier by Wright (1940), who showed that natural selection can operate most efficiently in small populations, and here also genetic drift, or chance change in genetic composition, is most likely to have significant effects. This conclusion might seem to contradict the other idea of Wright that was noted earlier—that evolution is likely to occur rapidly in certain large populations. From this seemingly paradoxical pairing of conclusions emerged the idea that rapid evolution within small subpopulations could accrue to the benefit of a

FIGURE 2-7
An adaptive landscape of the type proposed by Wright
(1932). The landscape represents genetic composition,
not geographic position. Peaks (+) represent highly
adaptive potential gene combinations, and valleys (−)
represent maladaptive potential gene combinations. The
landscape as well as the genotype may shift through time.

full-fledged species. This belief, which was introduced earlier in the chapter, is
still held by Wright:

> The situation is much the most favorable for evolution where there can be selection
> among interaction systems. This can occur, within widely ranging species, divided
> into small populations, sufficiently isolated to permit wide stochastic deviations in
> numerous loci but not so isolated as to prevent excess diffusion from those centers
> that happen to have acquired the most adaptive interaction systems.... (Wright,
> 1977, p. 471.)

Mayr, in contrast, has focused upon isolation as the key to rapid evolution.
Clearly, successful speciation is a rare event in the history of a lineage. To take
the simplest case, if constant diversity is maintained, an average lineage will bud
off only one descendant species before being terminated. Many temporary isolates become reunited with parent populations before diverging into new (reproductively isolated) species. Also, Mayr has estimated that something like 99
percent of permanently isolated small populations with the potential to form new
species may go extinct without doing so. In effect, these populations fail to cross
an adaptive valley (Figure 2-7). It is also important to appreciate that Mayr's
scheme makes no claim that all species arise in association with significant

change. Some new species are quite similar to their parent species. Nearly identical species that are newly recognized (i.e., that were formerly not recognized as being distinct from one another) are called **sibling species.**

Thus, Mayr's argument is that some fraction of speciation events produce rapid morphologic divergence and that these account for most net evolutionary change in the history of life. In effect, Mayr focused quantum evolution at the level of the species; and Grant (1963), who has also contributed much to the study of this phenomenon, coined the useful phrase **quantum speciation.** The nature of quantum speciation is not yet well understood. Only Mayr's original analysis has been outlined in the preceding paragraphs. Additional views on the subject will be reviewed in Chapter 6, after the importance of the phenomenon has been assessed.

Another point that deserves clarification relates to the timing of rapid divergence that is alleged to be associated with some speciation events. Does most evolutionary change associated with quantum speciation occur before or after reproductive isolation has been completed? The question is potentially difficult to answer, because if reproductive isolation of a recent population from the parent species has not yet been achieved, it is not possible to predict when it will be. Similarly, once reproductive isolation has been achieved, its timing cannot be reconstructed from geologic evidence. In fact, there will often be no brief moment at which reproductive isolation can be said to have been achieved. Some distinct species can hybridize to a degree, though hybrids may be selected against, leading to further divergence of the species and development of sterility barriers to their interbreeding. Because of these sorts of problems, consideration of the temporal relationship between rapid divergence and reproductive isolation will be deferred to Chapter 6.

For the present, we can define **quantum speciation** simply as speciation in which most evolution is concentrated within an initial interval of time that is very brief with respect to the total longevity of the new lineage that is produced. Implicit in this concept is the idea that during the rapid, early phase of evolution, the seminal population has not yet expanded from its small, initial population size.

It will be useful at this point to place quantum speciation in a more general perspective and to outline how its importance will be elucidated in the chapters that follow. In Figure 2-8 quantum speciation (diagram D) is contrasted graphically to other hypothetical modes of rapid morphologic transformation. Diagrams A and C of this figure depict dramatic acceleration of evolution within a single lineage. In other words, an entire established lineage undergoes one or more sudden morphologic shifts without branching. In A the transition occurs at large population size. In C the lineage is constricted, or bottlenecked, during the rapid shift; this mode of change resembles quantum speciation, in that a new species is formed from a small population, but differs in that no branching occurs. Modes C and D, in which population size is reduced during rapid

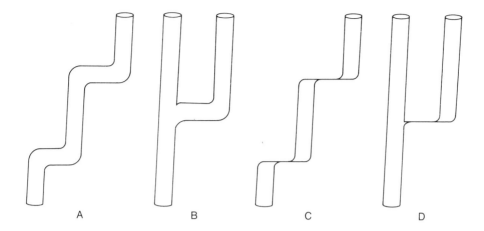

FIGURE 2-8
Schematic representation of possible modes of rapid evolutionary transformation. Population size is proportional to thickness of bar. A: Stepwise evolution in a large, established lineage. B: Rapidly divergent speciation by way of a large population. C: Stepwise evolution by bottlenecking of an established lineage. D: Rapidly divergent speciation by way of a very small population (quantum speciation, as defined in this book).

evolution, can be viewed together as representing quantum evolution, as originally defined by Simpson (1944).

In the following pages, arguments will be marshaled against the general significance of mechanism A of Figure 2-8 (Chapters 3 and 4). Mechanism B, in which rapid branching occurs at large population size, will not be treated explicitly because the theoretical and empirical arguments directed against the possible importance of mechanism A apply also to B: Large populations—the only kind displayed in A and B—on theoretical grounds cannot be expected to evolve rapidly (Chapter 3), and fossil evidence demonstrates that the evolution of large populations is, in fact, very sluggish (Chapter 4).

Because it proceeds by way of a small, ephemeral population, quantum speciation (diagram D) is almost impossible to observe or fully document, but the likelihood of its occurring is supported by circumstantial evidence from numerous evolutionary disciplines (Chapter 3). Furthermore, we are forced by the following kinds of evidence to invoke quantum speciation as the dominant source of macroevolutionary transition: (1) Fossil data demonstrate that large-scale evolution (transition between higher taxa) proceeds very rapidly, and this condition implies the existence of one or more mechanisms of rapid transformation (Chapter 5). Given the evidence that evolution within large, established species is very slow (Chapter 4), we can invoke only bottlenecking (C) or quantum speciation (D). As inferred from evidence on rates of extinction, the bottlenecking mechanism is too rare to be of general significance (Chapter 6). This leaves quantum speciation as the only alternative: It must represent the prevailing mode of large-scale transition. What we know of the genetics of quantum speciation will be discussed in Chapter 6.

It is important to understand that diagrams A and C of Figure 2-8 represent somewhat extreme positions in a spectrum of possibilities, as do diagrams B and D. We might conclude that if large populations, like those in A and B, evolve slowly, and if small populations sometimes evolve rapidly, then populations of intermediate size may, on the average, evolve at intermediate rates. This is presumably the case; yet I will offer evidence that few major transitions occur (few higher taxa originate) except in quite small populations.

THE STATUS OF QUANTUM
SPECIATION IN MODERN BIOLOGY

The notion that established species evolve relatively slowly met with little acceptance in the 1950's and 1960's. Today it enjoys greater support, but remains the subject of considerable controversy. Perhaps the boldest recent attack on neo-Darwinism by a prominent biologist is that of Løvtrup (1976), whose ideas on rapid morphologic transformation by simple genetic changes will be cited in Chapter 6. A sampling of professional opinion at the time of this writing has revealed a remarkable diversity of viewpoints on the relative merits of the gradualistic and punctuational models. Furthermore, widely divergent individual opinions exist as to the consensus among all workers. Some suggest that the gradualistic model is a straw man—that nearly everyone has long understood evolution to be episodic and concentrated in speciation. Others staunchly defend a dominant role for phyletic evolution. Many writings on the subject seem self-contradictory.

Prevailing opinion can be assessed most objectively by examining fundamental doctrines of modern evolutionary biology. Here, even in publications of the 1970's, there are many manifestations of a predominantly gradualistic posture. A measure of the initial failure of the punctuational view to gain widespread support is the rarity with which the idea of quantum speciation is favored, or even mentioned, in books of the 1960's purporting to summarize the nature of organic evolution. Most such books discuss speciation as a branching phenomenon that merely establishes new evolutionary pathways without accelerating rates of change. I will now single out four specific fields of endeavor that in recent years seem to have offered particularly strong evidence of the weak support among biologists for the punctuational view: the analysis of the function of sexuality; the study of population genetics, in general; the interpretation of human ancestry; and the investigation of coevolutionary relationships among species.

The Function of Sexuality

Sexual reproduction is so widespread in the organic world that its role has long been the subject of special interest. A basic idea of the Modern Synthesis is that sexual modes of reproduction have two functions: They shuffle genetic material, providing for the frequent origin of useful genetic combinations, and they spread useful new mutations and gene combinations throughout populations. These ideas were treated in some detail by Fisher (1958) and appear in nearly all major treatises on evolution. Often the implication is that sexuality is of benefit not to individuals, but to populations or entire species, which must evolve rapidly to survive.

So widespread is sexuality that the traditional view of its function implies that this mode of reproduction is not simply an asset but a near necessity in what is seen as a universal, never-ending race of phyletic evolution against evolving competitors and predators and changes in the physical environment. Birky and Gilbert (1971), for example, describe conversations with one worker who "argued that sexual reproduction was so essential that even bdelloids [reputedly asexual rotifers] must indulge in some form of exotic sex."

In opposing the gradualistic model, primarily because of the demonstrable slowness of phyletic evolution, I will also challenge the traditional argument for the prevalence of sexuality. The fossil record shows that phylogeny does not represent a severe race of phyletic evolution. Furthermore, there is evidence in the modern world that asexual "species" of higher life survive quite successfully. An alternative argument for the dominance of sexuality, consistent with the punctuational model, will be developed in Chapter 8. The central thesis will be that sexuality is a requisite primarily for speciation, not for rapid phyletic evolution. The germane point in the present context is simply that the traditional explanation, which until recently has enjoyed nearly universal support, reflects a pervasive philosophy of gradualism.

Population Genetics

In a very direct way, the focus of population genetics since early in this century has reflected a strong general preference for the gradualistic model. In some instances, the preference has been stated explicitly. An example is the opinion of Wright, expressed in the quotation on page 25, that the most favorable conditions for effective natural selection are to be found in large, complex populations. This statement is equivalent to saying that the most rapid evolution is phyletic in nature.

In complementary fashion, population geneticists have commonly given speciation a gradualistic interpretation. Lewontin (1974, Chapter 4), for example, examined and rejected Mayr's phylogenetic scheme that lays great importance to rapidly divergent speciation. Dobzhansky (1972) referred to the slow divergence of new species as the "usual, and by now orthodox, view." Quoting this phrase, Nei (1975) expressed agreement with the gradualistic characterization of speciation. In assessing the status of evolutionary population genetics, Lewontin (1974, p. 159) wrote: *"We know nothing of the genetic changes that occur in species formation."* (Italics for emphasis are in the original.) This lack of knowledge reflects a traditional lack of effort. Even in the expanded version of his classic book on population genetics, R. A. Fisher (1958) devoted only three and one-half pages to the subject of speciation.

There are now signs that within the general field of genetics the tide of gradualism is receding. Until about 1970, the writings of Theodosius Dobzhansky, the most prominent experimental geneticist of the mid-twentieth century, were strongly gradualistic (for example, see Dobzhansky, 1970). Before his death, however, Dobzhansky came to entertain the idea that quantum speciation might be important. "New excitement in a new field" was his subtitle for a review of the nature of species and speciation in *Drosophila* (Dobzhansky, 1972). Dobzhansky's enthusiasm was kindled in large part by the studies of Hampton Carson on the rapid formation of species of *Drosophila* in Hawaii. These findings will be reviewed here in Chapter 6, as will the importance of regulatory genes (Britten and Davidson, 1969, 1971; Wilson *et al.*, 1977); these genes seem likely to play an important role in rapidly divergent speciation, but are not easily studied by traditional techniques. New ideas that chromosomal alteration may play a major part in rapid divergence will also be discussed. A few of these ideas, though certainly less extreme, are reminiscent of the earlier schemes of De Vries and Goldschmidt. The strong general rejection of these geneticists' radical ideas on the instantaneous origins of new higher taxa seems to have imparted a momentum that was partly responsible for carrying population genetics largely into the province of gradualism.

Human Ancestry

The gradualistic view has perhaps been graphically expressed most often where biology, anthropology, and paleontology converge: in the study of human origins. It has been standard practice in reconstructing the phylogeny of the Hominidae to place *Homo sapiens* at the Recent terminus of a continuous lineage tracing back through at least two species, and perhaps three, including *Australopithecus africanus* or a close relative. Typical reconstructions are presented in Figures 2-9 and 2-10. Debates over human phylogeny are closely tied to taxonomic questions, the fates of which are at the mercy of a particularly depauperate fossil record. As will be shown in Chapter 4, the record has recently yielded material that shows reconstructions like Figures 2-9 and 2-10 to be untenable. At present, however, our focus is on the history of the subject.

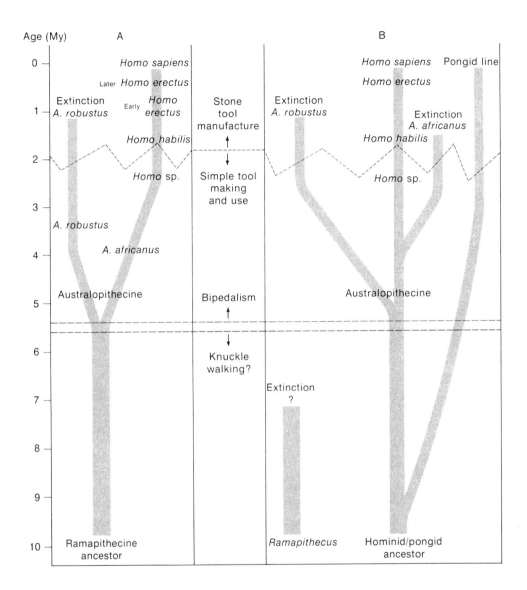

FIGURE 2-9
Two alternative, gradualistic reconstructions of hominid phylogeny. (From Clark, 1976; courtesy Glynn Ll. Isaac and Elizabeth R. McCown: *Human Origins*, copyright © 1976 by W. A. Benjamin, Inc.)

The belief that *Homo sapiens* should stand atop a solitary lineage was a natural outgrowth of our anthropocentric heritage, which portrayed humans as unique figures, closest to God. In the nineteenth century, our species preserved its superiority among earthly beings by conversion of the rigid *scala naturae* to a ladder of evolution—we still occupied the top rung. Still, the shocking impact of

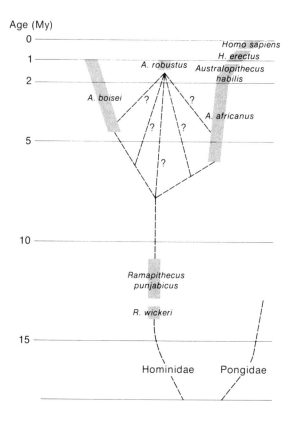

Age (My)

FIGURE 2-10
Gradualistic reconstruction of hominid phylogeny.
(From Buettner-Janusch, 1973.)

evolution on the traditional prejudice is perhaps reflected in Darwin's choice of a title for his book on human evolution. The first clause of the title is *The Descent of Man*.

The "single lineage" theme of hominid evolution culminated in the suggestion that trends in the hominid line of descent were not merely continuous, but also linear, or orthogenetic, in the descriptive sense of the term (Weidenreich, 1947). Beginning in the 1950's, however, even the less extreme forms of the "single lineage" idea were challenged by Louis Leakey and others, based on the discovery of fossil evidence for what was claimed to be a great variety of late Cenozoic hominids. This challenge met with opposition that, in retrospect, appears to have represented an overreaction fueled by taxonomic disagreement. Leakey was inclined to oversplit taxa (see, for example, Leakey, 1959), and other workers were even more guilty for the undue proliferation of new names, but in rejecting the introduction of names like *Zinjanthropus*, opponents of oversplitting went so far as to reaffirm the idea of a single line of descent (see, for example, Simons, 1963; Wolpoff, 1971). Defense of the "single lineage" view has, in part, been based on an early suggestion of Mayr (1950) that predated his full assertion of the punctuational scheme. Mayr noted that the competitive exclusion principle of ecology might oppose the notion that two hominid species could coexist.

Because it requires that all evolution be phyletic, the "single lineage" idea can only be entertained within an overview that is gradualistic in the extreme. Even after it was widely agreed that during the Pliocene two or more species of *Australopithecus* lived contemporaneously in Africa, the gradualistic model continued to prevail. The reaction to new fossil discoveries was simply to postulate the existence of more than one line of gradual descent (Figures 2-9 and 2-10). Gradualistic discussions of human evolution dominated the literature well into the 1970's (see, for example, Buettner-Janusch, 1973, pp. 274–275; Washburn and Moore, 1974, p. 119; Campbell, 1974, pp. 98–111; Clark, 1976, pp. 16–17; and Tobias, 1976, pp. 403–409).

As will be discussed in Chapter 4, the erosion of this viewpoint began with the discovery that hominid species traditionally assigned to a single phyletic succession leading to *Homo sapiens* (Figures 2-9 and 2-10) actually have overlapping stratigraphic ranges. The gradualistic scheme is further—and perhaps fatally—damaged by the observation that hominid species existed for extremely long intervals of geologic time. This observation refutes the idea that phyletic evolution proceeded rapidly; this is a striking revelation, because it remains obvious that the overall pace of change in human evolution was extremely fast. This argument will be developed more fully in Chapter 4, but its logic can be quickly grasped by evaluation of Figures 2-9, 2-10, and 2-11. It is widely agreed that the genus *Homo* evolved from *Australopithecus africanus*, a slightly more gracile (slender) form than the species depicted in Figure 2-11. The fossil record now indicates that *A. africanus* survived for perhaps as long as 3 My without changing enough for its lineage to be deemed a new chronospecies. In relying on phyletic evolution, a gradualistic interpretation would require that this creature, after this long period of virtual evolutionary stagnation, underwent abrupt phyletic transformation into the genus of man. The more palatable alternative is that *Homo* is not a lineal descendant of *A. africanus* but a dramatically new form that budded off by way of a small population, while the parent species persisted with little change. As will be shown in Chapter 4, the great longevity of hominid species in general seems to force this view upon us. It will, in fact, be shown that phyletic evolution has operated more slowly in the Hominidae than in an average family of mammals existing at the same time.

Coevolution

The term **coevolution** describes the complementary evolution of species that are intimately associated with one another. Some of the most striking examples

FIGURE 2-11
A reconstruction of the Plio-Pleistocene hominid species *Australopithecus boisei*. (Painting by Jay H. Matternes, courtesy of the National Museums of Kenya.)

involve relationships between flowering plants and their insect pollinators. The term "coevolution" attained popularity after appearing in the title of an important paper by Ehrlich and Raven (1964). It has since become customary in many circles to view species that today are found to have interlocking adaptations as having arrived at this condition through a process of gradual, correlated change (Leppik, 1957; Janzen, 1966; Southwood, 1973). For example, the idea here would be that if an insect species displays complex behavioral or structural adaptations for feeding upon the nectar of a particular plant species, and if the plant species displays special biochemical or structural adaptations to attract the insect species and to ensure that it transports pollen, then these interwoven adaptations will have evolved together over a long period of geologic time. Another example would be coevolution of a predator-prey system, in which a species of predator becomes more and more adept at capturing prey, and its chief prey species becomes more and more adept at avoiding or warding off the predator. The kind of phyletic race envisioned here is like the one that is assumed in the traditional analysis of sexuality, except that the coevolutionary race is usually seen as being a two-species event.

Thus, when the punctuational model is described to biologists, a common response is that the widespread occurrence of coevolution offers evidence contradictory to the model. It is little recognized that Ehrlich and Raven (1964, p. 605), who popularized the term "coevolution," described their concept as "a stepwise pattern of coevolutionary stages." Earlier workers such as Grant (1949) and Baker (1963) have also emphasized the saltational nature of much that is now called coevolution. Some of their interpretations will be presented in Chapter 6. For now, it will be sufficient to note that the widespread designation of coevolution as a slow phyletic process reflects a strong current of gradualism in modern biology.

THE STATUS OF QUANTUM SPECIATION IN PALEONTOLOGY

During the present century, the idea that adaptive innovations arise by rapid speciation has appeared sporadically in paleontology, without taking hold. Otto Schindewolf (1936, 1950) was a prime mover here, but as described earlier his views were extreme, in part reflecting the influence of DeVries and Goldschmidt. Schindewolf believed that a single *Grossmutation* could instantaneously yield a form representing a new family or order of animals. This view engendered such visions as the first bird hatching from a reptile egg. However unacceptable his explanations may have seemed, Schindewolf at least confronted the failure of the fossil record to document slow intergradations between higher taxa. As discussed above, Simpson addressed the same problem in 1944 with his less heterodox idea of quantum evolution. In his second book on large-scale evolution, Simpson (1953) cited the work of Schindewolf more frequently than that of any other author, but with strong expressions of disagreement.

Perhaps the extremism of Schindewolf's two major treatises published in 1950 elicited a gradualistic reaction. In any event, there are few vestiges of punctuational thinking in Anglo-American paleontologic literature published between 1950 and the early 1970's. The conventional gradualistic view of human evolution, which was discussed earlier in this chapter, falls partly within the sphere of paleontology and represents a prime example. There seems, however, to have simmered beneath the surface a measure of discontent with the Modern Synthesis. According to Olson (1960), who questioned the monolithic certainty of the synthetic theorists, most of the skeptics were morphologists and paleontologists:

> There exists, as well, a generally silent group of students engaged in biological pursuits who tend to disagree with much of the current thought but say and write little because they are not particularly interested, do not see that controversy over evolution is of any particular importance, or are so strongly in disagreement that it seems futile to undertake the monumental task of controverting the immense body of information and theory that exists in the formulation of modern thinking. It is, of course, difficult to judge the size and composition of this silent segment, but there is no doubt that the numbers are not inconsiderable. Wrong or right as such opinion may be, its existence is important and cannot be ignored or eliminated as a force in the study of evolution. (Olson, 1960, p. 523.)

Olson, of course, had in mind less extreme dissenters than Schindewolf. From this group, however, there emerged no punctuational scheme.

It is hardly surprising that paleontologists failed to adopt the suggestion in Mayr's paper of 1954 that rapidly divergent speciation by way of small populations could account for numerous gaps in the fossil record and for the sudden fossil appearance of evolutionary novelties. After all, this paper, which in effect proposed quantum speciation, was largely neontologic in nature and yet was even ignored by nearly all biologists. Unfortunately, even today Mayr's punctuational interpretation of the fossil record is not duly acknowledged in the paleontologic literature.

Thus, until very recently, modern paleontologists were trained in an atmosphere of gradualism. A particularly striking example of the prevailing view is the conclusion of Durham (1969; 1978) that metazoans must have originated many hundreds of millions of years before the Cambrian in order to have attained the high levels of complexity exhibited by Cambrian fossils. Assuming phyletic evolution to prevail, he noted that an average species of Cenozoic marine invertebrate has survived for something like 6 My. Applying this figure to early life, he concluded that hundreds of millions of years would have been required for phyletic evolution to produce a complex echinoderm. In effect, a large number of species would have had to be stacked end-to-end for phyletic evolution to achieve the transition.

The punctuational idea emerged as a more visible alternative to English-speaking paleontologists with the publications of Eldredge (1971) and Eldredge and Gould (1972). It is both interesting and surprising that, unkown to Americans, this view had previously gained support in the paleontologic community of the Soviet Union (Ruzhentsev, 1964; Nevesskaya, 1967; Ovcharenko, 1969). It was the initial view of Eldredge and Gould (1972, p. 99) "that the data of paleontology cannot decide which picture [punctuational or gradualistic] is more adequate ... [but] that the picture of punctuated equilibria is more in accord with the process of speciation as understood by modern evolutionists." More recently, these authors have countered many arguments of gradualists, adopting a more optimistic view of the potential role of paleontology (Gould and Eldredge, 1977). As of this writing, the controversy between punctuationalists and gradualists persists within paleontology, and any assessment of its status would be outdated before publication. Gould and Eldredge (1977) have provided an extensive review of recent gradualistic writings. It is safe to say, however, that no gradualist has provided a general paleontological argument based on comprehensive data.

SUMMARY

An important issue of evolutionary biology is whether most net evolutionary change in the history of life occurs in association with the branching of lineages (speciation) or by the gradual transformation of well-established species (phyletic evolution). The first alternative represents the punctuational model of evolution and the second, the gradualistic model. In subsequent chapters, evidence will be presented that the speciational component of evolution dominates, which is to say that the punctuational model is valid. Phyletic evolution accounts for enough change in many lineages that intergrading chronospecies are recognized, but, it will be argued, such evolution seldom produces major morphologic transitions. This view does not necessarily imply that most entities recognized as *species* arise by splitting of lineages, because it is possible that a small number of markedly divergent splitting episodes (quantum speciation events) account for more net change than a large number of gradual transitions between chronospecies.

The Modern Synthesis of the present century, like Darwinism of the last century, had a heavily gradualistic orientation. That the same persuasion has dominated modern biology is shown by: (1) the gradualistic nature of traditional explanations for the prevalence of sexual reproduction among higher organisms, (2) the general focus of theoretical population genetics, (3) the traditional representation of human evolution as having followed a single phyletic pathway from *Australopithecus africanus* to *Homo sapiens,* and (4) the common characterization of coevolution as a process of gradual transformation, whereby evolutionary changes within interacting species track each other over long periods of time.

Until recently, the punctuational model has been entertained only sporadically within paleontology. Just as the extreme punctuational schemes of the geneticists De Vries and Goldschmidt seem to have caused many biologists, and especially geneticists, to turn their backs on the possible significance of quantum speciation, the extreme punctuational scheme of the paleontologist Schindewolf may have engendered an overreaction among students of the fossil record, in favor of the gradualistic model. Only within the past few years has the possibility of a dominant role for quantum speciation been widely reconsidered.

3

Diverse Lines of Evidence

Some distinctive living species clearly originated in the very recent past, during brief instants of geologic time. Thus, quantum speciation is a real phenomenon. Chapters 4 through 6 provide evidence for the great importance of quantum speciation in macroevolution (for the validity of the punctuational model). Less conclusive evidence is as follows: (1) Very weak gene flow among populations of a species (a common phenomenon) argues against gradualism, because without efficient gene flow, phyletic evolution is stymied. (2) Many levels of spatial heterogeneity normally characterize populations in nature, and at some level, the conflict between gene flow between subpopulations and selection pressure within subpopulations should oppose evolutionary divergence of large segments of the gene pool; only small populations are likely to diverge rapidly. (3) Geographic clines, which seem to preserve in modern space changes that occurred in evolutionary time, can be viewed as supporting the punctuational model, because continuous clines that record gradual evolution within large populations represent gentle morphologic trends, while stepped clines seem to record rapid divergence of small populations. (4) Net morphologic changes along major phylogenetic pathways generally represent such miniscule mean selection coefficients that nonepisodic modes of transition are highly unlikely. Quantum speciation or stepwise evolution within lineages is implied. (5) The known fossil record fails to document a single example of phyletic evolution accomplishing a major morphologic transition and hence offers no evidence that the gradualistic model can be valid.

Evaluations of overall genetic distance, which ignore the fact that large evolutionary steps result from a very small number of regulatory genetic changes, have little bearing on the distribution of morphologic changes within phylogeny.

INTRODUCTION

Miscellaneous observations and inferences offer partial testimony to the relative importance of the phyletic and speciational components of evolution. Although it will be the purpose of this chapter to review these kinds of evidence, at the outset a negative note must be sounded. Scattered bits of data pertaining only to the pace of evolution within particular lineages or particular speciation events may tend to favor one point of view, but nonetheless fail to provide a comprehensive picture of the distribution of rates of evolution within phylogeny. Even taken collectively, these kinds of data tend to leave us with no more than a piecemeal approach to the problem.

Even a small set of lineages or speciation events may form a biased sample. For example, the employment of *Drosophila*, the genus of fruit flies, as the subject of a large proportion of all genetic studies has introduced a substantial bias to the study of evolution. *Drosophila* displays a striking lack of morphologic flexibility (Dobzhansky, 1956 and elsewhere). The genus contains over 1,000 species (Mayr, 1969b), and many of them belong to sets of nearly identical sibling species. Restricting one's studies to *Drosophila* diminishes the possibility of addressing the nature and incidence of quantum speciation.

What a case-by-case, piecemeal approach can provide is valuable evidence that certain kinds of things can happen, even if their frequency remains in question. If, for example, we establish that a living species has originated within the past few thousand years, we have direct evidence that speciation can occur during an interval that is brief on a geologic scale of time. On the other hand, if we find examples of rapid phyletic evolution in the fossil record, then we have evidence that the gradualistic model might be valid.

It should also be recognized that even when a comprehensive picture is unattainable, the accumulation of discrete examples can eventually form a strong statistical argument if an enormous quantity of evidence is found to weigh heavily on one side of an issue. Even if the evidence is known to be biased in some ways, a few exceptions to the prevailing kind of observation might be expected to appear. If none do, in time, a case begins to build against whatever idea exceptions would favor. As we shall see, this kind of argument may be emerging with respect to the general slowness of phyletic evolution; for although many phyletic lineages are well documented by fossil data, they do not display evolution sufficiently rapid to have produced higher taxa during intervals as brief as those comonly demanded for such transitions by other fossil taxa.

EVIDENCE OF RAPID SPECIATION

Is quantum speciation a real phenomenon or a hypothetical one conjured up to explain away certain dilemmas of evolutionary biology and paleobiology? Occasionally, information about the late Cenozoic physiographic history of an area reveals the nature of the speciation event that produced a modern species. We can begin to answer our question by examining this kind of evidence. We have available so few excellent examples that we are confined to a piecemeal approach. The opportunity is not to prove or to refute either the punctuational model or the gradualistic model but to demonstrate that the punctuational model can be valid—that quantum speciation can occur. To do so, an example must satisfy three criteria:

1. For an example amenable to dating, the process of divergence, normally by partial or complete physical isolation, must have begun only a short time ago (in the order of 10^3 to 10^4 years or less).

2. The population that has diverged must now be reproductively isolated from the ancestral species. Because we are considering living species, this condition is often testable.

3. The morphologic divergence must have been substantial (a quantum step).

With respect to the first condition, our present vantage point in geologic time is particularly fortunate. Biogeographic patterns have fluctuated markedly with the waxing and waning of Pleistocene continental glaciers. Many species have clearly arisen from populations isolated by geographic changes associated with glacial movements and concomitant shifts of climate.

A number of case histories satisfy, or seem to satisfy, only the first of the conditions listed here. In them, divergence began very recently, but only subspecific differences are recognized. One of the most striking is the origin of new subspecies of mammals on the island of Newfoundland (Cameron, 1958). These divergent populations have apparently been isolated for only about 12,000 years. Because a water barrier had to be crossed for their evolution from mainland ancestors, they were presumably founded by small numbers of individuals. It is quite possible that most of their divergent evolution was concentrated in an initial interval much briefer than 12,000 years, while populations were even smaller than they are today. Of 14 species represented, 10 have evolved into new subspecies. The Newfoundland beaver, *Castor canadensis caecator*, is almost so distinct as to warrant placement in a new species. Elsewhere in the world, the Faeroe Island house mouse, *Mus musculus faeroensis*, is so distinct that some authors have placed it in a new species (Huxley, 1942, p. 195), yet it was introduced not much more than 250 years ago! Mayr (1963, pp. 346–349; 1970, pp. 579–580) and Huxley (1942, pp. 194–196) have reviewed other examples. It should be appreciated that taxonomists seldom actually test reproductive incompatibility of populations considered to be separate species. Tests in captivity are indecisive because unnatural surroundings can alter behavior, and tests in nature are problematic because to be conclusive they require relocation of individuals without behavioral or ecological disturbance. On the other hand, if the populations in question are **sympatric** (occupy the same area), an absence of interbreeding can often be observed directly.

Even if ancestral and descendant populations are **allopatric** (have nonover-

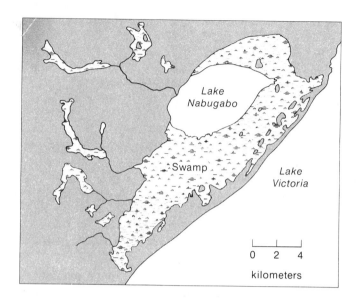

FIGURE 3-1
Map of Lake Nabugabo, Uganda, at the margin of Lake Victoria.
(From Greenwood, 1965.)

lapping geographic distributions), obvious differences in reproductive adaptations, including time of breeding and mating behavior, can attest to incompatibility. Particularly noteworthy from this standpoint, and for recency of occurrence, is the origin of five cichlid fish species of the genus *Haplochromis* in Lake Nabugabo, Uganda (Greenwood, 1965). This lake is the small remnant of a slightly larger body of water (Figure 3-1). The history of Lake Nabugabo is rather well known. The lake formed by growth of a longshore bar across an embayment of Lake Victoria. By radiocarbon analysis, the isolating event has been dated at roughly 4,000 years before the present. Among the fishes of Lake Nabugabo are five cichlid species that differ from all species of Lake Victoria. Each of the five species endemic to Nabugabo resembles a species in the parent lake, but differs as much from this apparent ancestral species as many species within Lake Victoria differ from each other. Furthermore, a unique male coloration has arisen in each of the Nabugabo species. This characteristic seems to confirm the attainment of reproductive isolation from counterparts in Lake Victoria, because ethological studies have shown male coloration to be an important feature of species discrimination in reproduction within the Cichlidae. Here, then, we have clear examples of speciation in less than 4,000 years. In fact, the morphologic differences now observed may have originated in considerably less time, but even 4,000 years is a miniscule span, in a geologic perspective.

Even tighter bracketing for an interval of speciation has been established for banana-feeding species of *Hedylepta*, a genus of Hawaiian moths, (Zimmerman, 1960). About 1,000 years ago, Polynesian immigrants introduced the banana plant to the Hawaiian Islands. Previously, no banana-like plants were found in the general vicinity. Five species of *Hedylepta*, and others not yet described, now form a clade of banana-feeding moths endemic to Hawaii. These species are most closely related to palm-eating species of the genus, only one of which occasionally feeds on bananas; but they themselves feed on nothing else. It seems evident that these species have arisen within the last 1,000 years, during which time bananas have spread widely throughout Hawaii, owing to human influence.

The planktonic copepod *Cyclops dimorphus* seems to have appeared in less than 30 years! This species was described by Kiefer in 1934, based on specimens from the Salton Sea, a large man-made body of water in California formed between 1905 and 1907. Johnson (1953) has redescribed the species and confirmed its distinctive nature.

Some of the examples of the previous paragraphs illustrate rapid speciation, but speciation that has been only weakly divergent. In other words, they do not satisfy the third criterion listed above. There are examples of more marked divergence that possibly, and in some cases definitely, represent quantum speciation. Auffenberg and Milstead (1965) have reported a likely instance of quantum speciation within *Terrapene*, a genus of box turtles. *Terrapene coahuila* now occupies springs and ponds in a small area with interior drainage in central Mexico. The species *Terrapene carolina* extends from the southeastern United States to within only several hundred miles of *T. coahuila* and is its nearest relative. It seems evident that *T. coahuila* arose from *T. carolina*, which spread well into Mexico during glacial intervals of the Pleistocene. *T. coahuila* seems to have evolved from a small relict population that was isolated by increasing aridity during an interglacial stage. It is a distinctive form and the only aquatic member of its genus, so that its origin seems indeed to have been a quantum event. Auffenberg and Milstead note that certain of its aquatic adaptations resemble characters found within the gene pool of *T. carolina*. These characters were apparently accentuated and fixed rapidly in the evolution of the peripheral population.

Barombi Mbo of West Cameroon, a crater lake only 3.5 kilometers in diameter, offers particularly striking evidence of quantum speciation (Trewavas *et al.*, 1972). This lake is apparently of late Quaternary age, perhaps being only a few hundred thousand years old. Its sudden volcanic origin precludes an ancient origin or complex history. Of 17 cichlid species living in Barombi Mbo, 12 are endemic and 7 of these are cichlids that belong to 4 endemic genera. So distinctive are these genera in comparison to species of neighboring faunas that their taxonomic origins are not entirely clear.

One might wish to question the time of divergence of some of the distinctive forms discussed above or to disparage the degree of divergence produced by those speciation events for which timing is well established. The pupfishes of Death Valley (Miller, 1950, 1961) seems invulnerable to both forms of skepti-

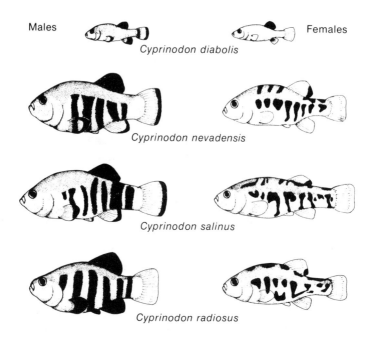

Males   Females
Cyprinodon diabolis

Cyprinodon nevadensis

Cyprinodon salinus

Cyprinodon radiosus

FIGURE 3-2
Four species of pupfishes that have evolved within the last 20,000 to 30,000 years in streams and thermal springs of the Death Valley region. (From J. H. Brown, "The Desert Pupfish," Copyright © 1971 by Scientific American, Inc. All rights reserved.)

cism. During the Pleistocene Epoch, lakes and large streams occupied lowland areas of the Great Basin of the western United States. Today there remain only small remnants of these large bodies of water. In Death Valley, isolated pools and hot springs now harbor small, relict populations of fishes, including members of the pupfish genus *Cyprinodon* that are assigned to four species (Figure 3-2). Of these, *C. diabolis* is the most distinctive. Its single population, which is confined to a thermal spring called Devil's Hole, has never been observed to grow larger than 500 individuals. The physiographic history of the region indicates that the four species of *Cyprinodon* have arisen during the past 20,000 to 30,000 years (Brown, 1971), since the last glacial interval of the Pleistocene ceased to affect the area (Brown, 1971; Turner, 1974). Miller (1950) noted that knowledge of the history of the group has led to a taxonomic bias: If a systematist unfamiliar with the species were to classify them, he would probably assign them to two or three separate genera! Presumably, the ancestors from which the species arose resembled the three larger species (Figure 3-2) more closely than they resemble *C. diabolis*. Certainly, the genus-level divergence of *C. diabolis* during 20,000 to 30,000 years represents quantum speciation.

Large African lakes have fostered remarkable adaptive radiations that bear special consideration here. In these lakes, members of the Cichlidae and other families have formed what are sometimes called species flocks or species swarms. These are groups of related endemic species that have proliferated rapidly on a geologic scale of time. Among the most remarkable is the cichlid flock of Lake Victoria, which is extremely well studied. It contains about 170 species, all but three of which are endemic, yet the lake is only 500,000 to 750,000 years old (Fryer and Iles, 1972; Greenwood, 1974). The three nonendemic species extend some distance upstream along rivers connected to the lake. The endemic radiation of the lake has produced striking evolutionary divergence of feeding structures and feeding habitats, but for the time being, Greenwood (1974) has assigned all but seven of the endemic species to the ancestral genus *Haplochromis*. Five of the other seven are placed in mono-specific genera. On purely morphologic grounds, numerous genera should be recognized. The range of morphologies is illustrated in Figure 3-3. Many of the species may have arisen by quantum speciation, yet the possibility remains that the lineages they represent attained reproductive isolation 500,000 or 750,000 years ago, and that their present degree of morphological divergence has been attained by gradual divergence during this interval. Even so, it will be shown in Chapter 9 that an average late Cenozoic species of freshwater fishes in the United States has survived for about 5 My, making marked divergence during even 500,000 or 750,000 years relatively rapid.

Lake Malawi (formerly Nyasa), which lies to the south of Lake Victoria, is considered to have formed earlier, about 1 to 2 My ago (Fryer and Iles, 1972, p. 465). Thus, it is no surprise that Malawi contains more endemic genera (20 compared to 6), although Greenwood believes that the differing geologic histories and ecologic conditions of the two lakes may have exerted an influence here as well. The origin of 20 genera in Malawi during only 1 to 2 My is suggestive of quantum speciation, but not conclusive. Taken at face value, the numbers alone do not rule out the possibility of divergence by rapid phyletic evolution during the entire history of the lake.

In fact, Fryer and Iles favor a gradualistic origin for the cichlid flocks of large African lakes, going so far as to question geologic evidence for the youth of certain lakes because they have harbored so much evolutionary change (Fryer and Iles, 1972, p. 463). In contrast, Greenwood (1974) has offered a punctuational interpretation, and an argument can be made for his view, based on the presence today of species that are believed to be the ancestral forms, or close relatives of the ancestral forms, of the markedly divergent species. There exist in Lakes Victoria, Malawi, Edward, George, Albert, and Turkana (Rudolf) morphologically primitive species that resemble each other. As Greenwood (1974) points out, it is reasonable that, having survived, the ancestral lineages would be widespread, because the rivers that originally drained the lakes all formed part of the east African highland drainage system. Relict streams of this system contain a population tentatively assigned to the species *Haplochromis bloyeti*, which Greenwood believes to represent the anatomical state of ancestors of the Lake

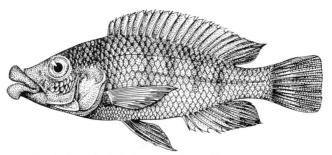

Haplochromis chilotes, a specialized insectivore ($\times \frac{2}{3}$).

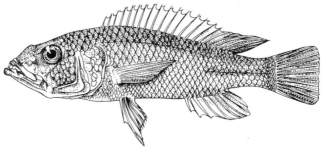

Haplochromis estor, a piscivore ($\times \frac{1}{3}$).

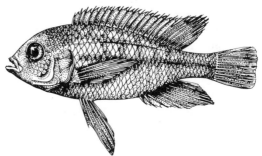

Haplochromis sauvagei, a mollusk eater ($\times \frac{2}{3}$).

FIGURE 3-3
Three species of cichlid fishes that have evolved within Lake Victoria, Uganda. (After Greenwood, 1974.)

Victoria flock. Greenwood favors the idea that the endemic cichlids of Lake Victoria are largely monophyletic, yet several species now living in the lake share the primitive characters of *H. bloyeti*. *H. callipterus* is a similar form, which lives in riverine habitats and in Lake Malawi; Fryer and Iles (1972) believe that it represents the ancestral condition of the endemic cichlids of Malawi.

A gradualist claiming that phyletic evolution has been responsible for most divergence in the great lakes of Africa would be hard pressed to explain the widespread persistence of primitive species. How can ancestral forms remain unchanged when they are supposed to have evolved phyletically into more specialized forms? The gradualist's only option would be to claim that some lineages have evolved incredibly rapidly while others have remained static. Why such a disparity in rates should occur remains unexplained. The general belief that the bulk of the Victoria species flock is monophyletic poses a real problem here. It is hard to believe that enough primitive species of similar morphology existed that some could have been transformed phyletically into strikingly different animals while others remain largely unchanged. The punctuational model provides a much less perplexing framework for understanding the present condition. Its basic tenet that lineages in general evolve slowly provides a ready explanation for the persistence of ancestral forms. If chronospecies survive for a million years or more without undergoing enough phyletic evolution to be considered new chronospecies and without being eliminated by termination of their lineages, we could expect the current species flocks of Lake Victoria and Lake Malawi to comprise both primitive species that happened to survive and specialized species that have budded off in stepwise fashion from the primitive forms. This composition is precisely what we now observe.

As has been noted above, the fossil record offers little promise for extensive documentation of quantum speciation. Even so, one possible example has been reported (Ovcharenko, 1969). The two brachiopod species *Kutchithyris acutiplicata* and *K. euryptycha* are well represented in Jurassic deposits of the Soviet Union. In one small area, in a marly bed only 1 to 1.5 meters thick, transitional forms are found. The lower part of this bed contains only *K. acutiplicata*, and the upper part, only *K. euryptycha*. In the middle, there occur transitional forms, most of which are concentrated in a layer only about 100 centimeters thick. While the exact duration of the apparent speciation event is unknown, it must have been brief, and divergence was substantial. The species are distinct enough to have been regarded by one early worker, albeit a taxonomic splitter, as belonging to separate genera.

It should be clear from the various examples discussed above that rapidly divergent speciation does occur. However it may be brought about and whatever its importance may be in the history of life, quantum speciation is a real phenomenon. Possible mechanisms by which it is accomplished and the role that it plays in macroevolution will be examined in Chapters 5, 6, and 7.

STABILITY AND GENE FLOW

As was discussed in the previous chapter, Mayr (1954) suggested that various subpopulations of a large species must be subjected to differing selection pressures and that gene flow among the subpopulations may represent a significant stabilizing force, opposing the rapid evolution of the entire species in any particular direction. More recently, some support has been withdrawn from this idea (Mayr, 1970; Eldredge and Gould, 1972) and the unity of the genotype (page 24) has been considered the primary source of species stability. This shift of emphasis reflects evidence, partly summarized by Ehrlich and Raven (1969; see also Endler, 1973; 1977) that there is only limited gene flow among the subpopulations of many species in nature. The alleged problem here is that there is insufficient gene flow to confer stability.

This problem may be more apparent than real (Stanley, 1978). Even localized gene flow should contribute greatly to the stability of a species. Inevitably, a unique set of selection pressures will bear on each deme of a species, depending on the nature of local food sources, competitors, predators, and conditions of the physical environment. If, however, every deme exchanges genetic material with a few others, even at a modest rate, stability should be conferred. All that should be necessary is for modest gene flow to occur among neighboring demes. It is immaterial how long is required for a new mutation or gene combination to diffuse throughout the *entire* range of the species. An important point here is that spatial clustering of individuals occurs on several scales within most species. An example is depicted in Figure 3-4. Individual demes at any level will occupy varied environments and experience differing selection pressures. There will normally exist one or more levels of clustering at which the conflict between local selection pressures and gene flow will tend to stymie phyletic evolution.

The general point of the preceding paragraph can be illustrated by concrete examples. Since the Isthmus of Panama formed about 3.1 My ago (Keigwin, 1978), marine populations of many species on either side have not diverged appreciably. This fact should not be surprising if, before division, each species was distributed as semi-isolated demes and if phyletic evolution was braked by gene flow among neighboring demes. Dividing each large species in half would not eliminate the stabilizing mechanism, but would simply separate its operation into two parts.

At the other end of the geographic scale, we can consider the implications of the local selection pressures that have been observed to operate on freshwater zooplankters, like *Bosmina longirostris*, a typical cladoceran crustacean. In Union Bay, a temperate embayment of Lake Washington in the American Northwest, predation pressure maintains a spatial dimorphism within populations of *B. longirostris* (Kerfoot, 1975). Inshore populations are heavily preyed upon by fishes, which select the individuals that are visually most conspicuous—those with long mucrones and antennules (Figure 3-5). Offshore populations are heavily preyed upon in a different manner, by grasping copepod crustaceans, which find it relatively difficult to attack individuals with long mucrones and antennules and therefore selectively remove animals with short features. The result of the two modes of selective predation is a dimorphism, proven to be underlain by genetic differentiation, in which populations tend to be short-featured inshore and long-featured offshore. *Bosmina longirostris* is broadly distributed, occurring

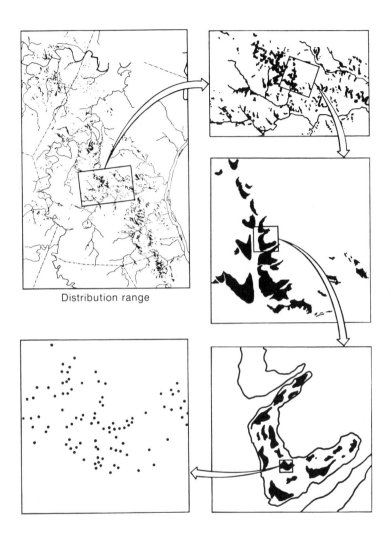

FIGURE 3-4
Levels of clustering in the distribution of the plant *Clematis fremontii.*
Dots (lower left) depict individual plants. (Courtesy of Ralph O.
Erickson.)

in temperate and tropical zones of Europe, Africa and the Americas. In Gatun
Lake, Panama, and perhaps elsewhere, heavy fish predation results in a decreased mean diameter of eye pigment in *B. longirostris* (Zaret and Kerfoot,
1975). The pigment serves as a target for visually searching fishes. Like other
cladocerans, *B. longirostris* commonly also occupies ephemeral bodies of water in

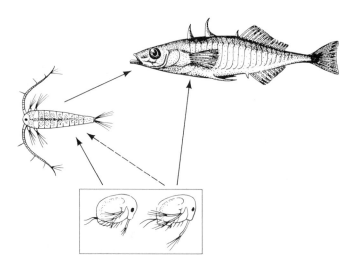

FIGURE 3-5
Diagrammatic representation of the impact of predation on the
morphology of *Bosmina longirostris*. The predatory copepod *Epischura* feeds preferentially on short-featured forms of *Bosmina* (enclosed in box). The stickleback fish feeds preferentially on long-
featured forms of *Bosmina* and upon *Epischura*. (After Kerfoot,
1975.)

which predation is lacking. In any sizable region within its range, the species
occupies many bodies of water, each of which must harbor a unique predatory
fauna or none at all. There is normally substantial dispersal and gene flow over
moderate distances, as shown by the enormous geographic range of the species
and by the great capacity for dispersal that it shares with other cladocerans by
virtue of the production of resting eggs, which withstand freezing, desiccation,
and transport by wind. The species rapidly invades newly formed bodies of
water. The complex conflict between divergent local selection pressures and
gene flow must result in very slow net evolution within any region. Therefore,
even if there is little gene flow across enormous barriers—say between Africa and
North America—it is difficult to imagine the complex population of one continent diverging rapidly from that of another. Similarly, gene flow may be extremely slow from end to end across a long chain of populations, as between
Washington State and Panama, yet regional genetic exchange within areas all
along the chain should prevent the rapid evolutionary divergence of any population. Certainly, the entire species cannot undergo rapid phyletic evolution.
    Thus, the direct observation that low rates of gene flow characterize many

living species (Ehrlich and Raven, 1969; Endler, 1973; 1977) becomes an argument against the gradualistic model (Stanley, 1978). Typically, a potentially useful new mutation or gene combination must diffuse very slowly through the populations of a species, and this condition precludes rapid phyletic evolution. (It is exceedingly unlikely that all subpopulations of a large species would coincidentally evolve rapidly in a particular direction, in parallel, without gene flow.) It is only when a small interbreeding subpopulation departs reproductively from the remainder of a species that a new feature, or a new constellation of features, can become fixed rapidly within an entire species (which the subpopulation soon represents). Then, as this subpopulation expands and develops a fragmented, heterogeneous distribution of its own, it too will become stabilized, to contribute major adaptive shifts only by casting off descendant species by way of small populations throughout which dramatically new features are able to spread.

In light of the foregoing discussion, it seems reasonable to suggest that in a typical species gene flow may be sufficiently weak over *long* distances that clines and genetically distinctive clusters of populations are common, but sufficiently strong over *short* distances that successful speciation is rare. Evidence for the rarity of successful speciation will be discussed in the following chapters.

CLINES AND RING SPECIES

We have, thus far, considered only evidence for the occurrence of rapid evolution in the process of speciation. Some workers might claim that the other component of large-scale evolution—the phyletic component—is shown to be of considerable significance by the presence of clines. A **cline** (Huxley, 1939) is a geographic gradient of variability in some character within a population. Clines are quite common in nature. Frequently, they parallel environmental gradients, as in the example depicted in Figure 3-6. In effect, a cline represents the result of two opposing processes: natural selection, which tends to adapt subpopulations to local environments, and gene flow, which tends to homogenize the total population. A cline preserves a temporal evolutionary change as a spatial gradient. The question, then, is whether the commonplace occurrence of clines shows phyletic evolution to be of great importance in nature.

There is general agreement that most smooth, continuous clines, like that shown in Figure 3-6, have arisen gradually, while all populations involved have remained in nearly continuous contact (Mayr, 1963, pp. 369 and 380). On the other hand, character gradients sloping so steeply as to represent abrupt geographic discontinuities are the subject of disagreement. The traditional view (Mayr, 1963, pp. 369 and 380–384) has been that such stepped clines must have

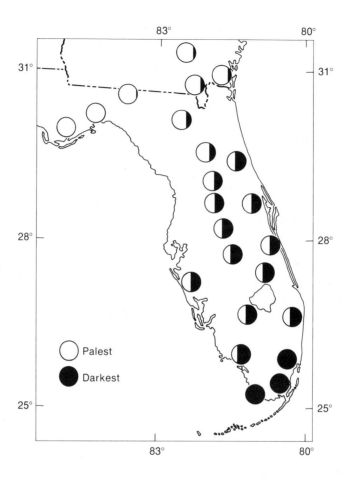

FIGURE 3-6
Cline in the coloration of the water rat *Neofiber alleni* in Florida and
Georgia. (From Burt, 1954.)

formed by the secondary contact of populations that diverged **allopatrically**
(while entirely isolated from one another). Endler (1977) has argued in favor of
an alternative mode of origin. One of his conclusions is that stepped clines can
originate as easily without geographic separation as with it. Another is that such
clines can exist only if different selective regimes obtain on either side of the zone
of contact. Thus, he suggests that many stepped clines have arisen by rapid
parapatric divergence, or divergence of a local population while it remains in
contact with the one from which it diverges. His conclusions are based on the
idea that gene flow is so weak among natural populations of a typical species that

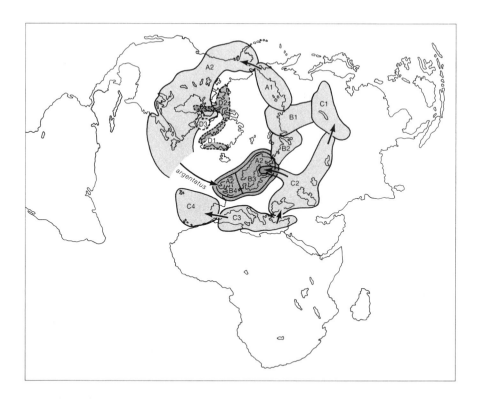

FIGURE 3-7

Distribution of a circumpolar ring species of gulls, the *Larus argentatus* complex. Letters denote species and subspecies; numbers identify distinct populations. The *argentatus* subspecies A, B, and C apparently evolved as Pleistocene isolates. D represents a separate species (*L. glaucoides*) that evolved in North America. After the Pleistocene, subspecies A spread from a north Pacific refuge across North America and into western Europe where it became sympatric with subspecies B, from which it remains distinct. (From Mayr, 1963.)

local selection pressures are generally strong enough to prevent homogenization by interbreeding.

What is important to the present discussion is not the relative frequencies of the two proposed modes of stepped cline formation, but the fact that in both the allopatric and parapatric modes, stepped clines are seen as forming rapidly on a geologic scale of time. In fact, Endler (1977, pp. 174–178) observes that it is difficult to assess the relative merits of the two models empirically, because they are considered to follow similar biogeographic patterns that entail rapid divergence of small populations.

Far from establishing a dominant role for phyletic transformation, the formation of clines may be quite compatible with the punctuational view. Smooth, uninterrupted clines (Figure 3-6) may indeed be the horizontal tracings of gradual evolution, but they seem invariably to be gentle. Seldom, if ever, do their end members represent a degree of adaptive separation equivalent to that produced by a single quantum speciation event—for example a genus-level transition. Rather, they represent minor variation in body size, color pattern, breeding season, or some other feature that, it will be suggested in the following chapter, represents the sort of fine tuning of adaptation that phyletic evolution may indeed accomplish. Similarly, stepped clines seem to be the geographic products of abrupt evolutionary transitions that have not quite represented full speciation events. It is not difficult to extrapolate from the mechanism of their formation to saltational speciation of the sort described earlier in this chapter. This fact has been fully recognized by both Mayr and Endler, whose writings will be cited again in Chapter 6 where quantum speciation will be considered in greater detail.

Ring species offer similar evidence. A ring species is a circular chain of partly intergrading subspecies (Figure 3-7). It documents the spread of populations around an inhospitable region. The oldest and youngest populations, which overlap, cannot interbreed. At first glance, ring species might be regarded as exemplifying extensive phyletic evolution; but quite to the contrary, Mayr (1942, 1970) has cited them as offering strong evidence of the importance of geographic isolation to rapid evolution and speciation. The rationale here is that in nearly all ring species there seems to be evidence of major spatial gaps, in the past if not in the present, indicating that subspecies arose by incomplete geographic speciation. The accumulation of evolutionary steps from subspecies to subspecies now prevents the terminal members from interbreeding. A parapatric origin of the subspecies, in accord with Endler's model for the origin of stepped clines, would be comparably punctuational in character.

GENETIC DISTANCE

In the decade following the pioneering work of Harris (1966) and Hubby and Lewontin (1966), the technique of gel electrophoresis was widely employed to evaluate genetic differences among many different groups of organisms. Basically, electrophoresis provides for detection of the products (primarily polypeptides and proteins) of various genetic alleles. Unfortunately, this approach has turned out to have limited value in the study of large-scale evolution. A measure of its restricted utility is illustrated by the fact that electrophoresis, used in combination with immunological techniques and other methods, has revealed that the average human polypeptide is 99 percent identical in composition to the equivalent polypeptide of a chimpanzee (King and Wilson, 1975). Certainly, we are entitled to question the macroevolutionary significance of studies showing that estimated degree of genetic divergence of two species correlates better with longevity of phyletic separation than with number of intervening speciational steps (Avise, 1977). We have no evidence that the measured genetic changes are of a kind that yields substantial morphologic transition. Even if adaptive, they may provide only the fine tuning of lineages.

A basic problem here seems to lie in the strong influence of **regulatory genes** on morphogenesis. These genes generally influence the expression of other genes without yielding enzymes or other products detectable by electrophoresis. The part they may play in macroevolution will be discussed more fully in Chapter 6, which will include, as a prime example, the origin of the giant panda, *Ailuropoda*. So distinctive are the superficial morphologic features of this animal that for years there was debate as to whether it should be assigned to the raccoon family or the bear family. Through detailed anatomical studies, Davis (1964) conclusively showed it to be a derivative of the bears, but one so distinctive in form as to represent a monogeneric subfamily or even a separate family. Even so, the singular features of *Ailuropoda* have evolved by means of extremely few genetic alterations, with these apparently being changes in regulatory genes that have operated pleiotropically upon integrated morphogenetic systems. Biochemical analysis confirms the close affinities between bears and the giant panda, but fails to reflect the enormous degree of morphologic divergence of the latter (Sarich, 1976), just as it fails to reflect the great phenotypic difference between humans and chimpanzees.

Up to the present time, attempts to compare the genetics of phyletic evolution and speciation have been hampered by a poor understanding of epistatic interactions among genes and, in particular, of the operations of regulatory genes (Ayala, 1975, p. 62). Lewontin (1974, p. 12) has written, "While population genetics has a great deal to say about changes or stability of the frequencies of genes in populations and about the rate of divergence of gene frequencies in populations partly or wholly isolated from each other, it has contributed little to our understanding of speciation...." Bush (1975, p. 339) has concluded that "at the molecular level, proteins used in the current flush of allozyme studies ... have almost no bearing on speciation."

The importance of regulatory genes and of epistatic interactions among genes pose serious problems for mathematical genetics, yet these problems have not been universally acknowledged. Nei (1975, p. 175) recently wrote, "From the standpoint of genetics the most appropriate measure of genetic distance would be the number of nucleotide or codon differences per unit length of DNA." This statement raises a question as to just what should be the "standpoint of genetics." With a graphic simile, Mayr (1970, p. 321–322) has expressed the view to which I subscribe:

> Indeed, it is becoming increasingly evident that an approach that merely counts the number of gene differences is meaningless, if not misleading.... Nor can species

differences be expressed in terms of genetic bits of information, the nucleotide pairs of DNA.... That would be quite as absurd as trying to express the difference between the Bible and Dante's *Divina Comedia* in terms of the difference in the frequency of the letters of the alphabet used in the two works.

And Lewontin (1974, p. 20) expresses this overview with equal force:

To concentrate only on genetic change, without attempting to relate it to the kinds of physiological, morphogenetic, and behavioral evolution that are manifest in the fossil record and in the diversity of extant organisms and communities, is to forget entirely what it is we are trying to explain in the first place.

Genomic components have significance only in terms of phenotypic expression. A bear probably has been transformed into a panda by a few genetic alterations, but the result is an enormous amount of adaptive change, not a little. The notion that rates of evolution ideally should be measured by genomic rather than by morphological parameters (Schopf *et al.*, 1975) excludes from consideration the phenotype, upon which selection operates. We desire to understand the genetic mechanism of major evolutionary transformations of the sort that occurred in the origin of the giant panda, but the kinds of genetic information to be sought can be gleaned only through study of phenotypic change. Davis (1964) distilled from the anatomy of the giant panda a few fundamental genetic questions, the answers to which, as will be discussed in Chapter 6, would shed much light on the nature of large-scale morphogenetic transformation.

PHYLETIC RATES OF PHENOTYPIC EVOLUTION

Whether fast or slow on a geologic scale of time, the pace of phyletic evolution is too sluggish to be measured comprehensively anywhere other than in the fossil record. In a particularly innovative paper, Lande (1976) evaluated net phenotypic change for a number of large-scale evolutionary transitions in well-studied Cenozoic taxa of mammals, including horses and oreodonts. The pattern of phylogeny accounting for these typical mammalian trends was not considered in detail, only the morphologic starting and end points. Lande's important finding was that net rates of conspicuous phenotypic change were typically so slow that, if evolution were continuous and constant in rate, they could have resulted from only about one selective death per million individuals per generation! Such selection pressures are so weak that Lande concluded that the analysis of rates alone offers no reason for excluding random drift as the primary agent of change. Such drift, he noted, could have taken place within the main populations of species or within isolated populations.

Viewed in another way, the calculations of Lande seem to deliver a severe blow to the gradualistic model. So many chance processes operate in nature that the probability is vanishingly small that an infinitesmimal selection coefficient will be maintained for 10 to 50 My, the approximate range of time intervals considered. We can invoke selection as the primary source of change only if its operation is highly episodic and therefore strong enough, over short intervals, to be realistically envisioned. Thus, the calculations are quite compatible with the punctuational model, a corollary of which is that stabilizing selection pressures prevail within large, established species. Genetic drift in small populations would not, of course, be ruled out *a priori* for a punctuational scheme.

The preceding calculations would seem to leave the gradualist only two recourses. One would be to claim that selection pressures fluctuate so markedly within lineages that phyletic evolution proceeds in a stepwise fashion. The most likely possibility here would be that lineages are periodically constricted to small population size, allowing for local selection pressures to operate intensively and consistently for brief intervals. Arguments against this possibility will be presented in the following chapter. The alternative would be to invoke genetic drift in single lineages, whether constricted or not, as the primary source of conspicuous phyletic trends. Our knowledge of adaptive advancement within numerous taxa makes this an absurd proposition.

The inference that evolution is episodic seems to be borne out by the study of discrete lineages. For some time it has been appreciated by at least some paleontologists that many recognized lineages exhibit very little morphologic change (Matthew, 1910; Zeuner, 1958 pp. 391–393, and other editions; MacGillavry, 1968). Rates of morphologic change have been measured for a number of apparent lineages. The word "apparent" is critical here. One of the problems lies in demonstrating that two or more populations being compared do indeed belong to a single lineage. Gould and Eldredge (1977) recently reviewed numerous alleged examples of phyletic evolution. For each one, they concluded either that the lineage in question has not been well documented or that the rate of evolution displayed is extremely slow. At present, one of the most thoroughly documented examples of phyletic trends is within the Permian foraminiferan *Lepidolina multiseptata* (Ozawa, 1975), as illustrated in Figure 3-8. Anticipating the approach to be taken in the following chapter, let me point out that this species survived for a very long interval of time—some 15 My—so that, in the context of large-scale evolution, the degree of morphologic transformation represented in Figure 3-8 is trivial.

Perhaps the variety of net morphologic change most commonly observed in lineages is change in body size or its equivalent. Two examples are illustrated in Figures 3-9 and 3-10. Certainly, change in size may be quite important ecologically. Hutchinson (1959) concluded that niche separation by feeding specialization is commonly achieved when one species exceeds a similar one in size by a factor of about 1.2. Even so, change of size by no means represents the sort of morphologic transition that achieves the crossing of a major adaptive threshold. Furthermore, the rates of change illustrated in Figures 3-9 and 3-10, like that shown in Figure 3-8, are very slow.

The question is whether, if we find only weak phyletic trends in the fossil

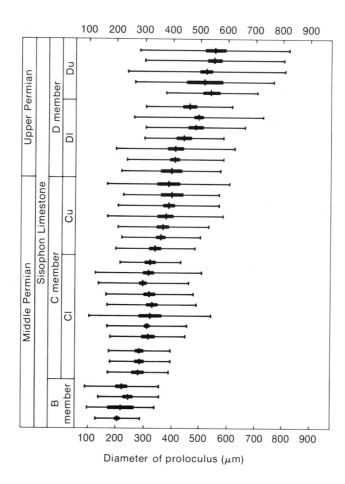

FIGURE 3-8
Evolutionary increase in the diameter of the proloculus (initial chamber) of the fusulinid foraminiferan *Lepidolina multiseptata*. For each population, the horizontal line represents the range of sizes, and the black rectangle depicts the 95 percent confidence limits for the mean, which is indicated by the vertical line. The six clusters of populations are in stratigraphic succession, but within each cluster, the vertical ordering of populations is arbitrary. (From Ozawa, 1975.)

record, we have conclusive evidence against phyletic gradualism. If we confine ourselves to a case-by-case consideration of apparent fossil lineages, is the fossil record complete enough to build such an argument? Were Lyell and Darwin correct in challenging the record's adequacy? My view is that apparent phyletic trends can be analyzed statistically and continually added to our store of exam-

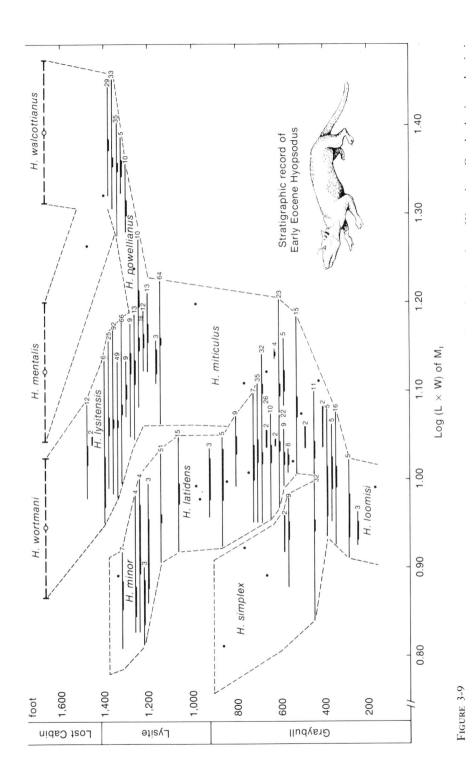

FIGURE 3-9

Changes in size within a phylogeny reconstructed for the Eocene condylarth *Hyopsodus* in northern Wyoming. On the horizontal axis is plotted the logarithm of the product of length and width (in millimeters) of the first lower molar (log L × W) of M₁. For each sample, the heavy bar represents the standard error and the light bar, the range of values. Numbers to the right indicate number of specimens in the sample; dots represent single samples. Open circles and horizontal dashed lines at the top are means and expected ranges for species poorly represented in the stratigraphic section but well studied elsewhere. (From Gingerich, 1974.)

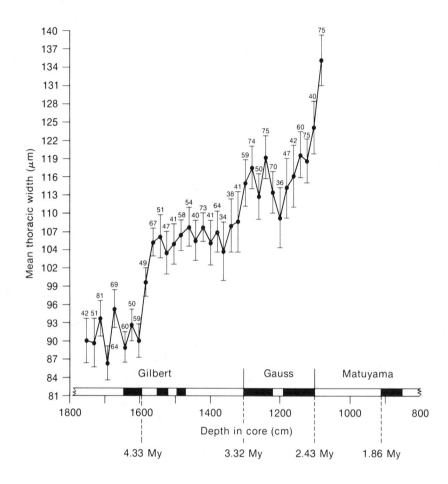

FIGURE 3-10
Increase in thoracic width of the Antarctic radiolarian *Pseudocubus vema* during an interval of about 2.5 My. Alternating black and white bars depict reversals of the earth's magnetic field that have been measured in deep-sea cores from which samples were obtained. (From Kellogg, 1975.)

ples, but even if only sluggish evolution continues to be recorded, it seems unlikely that in the near future the collective evidence will persuade advocates of the gradualistic model to abandon their views. As discussed at the start of this chapter, this kind of piecemeal approach, employing negative evidence, leaves open the possibility that our sample of data is biased. Perhaps more rapid evolution than we have discovered actually occurred in segments of time and space for which good fossil samples are lacking. I would agree with Gould and Eldredge (1977) that all examples brought forward to date in support of the gradualistic model either lack convincing documentation or represent very slow evolution. The case seems to be building, and it may be that in time an absence of unequivocal examples for rapid morphological change will be seen as forming a convincing argument. The primary approach to be followed here, in the following chapter, offers a more immediate test, making use of comprehensive taxonomic data rather than scattered morphologic information. Additional lines of evidence from the fossil record will be considered in Chapter 5.

SUMMARY

Though offering no final resolution, various kinds of fragmentary data shed light on the relative merits of the gradualistic and punctuational models. Some seem to weigh on the punctuational side of the controversy, and others at least establish that quantum speciation is a real rather than hypothetical phenomenon. For example, there is clear evidence that some events of divergent speciation have occurred since the withdrawal of Wisconsin ice sheets. Less direct evidence of quantum speciation is found in essentially monophyletic radiations of cichlid fishes in geologically young large lakes of Africa. Here primitive, seemingly ancestral lineages exist alongside strikingly deviant species, suggesting that the latter have evolved by stepwise branching, not by gradual transformation of ancestral stocks.

Evidence that intraspecific gene flow is very slow in nature seems to oppose the gradualistic model. Many large, isolated populations are known to have diverged slowly, probably because, on some scale, gene flow among neighboring subpopulations conflicts with divergent local selection pressures to prevent rapid unidirectional change within the entire system. It does not matter that new genetic features diffuse very slowly across the geographic ranges of many species. In fact, this condition represents evidence opposing the gradualistic view, because phyletic evolution requires that useful new genetic features spread throughout a species' gene pool. It seems likely that dramatically new adaptations can generally become fixed within a species only if they arise in a small, local population that is ancestral to the species.

Gradual, continuous clines seem to record change that is equivalent to phyletic evolution, but with selection being imposed by spatial, rather than temporal, environmental change. The gradients of such change are gentle, however, and the end-member populations display only minor morphologic separation. Stepped clines manifest the recent operation of more rapid evolution in small populations, suggesting the possibility of an important role for quantum speciation.

The field of genetics has shed little light on the nature of importance of quantum speciation, in part because the role of regulatory genes in large-scale adaptive transitions has only recently become apparent. Simple estimates of overall genetic distance between species reveal little about degrees and rates of morphologic divergence.

When we consider the *net* degree of morphologic change between some ancestral mammal species and their living descendants, the *mean* number of selective deaths per generation is extraordinarily small. Maintenance of continuous selection pressures as feeble as the measured means is virtually impossible. Excluding genetic drift from consideration because of the adaptive nature of the transitions, we must invoke either stepwise evolution within lineages or quantum speciation.

The fossil record has failed to yield unequivocal examples of rapid phyletic evolution that produced marked transformations in morphology. Whether this condition militates against the gradualistic model depends upon the quality of the record. In the following chapter, taxonomic, rather than morphologic, fossil data will be employed in order to undertake a comprehensive analysis of this question.

4

Longevity of Chronospecies and Rates of Large-Scale Evolution

The fossil record demonstrates that phyletic evolution (evolution within established species) proceeds very slowly. Especially in episodes of adaptive radiation, origins of higher taxa have generally been too rapid to be attributed to phyletic transformation. Particularly convincing evidence of this relationship is offered by Plio-Pleistocene mammals, for which it is possible to construct histograms representing chronospecies longevities and rates of phyletic evolution. The measured rates are much too slow to account for the origins of genera that appeared during the Pleistocene. Most of these, and in fact most genera of animals in general, must have formed rapidly, by divergent speciation. If typical for genera, this mode of origin must characterize the origins of families and other still higher taxa. Phyletic evolution in the Hominidae (human family) has apparently been even slower than that for other groups of the Mammalia, implying a punctuational pattern in the ancestry of Homo sapiens. *The fossil record of marine invertebrates is also complete enough on an appropriate scale to document generally slow rates of phyletic transition. For both vertebrates and invertebrates, increase in body size may be the most conspicuous form of structural change accomplished by phyletic evolution.*

INTRODUCTION

The traditional employment of higher taxa, rather than species, as units in the study of large-scale evolution can in part be attributed to what is termed the species problem of paleontology: the taxonomic difficulty imposed by the temporal dimension of the fossil record. This added dimension makes the designation of species a highly subjective matter. Although this subjectivity has clearly deterred paleontologists from studying large-scale evolution at the level of the species, the seemingly problematic temporal dimension can actually be turned to our advantage. In particular, by studying chronospecies, we can assess rates of evolution within lineages. No such analysis can be undertaken through the study of higher taxa.

Several years ago, Zeuner (1959) wrote, "I am confident that a thorough investigation of the Pleistocene fauna will, in the long run, provide most valuable information concerning the evolution of new species." He took note of the great temporal durations of many Pleistocene species and drew from this observation the punctuational inference that a typical species may evolve rapidly near the time of its origin, and slowly thereafter. It is this line of reasoning that will be pursued in the present chapter. The basic approach will be slightly more oblique than some of the diverse avenues explored in the preceding chapter, but more comprehensive and therefore, I believe, more compelling. The techniques that follow are based on a simple deduction: If phyletic evolution proceeds at a rapid pace, the resulting lineages, if documented by fossil data, will necessarily be divided into chronospecies of short duration. Figure 2-1 illustrates this point. The assumption is that a chronospecies encompasses a particular range of morphologic variation. The degree of variation will, of course, depend on the species concept of the worker who is evaluating the lineages considered; but reliance on the taxonomic efforts of a single worker will provide internal consistency.

An attempt to assess the gradualistic model by measuring longevities of a small sample of chronospecies would suffer from the kind of deficiency attributed in the previous chapter to other approaches: It would represent a piecemeal effort. Suppose, however, that a nearly complete evaluation could be provided—that within a large segment of phylogeny the duration of nearly all chronospecies could be measured. Imagine that nearly all of these durations turned out to be long relative to intervals of time required for the origin of higher taxa known to have formed within the phylogeny. Then we would be forced to conclude that the major evolutionary changes producing the higher taxa were concentrated in speciation events (unless entire lineages are commonly transformed quite abruptly—a prospect that will be deprecated in the next chapter). This hypothetical approach represents the kind of analysis that will be undertaken in the present chapter. To place the analysis in perspective, it will be necessary to examine the phenomenon known as adaptive radiation.

THE NATURE OF ADAPTIVE RADIATION

As discussed extensively by Simpson (1944; 1953), most evolutionary change occurs during **adaptive radiation.** Adaptive radiation is the rapid proliferation of new taxa from a single ancestral group. In other words, rapid evolution usually comes in bursts, with not just one new kind of organism appearing, but with many new kinds diverging from an ancestral group. An example is depicted in Figure 4-1, where we see the pattern of diversification of the scleractinian corals, or hexacorals. This group arose in the mid-Triassic, about 215 My ago. Of the nine living superfamilies of corals, six had arisen by about the beginning of the Jurassic, after only about 20 My of evolution. Only two more superfamilies appeared during the Jurassic, one during the Cretaceous, and none during the entire Cenozoic.

It is often pointed out that our retrospective view leads us to push the origins of higher taxa backward in geologic time. The idea is that distinctive groups that deserve recognition as discrete higher taxa are frequently denied this recognition because they have evolved recently in geologic time. The concept of the clade is useful here. When we see that a clade has become distinctive because of its size (number of lineages) or other biologic properties, we tend to classify it as a discrete higher taxon. This kind of taxonomic decision has been made at the present time, T_2 in the hypothetical example depicted in Figure 4-2. Had we lived at an earlier time, such as T_1, we might have tended to lump the descendant clade within the higher taxon (A) from which it arose. The reason for this hypothetical consolidation would have been that the incipient clade was less conspicuous in containing fewer species. Its representatives might also on the average have been less different from those of the ancestral group than are its more numerous representatives at T_2. On the other hand, phylogenies like the one displayed in Figure 4-1 show that this kind of taxonomic bias often has only a minor influence on the placement of taxonomic boundaries in geologic time. It is true that if we were living in the earliest Jurassic we might not recognize as many as seven superfamilies of corals because many of the groups now recognized as superfamilies were then only incipient clades. On the other hand, by the Middle or Late Jurassic, these clades were large and discrete. The taxonomists who today recognize them as superfamilies would certainly also have done so then. Clearly, during the last 150 My, the Scleractinia simply have not undergone adaptive shifts as dramatic as the shifts that characterized the initial phase of their phylogeny. Much more time has expired, yet there have appeared very few body plans distinctive enough to be assigned to new superfamilies.

The early portion of scleractinian phylogeny typifies early adaptive radiation in general. In some instances, the apparent rate of initial adaptive radiation is even more remarkable. Trilobites are unknown before the beginning of the Cambrian, and yet, during the Early Cambrian, which lasted between 15 My and 30 My, about thirty families made an appearance (Harrington, 1959). The ammonoids nearly became extinct at the end of the Permian. Renewed radiation

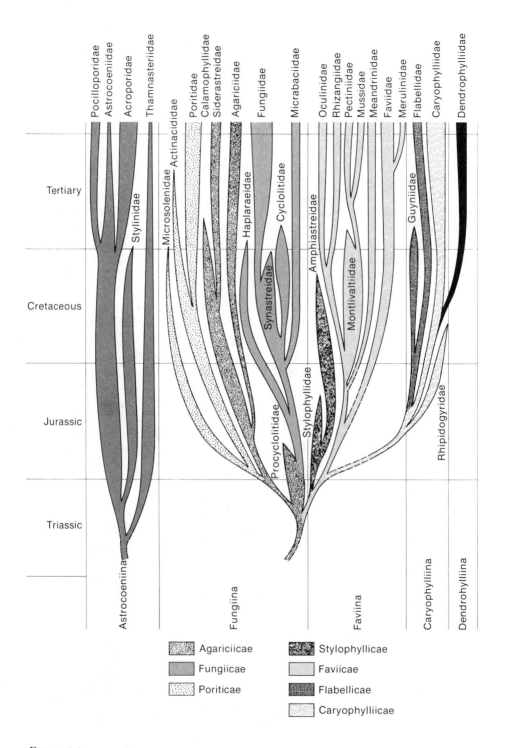

FIGURE 4-1
Pattern of adaptive radiation of scleractinian corals (hexacorals). Five suborders contain seven superfamilies, which are depicted by patterns. Families are labeled at the top of the diagram. (From Wells, 1956).

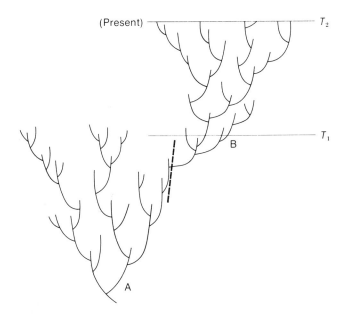

FIGURE 4-2
Pattern of phylogeny in which one clade emerges from another.
In retrospect (time T_2), the two clades are seen as being distinct,
and the phylogeny is divided at the position of the heavy,
dashed bar into taxa A and B. A taxonomist living at time T_1,
however, would have recognized only a single clade and would
have grouped the entire phylogeny that had developed by that
time into a single taxon (A).

from perhaps only two or three surviving genera produced numerous new families during the Early Triassic, an interval of perhaps just 7 or 8 My (Figure 4-3).

When confronted with the abrupt appearance of many new families of a
higher taxon, some workers have sought an explanation in the imperfection of
the fossil record. The most general problem to which this approach has been
taken is the apparently sudden appearance in the fossil record of the diverse
classes of invertebrates that characterize the Cambrian system. In fact, it was
Darwin, the first worker seriously faced with the "Cambrian problem," who
initially hypothesized a long Precambrian interval of diversification for which we
have no fossil record. In the absence of direct evidence for such an interval, the
magnitude of the problem seemed so great that he wrote in *On the Origin of Species*
(1859, p. 308) "The case at present must remain inexplicable; and may be truly
urged as a valid argument against the views [evolution by natural selection] here
entertained." As noted earlier (page 36), Durham (1969) recently addressed the
problem with more concrete data. Noting that an average Cenozoic species of

68

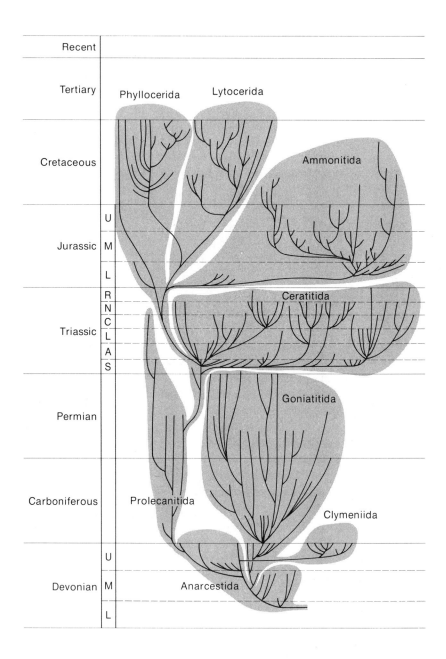

FIGURE 4-3
Phylogeny of the Ammonoida, showing the rapid appearance of new families in the earliest Triassic, following the mass extinction at the end of the Permian. Each line in the phylogeny represents a particular family. The Early Triassic, or Scythian, is identified by the letter *S*. (From Teichert, 1967.)

marine invertebrates seems to have existed for about 6 My, he suggested that the complex nature of Cambrian taxa like echinoderms points to a very long Precambrian interval of evolution. He argued that numerous species would have to be arrayed end-to-end to produce such advanced animals from simple ancestors. This reasoning, of course, represents a gradualistic position of the most extreme sort.

In fact, as hypothesized by Cloud (1948), the evidence is now mounting that most of the major fossil groups of the Cambrian arose by rapid evolution. It appears that the main radiation spanned an interval lasting in the order of 100 My, during the latest Precambrian and Early Cambrian. A long earlier history for Cambrian taxa seems at present to be without support (Stanley, 1976). In the first place, fossil assemblages consisting of the imprints of soft-bodied creatures (mostly coelenterates), including the famous Ediacara fauna of Australia, have been found in many areas of the world, but are never older than latest Precambrian. Also, trace fossils (tracks, trails, and burrows) extend back into the latest Precambrian, but increase in abundance and diversity toward and into the Cambrian, apparently recording the initial adaptive radiation of the dominant multicellular groups of the Paleozoic. Even skeletal taxa appear, not simultaneously at the base of the Cambrian, but sequentially. Skeletonization was merely one aspect of the general radiation. The picture now taking form is of a rapid, but not instantaneous, adaptive radiation.

A parallel example is provided by the seemingly sudden appearance of the angiosperms (flowering plants) in the Cretaceous. This, too, was a source of great concern to Darwin, who referred to it as an "abominable mystery." For years, many paleobotanists invoked a long, imaginary pre-Cretaceous history, for which no fossil record was definitely known, to explain the presence of diverse angiosperm floras in the latter half of the Cretaceous. Recently, Doyle and Hickey (1976) have presented evidence that the angiosperms actually diversified during an Early Cretaceous interval of only about 10 My. Their study of the Potomac group of the Atlantic Coastal Plain of North America has revealed a rapid increase in morphologic diversity accompanied by net trends from simple, archaic features of leaf and pollen morphology to more advanced features (Figure 4-4).

For other examples of the sudden fossil appearance of taxa, we cannot even reasonably begin to entertain the hypothesis of a long, unrecorded interval of diversification. One of these is the appearance of most orders of Cenozoic mammals early in that era (Figure 4-5). Nearly all of the higher mammalian taxa of the Paleocene and Early Eocene evolved from a small number of primitive representatives of the latest Cretaceous (see page 139). Among the new mammalian taxa of the early Cenozoic were the order of bats (Chiroptera) and the order of whales (Cetacea); the divergent nature of both of these orders underscores the point that rates of large-scale evolution were very rapid.

If we abandon the notion of a long unrecorded history for taxa that display high diversities soon after they appear in the fossil record, we are left with the

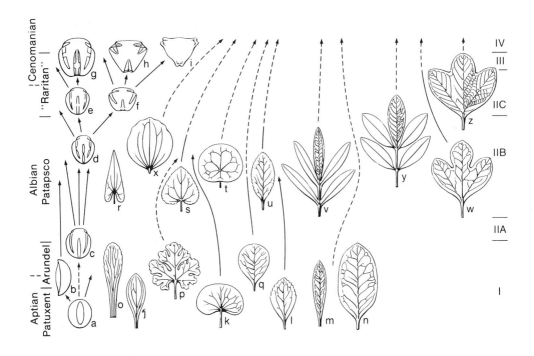

FIGURE 4-4

Increase in the diversity and morphologic complexity of pollen (a to i) and leaves of early angiosperms in the Lower Cretaceous Potomac Group of the Atlantic Coastal Plain of the United States. The time interval represented is about 10 My. (From Doyle and Hickey, 1976.)

idea that evolution must somehow be greatly accelerated during adaptive radiation. Two hypotheses can be advanced to explain the occurrence of such high rates of evolution. One is gradualistic and the other, punctuational.

The gradualistic hypothesis is that at times of adaptive radiation, phyletic evolution is greatly accelerated. As discussed earlier (page 65), this view was expressed by Simpson in 1953. One of Simpson's most important earlier contributions was the recognition that entry into a distinctive new adaptive zone (set of niches) is commonly triggered by the origin of a key adaptation. His interpretation was that, "it is, normally, phylogenetic [phyletic, as defined here] progression rather than speciation that leads into the new zone" (Simpson, 1953, p. 350; see also p. 354). The very establishment of the phrase "adaptive radiation" seems to reflect the historical prevalence of this view, because the word "radiation" refers to the spreading of linear rays from a point source. Darwin's original sketch of phylogeny, in fact, depicts such a pattern (Figure 2-6).

The punctuational hypothesis is that rates of evolution are accelerated by rapidly divergent speciation. Either or both of the following factors may be

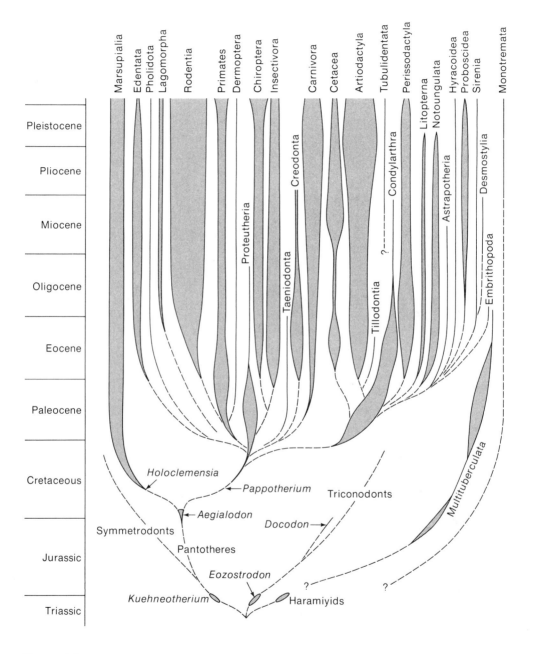

FIGURE 4-5
Stratigraphic occurrence of major taxa of the Mammalia, showing the appearance of nearly all orders of placental mammals (Edentata through Embrithopoda) during the Paleocene and Early Eocene. Vertical scale is nonlinear. (From Gingerich, 1977.)

responsible: (1) rates of speciation may be greatly accelerated, with speciation being the locus of most evolution, or (2) an unusually large percentage of speciation events may be markedly divergent, as a new adaptive zone is occupied. In other words, both number of speciation events and degree of divergence per event must be considered. These two factors will be evaluated in Chapters 5 and 6.

For the present, our goal will be to test the gradualistic hypothesis that adaptive radiation is characterized by the dramatic acceleration of phyletic evolution. To examine this alternative, let us consider the extreme case by assuming that all evolution in a higher taxon is of the phyletic variety: that there is no speciational component. Let us further imagine that we could measure something approaching a maximum rate of morphologic change within established lineages of the taxon. Then we would have established an upper limit for rate of large-scale evolution within the taxon. As shown in Figure 4-6,A, this hypothetical condition can be represented graphically in two dimensions for a single morphologic variable. Speciation changes the direction of evolution by forming independent lineages, but accomplishes nothing else. Phylogeny cannot move beyond the triangular area bounded by dashed lines. Overall rate of evolution will be high only if maximum rate of phyletic evolution is very high and if rates approaching the maximum are fairly common. If a speciational component is added, as in Figure 4-6,B', more rapid net rates of change are permitted; then very rapid change will occur if rate of speciation is high or if average amount of change per speciation event is large (or if both of these conditions obtain).

What we seek is an understanding of the anatomy of phylogeny, as discussed in Chapter 2. Does the pattern of phylogeny in adaptive radiation more closely resemble Figure 4-6,A or Figure 4-6,B? As already discussed, an intermediate shape is inevitable, but the question is whether both the speciational component and the phyletic component of evolution play an important role in adaptive radiation, which is to say in transpecific evolution in general. The evaluation of the phyletic component that follows will indicate that it has played only a minor role, leaving us with the inference that divergence associated with speciation is responsible for most net evolutionary change in the history of life.

Subjecting the gradualistic model to what is called the test of adaptive radiation (Stanley, 1975a), I noted that an average species of Late Cenozoic mammals has lasted for at least a million years. This figure seems extremely large when one considers that most Cenozoic orders of mammals evolved within an interval of only about 12 My of the Paleocene and earliest Eocene (Figure 4-5). An average lineage spanning the interval would have contained only about 10 intergrading chronospecies of average duration. Thus, in order to invoke phyletic evolution as the primary source of the major evolutionary transition within the early Cenozoic Mammalia, we would apparently have to call upon an extraordinary intensification of the selection pressures normally acting upon established species. The difficulty here is that the great acceleration of phyletic evolution would have had to take place within a wide variety of unrelated taxa occupying

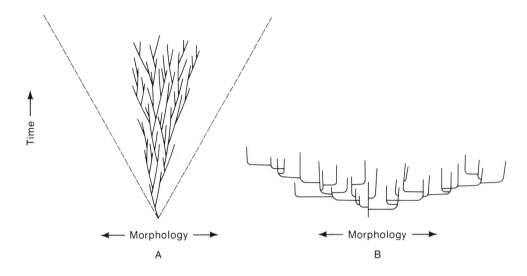

FIGURE 4-6
Idealized phylogenies showing how, if phyletic evolution is slow, a strongly gradualistic pattern (A) permits much slower diversification than a strongly punctuational pattern (B). In the gradualistic model, rate of diversification cannot expand phylogeny beyond the area between the dashed lines, which represent the maximum rate of phyletic evolution.

distinct adaptive zones in many different parts of the world. Why this should happen remains unexplained. It seems clear that the adaptive radiation was occurring because adaptive zones vacated by the dinosaurs were available for occupancy. The mere availability of niche space, however, could hardly be expected to cause rapid phyletic shift of adaptation. Why should all populations of any established species abandon their original niche because adjacent ecologic space is free for occupancy? Certainly, expansion of the original niche might be expected, but it is difficult to imagine that this could produce major adaptive shifts without fragmentation into new species. Far more likely would be the rapid invasion of adjacent ecologic space in association with divergent speciation. In fact, in the presence of vast expanses of ecologic space of the sort that seem to have existed in the terrestrial realm at the end of the Mesozoic Era, markedly divergent speciation events would be expected to occur in rapid succession. Thus, while the mammalian radiation is not easily accommodated within the gradualistic model, it is quite compatible with the punctuational view.

Objections may reasonably be leveled at the test of adaptive radiation as applied to early Cenozoic mammals. The conclusion that an average species of mammals has survived for at least a million years is based on data for the late Cenozoic. Subdivision of supposed lineages of the early Cenozoic has yielded similar durations for chronospecies (Gingerich, 1974; 1976), but such data are scanty and are based on limited skeletal information (primarily dentition). Also, we have referred only to average duration. A typical histogram of durations could be strongly skewed to the right so that, while the mean duration is quite long, the mode lies near the vertical axis, indicating the presence of a substantial percentage of short-lived species representing segments of lineages undergoing rapid phyletic transformation. These deficiencies can be overcome by considering only late Cenozoic mammals, many groups of which are also undergoing adaptive radiation, and by developing a comprehensive picture of chronospecies longevity. This, then, becomes the task of the following section.

DURATIONS OF
MAMMALIAN CHRONOSPECIES

We will focus here upon formal taxa, which can be substituted for morphologic variables in the measurement of rates of evolution (Simpson, 1944). In particular, we will analyze the durations of Plio-Pleistocene chronospecies of mammals. Into the diagnosis of chronospecies has gone an enormous amount of collective labor, which fortunately is easily converted to our purpose. A second advantage of using chronospecies as units of analysis is that they represent composite measures of morphologic variation. Many characters have been used to delimit them. Finally, a technique exists for generating a complete distribution of chronospecies durations (Stanley, 1978). Thus, it is possible to improve upon approaches that fail to sample phylogeny comprehensively (Chapter 3). It matters little that chronospecies, as segments of lineages, are delimited subjectively. Almost invariably, they are entities that exhibit only modest amounts of morphologic change.

If taxonomic data are to be employed meaningfully here, they must be internally consistent. Preferably, they should represent the species concept of a particular worker. Probably the most richly documented group of related phyletic lineages belongs to the Plio-Pleistocene Mammalia of Europe. For these, Kurtén (1968) has provided an authoritative taxonomic and stratigraphic evaluation. One measure of the excellence of the data is the fact that only about 12 percent of living species of European mammals are unrecognized in the fossil record of the Pleistocene. The few species lacking recognized records seem either to be recent immigrants to Europe or small, fragile forms unlikely to leave a conspicuous fossil record.

A histogram of chronospecies longevities for Plio-Pleistocene mammals will be derived from a particular kind of **survivorship curve.** Simpson (1944) pioneered in adapting the survivorship curve of population biology for use in the evaluation of paleontologic data. In population biology, this kind of curve depicts the decline through time of an average cohort, or set of individuals born at the same time. In other words, a survivorship curve is a reverse cumulative distribution of the longevities of individuals within a population. In the paleon-tologic application, taxa are substituted for individuals, and extinction takes the place of death. The following analysis is somewhat unorthodox. It evaluates the lengths of time that segments of lineages have survived before having undergone enough phyletic evolution to be considered new chronospecies. Extinction is by phyletic transition rather than by termination of a lineage.

In order to construct a survivorship curve for chronospecies of mammals, we will make use of a general bias in the taxonomic evaluation of late Cenozoic chronospecies: the practice of viewing living species as starting points for taxonomy (Stanley, 1978). A living population naturally displays a much more complete set of taxonomic characters than is represented by the skeletal remains of the fossil record. In the Recent, not only anatomical but also behavioral, ecologic, and geographic data are available for evaluation. Taxonomic paleon-tologists therefore tend to adopt authoritative classifications of living species, and European mammals are so well studied that there is widespread acceptance of species designations. The standard practice of a taxonomic paleontologist study-ing an extant lineage is to assign fossil populations to the living species as far back in geologic time as his species concept dictates, and then to draw a line demarcating the beginning of another chronospecies. Therefore, living species traced backward tend to represent chronospecies of *full* duration.

Taxonomic durations of extant species traced backward in time will vary according to rate of evolution in the lineages considered. Living chronospecies within a taxon, in effect, represent a cohort and can be used, collectively, to form a taxonomic survivorship curve. Consider all lineages existing at some previous time (T) that survive to the present day. The percentage of these extant lineages that are still assigned to living species at time T will, in effect, represent the percentage of "newborn" Recent chronospecies that "live" taxonomically back-ward in time to time T. This percentage, then, becomes a point for a survivor-ship curve.

The procedure just outlined will now be used to form a survivorship curve for mammals, and from this, a histogram of durations. The data for Plio-Pleistocene mammals of Europe are of such unusually high quality that it seems unlikely that comparable data are currently available for any other taxon. Even for the Mammalia, the procedure described above will be modified because of the fa-mous mass extinction of Würm and post-Würm times. (The Würm is the final stage of the Pleistocene, equivalent to the upper Wisconsin of North America.) The cause of this extinction remains problematic (Van Valen, 1969), but human activities may have been largely responsible. The beginning of the Würm will be used in place of the Recent as an endpoint for the survivorship curve. The composition of the Würm fauna is extremely well known, both from numerous collecting sites within the stage and from the occurrence of many species in both older deposits and the Recent fauna, so that this procedural adjustment intro-duces little error.

Figure 4-7,A is the survivorship curve for mammals. Where two chrono-

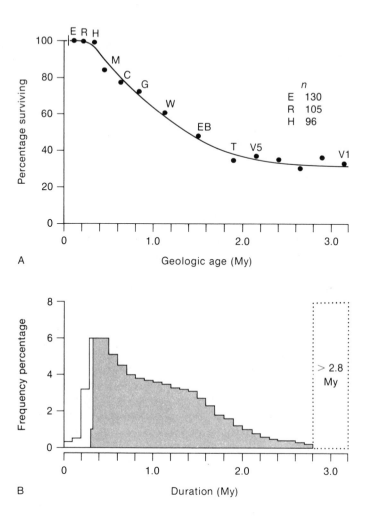

FIGURE 4-7
Longevities of chronospecies of Plio-Pleistocene mammals of
Europe. A: Survivorship curve for chronospecies. Each point represents all lineages in a geologic stage that survived to the beginning of
the Würm. The curve shows the percentage of these lineages that
survived to that point without having evolved enough to be consid-
ered new chronospecies. Also shown is the total number of lineages
(*n*) that survived to the Würm from each of the final three stages.
Dots are located at the midpoints of stages. Stage abbreviations:
E—Eem, R—Riss, H—Holstein, M—Mindel, C—Cromer,
G—Günz, W—Waalian, EB—Eburonian, T—Tegelen (Tiglian),
V5 through V1—divisions of the Villafranchian (midpoints arbitrarily placed at equal intervals within the stage). From Stanley, 1978.)
B: Histogram of durations derived from A. The left-hand region of
the histogram is unpatterned in order to show that only 1 percent of

species are considered to be ancestor and descendant but are currently separated by a stratigraphic gap, the boundary between the two has been placed at the midpoint of the gap. Figure 4-7,B is the histogram of chronospecies longevities that corresponds to Figure 4-7,A.

The depressed left flank of the histogram, reflecting the shoulder at the left end of the survivorship curve, attests to the rarity of rapid phyletic transition. The shoulder of the survivorship curve is drawn conservatively, to fall below the data point for the Holstein Stage. Thus, although the histogram constructed from the curve shows about 4 percent of chronospecies surviving less than 0.3 My, there is evidence that the actual percentage is even lower. The untreated data show only about 1 percent of living species going "extinct" backward in time between the Recent and the Holstein Stage, which represents an interval of about 0.35 My. This condition is displayed by leaving the appropriate left-hand portion of Figure 4-7,B unpatterned.

A valuable aspect of the plots shown in Figure 4-7 is that from them we can derive a histogram of rates of phyletic evolution. To understand how this is done, consider that, in taxonomic terms, the rate of evolution of a chronospecies that lasts 2 My will be 0.5 chronospecies/My. Thus, assuming that a uniform species concept is applied, rate of phyletic evolution for any chronospecies is the reciprocal of its longevity. Accordingly, a histogram of chronospecies longevities can be converted into a histogram displaying rates of phyletic evolution (Figure 4-8). Actually, the histogram (Figure 4-8,B) is derived from the survivorship curve (Figure 4-8,A) by converting the horizontal scale from durations into rates (reciprocals of durations) and dividing the converted scale into equal classes. Figure 4-8,B shows that mean rate of phyletic evolution for Plio-Pleistocene mammals was only 1.1 chronospecies/My. Extremely few lineages evolved at rates exceeding 4 chronospecies/My. In fact, this histogram, like Figure 4-7,A, is constructed using the conservatively drawn shoulder of the survivorship curve. As depicted by the data points used to draw the curve, only about 1 percent of living chronospecies are actually known to go "extinct" phyletically backward from the Recent during a span of 0.35 My. In other words, only about 1 percent of lineages evolved at rates as high as 2.9 chronospecies/My.

The survivorship curve and histograms display very sluggish phyletic evolution. That this is a general condition for late Cenozoic mammals is indicated by the relatively complete sample of data plotted. Today about 140 species exist in

chronospecies are actually known to have lasted less than 0.35 My; a larger percentage appear in the histogram because the survivorship curve is drawn conservatively, below the data point for the Holstein Stage. Dotted rectangle at right shows the percentage of chronospecies lasting longer than 2.8 My; inclusion of the actual ranges of these species would extend the histogram far to the right. (Data and phyletic interpretations from Kurtén, 1968. Time scale from Berggren and Van Couvering, 1974.)

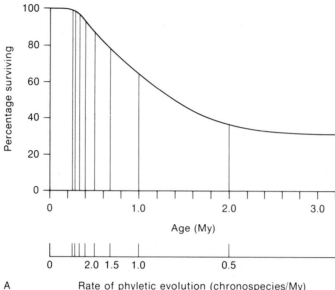

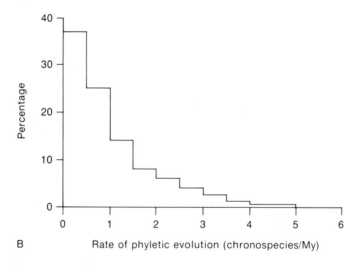

<small>FIGURE 4-8</small>
Derivation of a histogram (B) depicting rates of phyletic evolution
for Plio-Pleistocene mammals of Europe. This procedure is accomplished by partitioning the survivorship curve for chronospecies (Figure 4-7, A) into equal intervals in terms of rate of phyletic evolution (A). The rate of phyletic evolution represented by a
given chronospecies longevity equals the reciprocal of the longevity.

Europe. Restoration of species lost in the Würm and post-Würm extinction would raise this number to perhaps 150, which can be taken to approximate standing diversity for the Würm. Note that in Figure 4-7,A the number of lineages documented for the Eem is 130, or about 85 percent of the number that actually existed. The number of lineages plotted declines toward the right of the graph not simply because quality of preservation declines, but also because lineages are of finite duration and the extant number decreases with age. The critical shoulder of the curve, however, is documented by a large majority of the lineages that actually existed. The right-hand portion of Figure 4-7,A is probably elevated somewhat by a definite source of bias: the relative ease with which lineages can be traced backward when their early populations are assigned to living species. Lineages that have evolved more rapidly will be traceable only if documented by sequential populations, and a smaller fraction of such lineages will therefore be recognized to yield data for the right-hand portion of the curve. Still, the left-hand region of the curve is the critical part. Furthermore, the most rapidly evolving lineages seem to be those composed of large animals, like elephants, which tend to be easily preserved and discovered in a fossil state.

As discussed above, most living species of European mammals that lack a fossil record but are not recent immigrants are small, fragile forms. Furthermore, the six living species believed possibly or definitely to extend back into the Astian Stage (at least 3 My) fall within this category. They are the pygmy shrew (*Sorex minutus*), the common field mouse (*Apodemus sylvaticus*), two species of bats (*Myotis dasycneme* and *Miniopterus schreibersi*), the bank vole (*Clethryonomys glareolus*) and the steppe-pika (*Ochotona pusilla*) (Kurtén, 1968). The future will bring new data and revised interpretations, but it is most unlikely that the overall picture will change. The data now in hand are simply too complete to allow for substantial error.

Actually, Figure 4-7,A is in various ways biased against the conclusion that late Cenozoic lineages of mammals have evolved very slowly. For one thing, nearly all of the lineages upon which it is based survived to the Recent, which therefore represents the time of their taxonomic "birth." Termination of the curve at the start of the Würm instead of at the Recent abbreviates actual longevities of most chronospecies by 40,000 years. Also, relaxation of the basic premise—that the Recent represents the starting point for taxonomy—would lengthen the durations of certain chronospecies. Finally, while phyletic origins are well established for some species like *Ursus arctos*, the brown bear (Kurtén, 1958), other species thought probably to have originated phyletically must, in fact, have arisen by branching. In some instances, the time of branching must have preceded the assumed time of phyletic transition.

We might question whether the Pleistocene interval of mammalian phylogeny in Europe can be taken as representative of mammalian evolution in general. I believe that it can. Certainly, an epoch during which local climates and floras were fluctuating in the extreme is hardly a time when we would expect phyletic evolution to have stagnated in comparison to its rate during other intervals of the

Cenozoic. Selection pressures can hardly have subsided for the class as a whole. The Pleistocene intersects the phylogenies of genera and families at various stages of diversification and decline. Numerous geographic areas are represented, and varied adaptive zones. During the late Cenozoic, many taxa have obviously been in early stages of adaptive radiation. Examples are the Cervidae, the Bovidae, and various subgroups of rodents and carnivores. As will be discussed in Chapter 6, the giant panda, representing a new monogeneric subfamily or family, may well have arisen in the Pleistocene.

HUMAN EVOLUTION

The traditional gradualistic manner of reconstructing human evolution was reviewed in Chapter 2. The necessarily gradualistic view that no more than one hominid species has ever existed at one time has been refuted by concrete evidence. Even so, the idea has persisted that *Australopithecus africanus* and *Homo sapiens* belong to a single phyletic lineage (Figures 2-9 and 2-10). Only recently has the branching pattern indicated by new evidence for the temporal overlap of species led to the observation that human evolution is compatible with the punctuational model (Gould, 1976; Leakey and Walker, 1976). Gradualism is not ruled out *a priori*, because branching could simply have increased the number of directions taken by phyletic evolution without contributing rapid transitions.

In light of the kind of analysis undertaken in this chapter, new evidence of hominid fossil occurrence forms a strong case against gradualism (Stanley, 1978). The trait to be emphasized here is not coexistence, but longevity. All five recognized Plio-Pleistocene species predating *Homo sapiens* existed for intervals in the order of a million years (Figure 4-9). There is evidence that *Australopithecus africanus* survived much longer. For two reasons, all five ranges are underestimates of actual chronospecies longevity. First, the recognized species are not full chronospecies. This is obvious because some must have disappeared by lineage termination. (There are too few potential descendant species for all five species to have become "extinct" by phyletic transition.) Second, the historical trend in hominid taxonomy is toward extension of ranges by discovery of fossils at new sites: The more collecting that is done, the longer become the spans of these slowly evolving species. If we make the conservative assumption that all five species are actually full chronospecies, and if we ignore the uncertain range extensions depicted in Figure 4-9, we can calculate a maximum average rate of phyletic evolution. The total span of the five ranges is almost exactly 5 My, giving a phyletic rate of 1 chronospecies/My—virtually the same as the mean rate for other mammals (Figure 4-8). Given knowledge of full chronospecies longevities, downward adjustment of the hominid phyletic rate would depress it

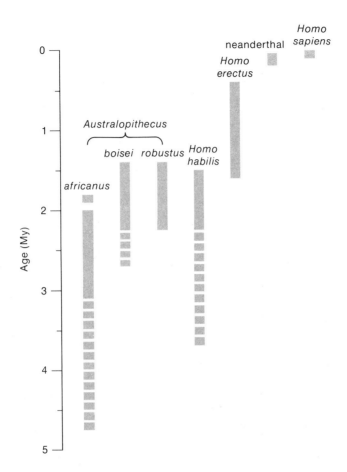

FIGURE 4-9
Stratigraphic ranges of Plio-Pleistocene species of the Hominidae. Broken portions of bars represent fossil material tentatively assigned to species. Taxa represented: *Australopithecus africanus, A. boisei, A. robustus, Homo habilis, H. erectus,* neanderthal man, and *Homo sapiens sapiens.* The species *Homo habilis* is a somewhat problematic taxonomic entity. Also, the robust australopithecines, *A. boisei* and *A. robustus,* may constitute a single species. Data from Tattersall and Eldredge (1977) and Leakey and Walker (1976). (From Stanley, 1978.)

below the average for mammals in general. This is a rather startling revelation, considering the extraordinarily high *net* rate of evolution that is the hallmark of human phylogeny. The only alternative to adopting a punctuational scheme would seem to be to claim that the record is woefully inadequate and that rapid evolution is embodied in lineages for which we have no record, but this could hardly represent a comfortable refuge. Stratigraphic ranges of recognized species are continually being extended, not only by periodic fossil discoveries, but also by the current trend toward taxonomic lumping. We seem forced to adopt the punctuational view that distinct lineages of hominids evolved rapidly by way of small populations. Once expanded into full-fledged species, these lineages were subjected to selection that was largely stabilizing rather than directional. In short, all evidence suggests that the traditional view of hominid evolution is wrong. Phyletic evolution has been unusually slow relative to that of other mammalian families, not unusually rapid.

THE ORIGINS OF MAMMALIAN GENERA

Another way in which the record of Pleistocene mammals testifies to the sluggish nature of phyletic evolution concerns the origin of genera. The genus lies but a single taxonomic step above the species, yet living genera of mammals tend to be quite distinctive, and taxonomy at this level, summarized by Walker (1975), suffers from relatively little controversy. If, as will be concluded, the phyletic component of evolution proves generally too weak to form a typical genus in the time called for by fossil documentation of generic origins, then phyletic evolution must also be too sluggish to account for the rapid appearance of families, orders, and classes in adaptive radiation.

A simple observation of great significance here is that, despite the detailed study of the Pleistocene mammals of Europe, not a single valid example is known of phyletic transition from one genus to another. The only example reported by Kurtén (1968) has disappeared with the unification of the elephantid genera *Archidiskodon* and *Mammuthus* because the phyletic transition in question occurred "without basic adaptive shifts" (Maglio, 1973, p. 51). Virtually all ancestral-descendant pairs of chronospecies of Pleistocene mammals that are now recognized (Kurtén 1968) differ from each other in only minor aspects of morphology. Many are separated by little more than body size. Genera, on the other hand, tend to possess distinctive morphologies that reflect a fundamental uniqueness of adaptation.

It seems reasonable to estimate that a minimum of eight or ten chronospecies transitions would, on the average, be required for phyletic evolution to turn one brand new genus gradually into another. In other words, this number of intergrading chronospecies would form a lineage segment that might be termed an average "chronogenus." The Pleistocene Epoch lasted about 1.8 My. Thus, to yield one chronogenus during the Pleistocene by way of eight to ten chronospecies, an average lineage would have had to represent rapid enough evolution for its component chronospecies to average only 200,000 years in duration. It is apparent from Figures 4-7 and 4-8 that virtually no Pleistocene chronospecies were undergoing phyletic transition as rapid as this. There is, in fact, no evidence at all for such high rates. Certainly, it is extremely unlikely that such unusually high rates would have been sustained in a single lineage for the duration of the Pleistocene. The obvious implication is that the phyletic component of evolution was too weak during the Pleistocene to be primarily responsible for the origin of more than a miniscule number of new genera. To test the relative efficacy of phyletic evolution more precisely, it is necessary to consider (1) the number of new genera that actually arose and (2) the number of phyletic pathways by which they might have arisen.

The number of genera that evolved in Europe during the Pleistocene cannot be determined accurately because genera must to some extent originate in pulses, but an approximation is possible. An average genus of mammals has existed for something like 8 My (Simpson, 1953; Van Valen, 1973). Since the Pleistocene lasted about one-fourth this long, we might expect between 15 and 20 of the 75 genera of terrestrial mammals that occupy Europe today to have evolved during the Pleistocene. Some, of course, must have originated on other continents, but others must have evolved in Europe or very nearby. As noted by Kurtén (1968, p. 253), few if any members of the modern Arctic fauna could have existed before the Pleistocene. The fossil record and, in some cases, adaptation to cold climates point to Pleistocene origins for genera like *Thalarctos* (the polar bear—see pp. 85–86), *Rangifer* (the reindeer), *Ovibos* (the musk ox), *Capreolus* (the roe deer), and *Microtus* (the common vole). To the number of extant genera that arose in Europe during the Pleistocene, we must add a few extinct forms, like *Megaloceros* (the Irish "elk"). It seems likely that altogether at least 15 genera originated in Europe. This minimum number can be compared to the number of continuous phyletic pathways that traversed the Pleistocene. As noted above, average standing diversity of mammals in Europe during the Pleistocene can be estimated at about 150 species. As illustrated in Figure 4-10, for any segment of

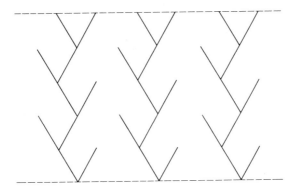

FIGURE 4-10
Hypothetical segment of phylogeny showing standing diversity of six species, with only three phyletic pathways spanning the interval represented.

phylogeny exhibiting branching and termination of lineages, standing diversity will greatly exceed the number of continuous nonoverlapping phyletic pathways. Kurtén (1968) identified 25 to 40 species that probably or definitely were terminated during the Pleistocene, and many additional lineages are likely to have gone extinct. It seems evident that no more than 100 phyletic pathways spanned the Pleistocene Epoch in Europe. Considering that very few contemporaneous chronospecies lasted for as short a time as 300,000 to 400,000 years (Figure 4-7,B), it seems evident that very few Pleistocene lineages could be divided into as many as five chronospecies. Kurtén (1968), in fact, recognized none. Clearly, only a minute fraction of the 100 or so composite lineages could have undergone rapid enough phyletic evolution to yield a new genus, and even fewer genera could have arisen within lineages that were terminated. Even clades that underwent substantial net evolution at this time were undergoing extremely sluggish phyletic evolution. We are left with the inference that most genera must have formed rapidly by way of divergent speciation events. Certainly, two or more speciation events may have contributed to some generic transitions.

The interval depicted in Figure 4-7,A represents about 5 percent of the Age of Mammals, and, as noted above, the environmental vicissitudes of the European Pleistocene could hardly have depressed selection pressures to levels below those normally operating upon established mammalian species. Sluggish evolution must typify mammalian phylogeny in general. The gradualistic model is contradicted.

The phylogeny of the Elephantinae (Maglio, 1973) provides a tangible illustration of the foregoing conclusions. This example is especially compelling because elephants are famous for having left an excellent fossil record and also for having undergone the most rapid phyletic evolution yet documented within the Mammalia. Their skeletons are both sturdy and conspicuous. Few unusual fossil finds of elephantines have been made in recent years, and Maglio regarded his phylogeny of the group as being relatively complete. Significantly, Maglio showed none of the three Plio-Pleistocene genera as having arisen by phyletic transition (Figure 4-11). Furthermore, rates of phyletic evolution within these genera, if projected backward, would require an absurdly long interval for derivation of each genus from the ancestral genus *Primelephas*, which seems itself to have arisen only about a million years before the three descendant genera appear in the record. Even if we ignore these points, *Primelephas* could have turned into only one of the three by phyletic evolution. At least two, and probably all three, must have arisen rapidly, by divergent speciation.

Two other points emerge from Figure 4-11. One is that, as discussed on page 17, the fraction of net large-scale evolution contributed by the speciational component of phylogeny does not necessarily correspond to the percentage of species that arise by branching. Figure 4-11 shows twelve species formed by branching and nine by phyletic transition. As seems to have occurred in elephantid phylogeny, a single event of quantum speciation such as one yielding

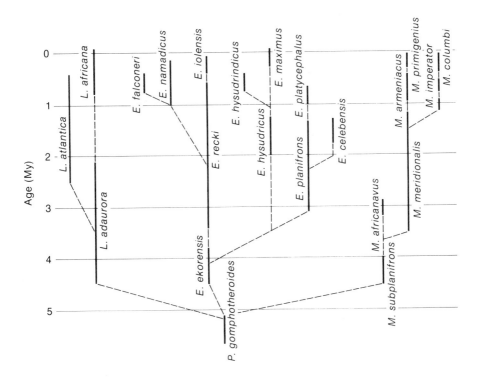

FIGURE 4-11
Phylogeny of the Elephantinae. Generic abbreviations: *P—Primelephas, E—Elephas, L—Loxodonta, M—Mammuthus.* Vertical alignment of chronospecies depicts apparent phyletic transition. (After Maglio, 1973.)

a new genus must often yield more net change than several successive chronospecies transitions.

The other point illustrated in Figure 4-11 is that relatively short-ranging chronospecies—ones representing the most rapid phyletic evolution—are often aligned with long-ranging ones. Species like *Elephas iolensis* and *Mammuthus primigenius,* which display phyletic evolution as rapid as is recognized anywhere in the entire fossil record, each descended from chronospecies that lasted well over 2 My! Thus, only a relatively small amount of total evolution occurred during the 3.5 My span of each lineage. If even here phyletic evolution moved only a short distance in 3.5 My, where can it be expected to have accomplished anything but minor transformation? Recall that, as shown in Figure 4-5, most orders of Cenozoic mammals arose during an interval only three or four times this long!

The polar bear (*"Ursus" maritimus*) may turn out to represent an excellent example of the recent origin of a genus by quantum speciation. It is adaptively distinct enough to have been assigned by some to the monospecific genus *Thalarctos*, rather than to the genus of its parent species, *Ursus arctos* (the brown bear). The genetic difference between the two species is not great because interbreeding yields fertile offspring in captivity (Thenius, 1953), but the polar bear maintains a discrete gene pool in nature and clearly fits the definition of a species (page 20). Furthermore, however few genetic changes may separate it from the ancestral species, it has diverged markedly in form and habit, being exclusively carnivorous, with a diet of seals, and possessing teeth that are specialized accordingly. It has clearly entered a new, semiaquatic adaptive zone. If morphology, in the broad sense, is to provide the criteria for taxonomy, as I believe it should, there seems every justification for placing the polar bear within a discrete genus. Hecht (1974) claimed that the origin of this animal reveals the efficacy of phyletic evolution. While the fossil record is not yet well enough known to settle the issue, there is a strong likelihood that the polar bear is actually the product of quantum speciation. An analysis by Kurtén showed that (1964) subfossil skull material from Scandinavia differs less from the morphology of *U. arctos* than do skulls of living polar bears, indicating that the polar bear has evolved substantially sometime within the last 20,000 years or so. Possibly, the species originated not much earlier. Except for its enormous size, a single bone from the lower Würm at Kew, England, resembles the ulna of a living polar bear and has been assigned to a distinct subspecies (Kurtén, 1964). The age of this specimen is perhaps 30,000 to 40,000 years; the possibility remains that it belongs to a nonaquatic bear of some unknown species or subspecies. No other fossil material showing affinities to *Thalarctos maritimus* is known, which is perhaps evidence of a recent origin because the semiaquatic habitat of the species should be conducive to preservation. As a potential product of quantum speciation, the polar bear is a tantalizing creature. At present, all that can be claimed with reasonable certainty is that it is adaptively distinctive; it is remarkable in having no known fossil record older than about 40,000 years, and perhaps none older than about 20,000 years; its parent species is still alive, indicating origin by phyletic branching; and its ecologic occurrence is marginal to that of its parent species (which has an enormous geographic range), suggesting origin of the polar bear via a small segment of the ancestral gene pool.

There is evidence that genera of birds normally originate in the manner suggested here for mammals. Mayr (1942, pp. 167–168) found evidence of such origins for several bird genera, and Bock (1970) outlined a stepwise sequence by which genera of the Drepanidae (honeycreepers) seem to have evolved in Hawaii.

The proposed rule that genera originate primarily through divergent speciation is, in accordance with the adage, expected to have exceptions. The alleged phyletic transitions between *Plesiadapus* and *Platychoerops* and between *Pelycodus* and *Notharctus* (Gingerich, 1976; Gingerich and Simons, 1977) are possible exceptions, though by no means definite ones. The populations claimed to bridge these genera of Eocene mammals are represented by very small samples, which

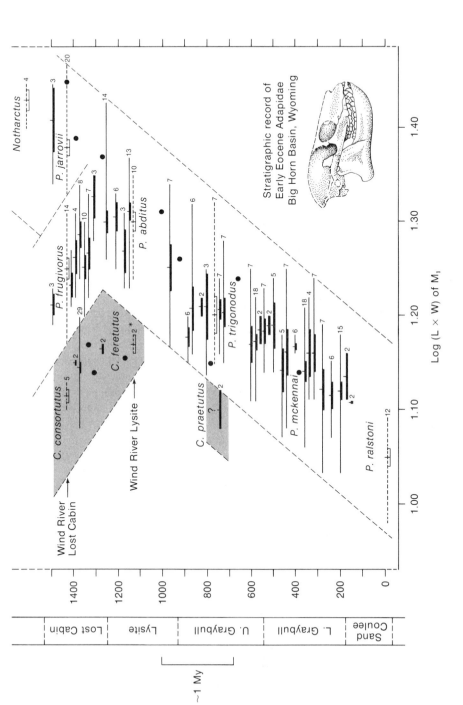

FIGURE 4-12
Phylogenetic interpretation of Gingerich and Simons (1977) for Early Eocene adapid mammals, suggesting the phyletic descent of *Notharctus* from *Pelycodus*. Symbols are the same as in Figure 3-9. Solid horizontal lines show sample ranges, vertical slashes show means, horizontal bars show standard errors of means, and small numbers indicate sample size.

exhibit suspiciously high dental variability to be assigned to single species and which are separated by stratigraphic intervals of 100,000 years or more (Figure 4-12). The marked molar alterations that have evolved in small populations of British voles during the past few thousand years and the complex patterns of geographic shifting that have accompanied this evolution (see page 165) raise the possibility that the purported lineages of Eocene mammals are artificially aligned bits and pieces of branching phylogenies.

THE ORIGINS OF INVERTEBRATE GENERA
AND THE ADEQUACY OF THE FOSSIL RECORD

There seems, over the years, to have arisen in paleontology a notion which, put cynically, claims that wherever in the fossil record one is looking happens to be where rapid phyletic evolution is not recorded. This idea resembles, in reverse, Locke's philosophical postulate that objects exist only when being observed. Embodied in the traditional indictment of the fossil record are two distinct allegations. One is that rapid evolution is confined to geographic areas that are unstudiable, or at least not yet studied. The other is that local sequences of sedimentary rocks harbor so many gaps in their temporal record that fossil data can almost never yield evidence of significant phyletic trends. The geographic argument was employed by Darwin, in fact, to explain the apparent absence of any record of the initial diversification of metazoans (see page 67). Darwin postulated that strata preserving the early record might happen to lie beneath our present-day ocean basins. More recently, Gingerich (1977), who adheres to gradualism but finds no evidence of large-scale phyletic transition in the early Cenozoic mammals of the Western Interior of the United States, has hypothesized that the major transitions occurred south of the border, in Mexico. Any possible argument of this kind for the Pleistocene of Europe has been precluded by the analysis presented earlier in this chapter: The totality of phyletic evolution in Europe was too slow to produce more than a miniscule number of genera, yet a number of genera arose in Europe. Human evolution now seems to be largely immune to the "poor fossil record" argument of phyletic gradualism. This argument also largely disappears for well-fossilized invertebrate animals, in general, when we consider that these creatures, unlike the vertebrates, have records that for major segments of geologic time are widely distributed in space and are also well studied. It is to the invertebrate record that we will now turn, to sharpen this point and also to evaluate the second traditional criticism of the fossil record: It is riddled with gaps that leave an artificial imprint of punctuation upon what is actually a gradualistic phylogeny.

The invertebrate record provides abundant evidence opposing the idea that gaps hide major phyletic trends. As Gould and Eldredge (1977) have put it, "stasis is data." Omission of this message from most discussions of the gradualistic and punctuational views would seem to represent the forest not being seen for the trees. Attention has been paid largely to lineages that seem to exhibit measurable phyletic evolution, but to these must be added the much larger number of lineages that display so little phyletic evolution that none has yet been pointed out: most of the lineages that are assigned to single species. In keeping with the theme of this chapter, it is the ranges of invertebrate species that we will now consider.

Myriads of invertebrate lineages have been traced through time by sequential sampling in the determination of stratigraphic ranges for species. The normal procedure here is to measure a stratigraphic section and to record the vertical distribution within that section of lineages, most of which are assigned to single species, but none of which need be. The final clause is critical. Biostratigraphers do not know, in advance, where a lineage is headed, in an evolutionary sense, when they trace it upward through a local section or through several sections that overlap in time to yield a composite stratigraphic range. Any one of the thousands of stratigraphic ranges of species that have been delimited *could* have been found to record sufficient evolutionary change within 5 or 10 My to display the transition from one family to another, yet no such change has ever been recorded. One might argue that such rapid change could only occur so rarely that it is not likely to be sampled, but, as with the analysis of mammals presented earlier in this chapter, the key lies in the origin of genera, which are far more numerous.

As discussed earlier, in singling out the genus for evaluation, I by no means intend to imply that it is equivalent from taxon to taxon, but it is the usual formal unit lying just one step above the species, and if lowly genera originate primarily by way of rapidly divergent speciation events, then it is in such events that most large-scale evolution must be concentrated. The examination of any well-fossilized invertebrate group will show that many family level transitions have occurred during intervals in the order of 50 My. Thus, documented rates of large-scale evolution are so high that, for phyletic evolution to have played a major role in large-scale transformation, phyletic transitions from genus to genus within about 5 My would have to be commonplace in phylogeny. (Obviously, only a fraction of rapid generic transitions would ultimately be carried far enough to produce a new family.)

In fact, only rarely has a lineage been found to yield what is considered to be a new genus. On the contrary, an average *species* of marine echinoids, bivalves, gastropods, or brachiopods has survived for at least 5 My without even evolving enough to be regarded as a new species (Durham, 1969; Stanley, 1975a). Figures 2-2, 3-8, and 3-10 illustrate examples from the invertebrate record. Additional evidence will be presented below and in Chapter 9. In some instances, mean species duration is so great that, even in the absence of a histogram of species durations, the previously discussed test of adaptive radiation (page 72) convincingly opposes the gradualistic model. By methods that will be discussed in the next chapter, it is possible to obtain crude estimates of mean species duration even within Cenozoic taxa that are less well fossilized than Pleistocene mammals of Europe. It will, for example, be shown in Chapter 9 that an average species of benthic Foraminifera has a range of approximately 30 My, yet more than 20 families of benthic forams arose during the Paleocene and Eocene, which spanned almost exactly this period of time! Few families of benthic forams are natural units, but if "families" are markedly polyphyletic, then even more than the apparent number of major transitions are indicated. A possible example of moderately rapid evolution of benthic forams and of the phyletic origin of a genus is the apparent transition between *Helicorbitoides* and *Lepidorbitoides* during the Upper Cretaceous (Gorsel, 1975). The only invertebrate taxon for which phyletic intergeneric transitions have been claimed with any frequency, however, is the Ammonitina, and as Hallam (1975) noted, taxa of this group are grossly oversplit, so that most lineages treated in this way should actually be assigned to single genera. In fact, many ammonite species exhibit an extraordinary degree of variability (review by Kennedy and Cobban, 1976). Morphotypes previously regarded as representing discrete genera have been found to occur, with intermediate forms, in single populations (Figure 4-13).

There is no denying the possibility that paleontologists have a natural reluctance to divide a lineage into two genera. This kind of bias cannot explain the extreme rarity of phyletic transitions, however. Most genera are erected in the absence of appreciable stratigraphic information, yet rarely has a genus, once erected, been found to intergrade phyletically with another as data have continued to accumulate.

The view that virtually all stratigraphic sections contain gaps that disguise a gradualistic phylogeny seems to me to reflect little thought about the nature of the stratigraphic column. To be sure, the column is riddled with hiatuses, but the critical question is largely one of scale. Certainly, in many regions we lack segments of record representing 10 My of time or more, but myriads of packages of rock to which we do have access were laid down during comparable intervals. Furthermore, these packages of rock, while containing gaps on a finer scale, typically offer numerous sequential samples of invertebrate populations separated by relatively short intervals of time. These sequential samples serve to establish both the continuity of lineages and the sluggish nature of phyletic evolution.

It is difficult for a neontologist to form a reasoned opinion about the adequacy of the fossil record for the study of evolution. Only observation of geologic outcrops, study of biostratigraphic data, and inquiry into sedimentary processes can supply a perspective. I will attempt here to illustrate briefly the evolutionary value of the record in a way that will be intelligible to the uninitiated reader. Earlier in this chapter, I summarized the evidence for the remarkably high quality of the Pleistocene record of European mammals, and now I will offer two examples in support of the preceding claims that the invertebrate record demands a punctuational interpretation.

The first example is of the Upper Devonian sedimentary sequence of New

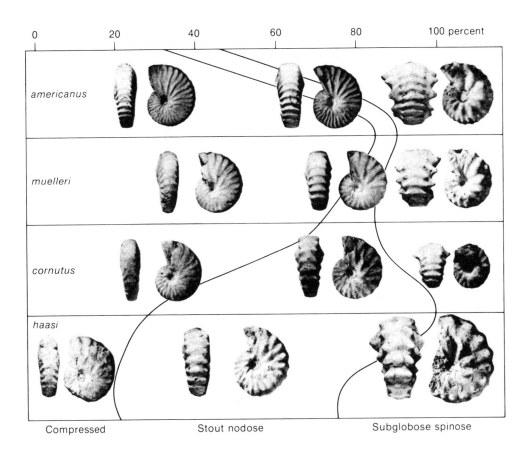

FIGURE 4-13
Diagram depicting the enormous variability within an Albian ammonite lineage of the genus *Neogastroplites*. The four chronospecies are defined on the basis of percentage of coexisting morphologic variants. In the past, members of a single population have been assigned to different genera. (From Kennedy and Cobban, 1976, based on Reeside and Cobban, 1960.)

York State, where, it happens, North American geology was born in the last century through the endeavors of James Hall. This example is presented here to illustrate how much information about evolutionary lineages is commonly revealed by fossils in single packages of rock. McAlester (1963) compiled stratigraphic ranges for bivalve mollusks of the Chemung Stage, which occupies the central position in the Upper Devonian sequence. A partial illustration of his results is shown in Figure 4-14. In all, he reported on about 40 species. Note that the ranges of most species span 5 to 10 My. The important point is that "the Chemung species show little or no evolutionary change throughout their known Appalachian occurrences" (McAlester, 1963, p. 1223). The species concept is

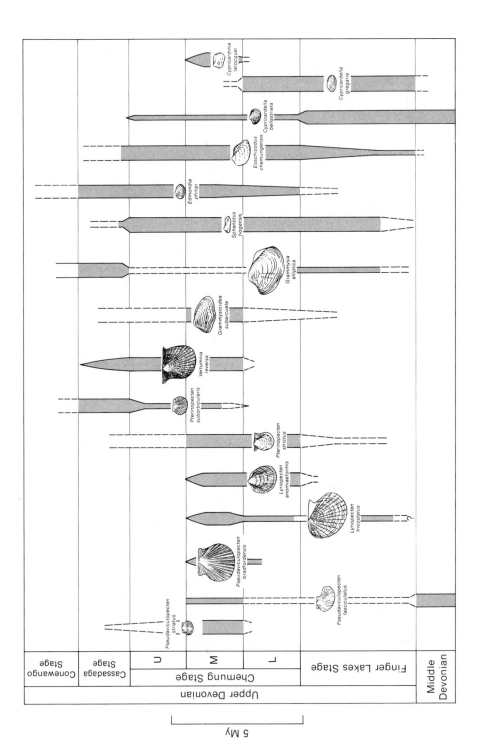

FIGURE 4–14
Stratigraphic ranges of species of bivalve mollusks recognized in the Chemung Stage of New York State. (From McAlester, 1963.)

hardly at issue here. Even if there were some erroneous lumping of similar species for lack of detailed morphologic information, gradualism would require evidence of change, as long as the fossil record is fairly complete.

Just how complete is the Chemung record? The outcrop belt for the Sonyea Group, the stratigraphic unit underlying the Chemung, resembles that of the Chemung and will be considered in the present context (Figure 4-15) because studies conducted on it serve to illustrate the detailed nature of the information provided by the Upper Devonian series of New York State in general. The outcrop belt is narrow, but stretches for hundreds of miles perpendicular to the ancient shoreline, which, as illustrated in Figure 4-15,A and B, lay to the east of the area where most sediment was depositied. Variation in thickness of rocks across the outcrop belt reflects not the depth of the ancient sea, but a higher rate of subsidence and sediment accumulation near the shoreline. Sediment was supplied by rivers flowing from the east. The Late Devonian (the time interval representing Upper Devonian sedimentary units) spanned 15 My, which means that the Sonyea Group, for which sampled sections are shown (Figure 4-15,D), spanned perhaps a little more than 2 My. (Note that the vertical bars in Figure 4-15,D depict vertical ranges of rock outcrops, not ranges of taxa.) The sections were sampled as part of a paleoecologic investigation (Sutton *et al.*, 1970; Bowen *et al.*, 1974). Occurrences of particular species are not depicted, but it is sufficient to point out that a species of the sort displayed in Figure 4-14 is represented by many samples throughout the three-dimensional body of rock. Occurrences of this kind, when collapsed to a single time axis, document lineages with many sequential and overlapping populations. During the deposition of Sonyea sediments, environments shifted and faunas migrated with them, leaving gaps in the fossil record at any single geographic position, but gaps bracketed by occurrences above, below, and at other localities. Not all sites from which fossils have come are shown in Figure 4-15,D, only those examined for paleoecologic analysis, and most Chemung species span a much longer interval of time than is represented in Figure 4-15,D. Some species made only brief appearances, but no species, however long its lineage, exhibits appreciable change. The same picture holds for brachiopods, which, owing to resistant shell structure, are even better preserved in the same rocks. There is no reason why the comprehensive data for Chemung lineages, and data like them from hundreds of other units of rock, should not be considered a good sample of phyletic evolution for major classes of invertebrates. Hallam (1976), summarizing data for the extremely well-studied Jurassic deposits of England, concluded that an average bivalve species spans 20 My! He found evidence of only very sluggish phyletic evolution. If phyletic evolution often moved at a rapid pace, many new genera should be found to have arisen phyletically during the Upper Devonian or Jurassic.

The preceding biostratigraphic example was chosen because it represents a relatively comprehensive picture of species ranges for diverse taxa found in one body of rocks. The second example represents one taxon as it occurs throughout the world. This taxon is the Spiriferellinae, a subfamily of brachiopods

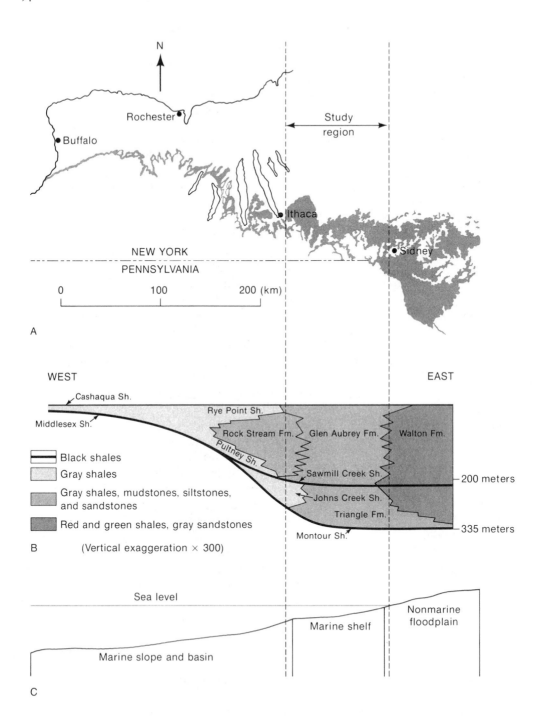

A

WEST   EAST

Cashaqua Sh.

Middlesex Sh.

Rye Point Sh.

Rock Stream Fm.   Glen Aubrey Fm.   Walton Fm.

Pultney Sh.

Sawmill Creek Sh.

                                                           — 200 meters

Johns Creek Sh.

Triangle Fm.

                                                — 335 meters

Montour Sh.

Black shales

Gray shales

Gray shales, mudstones, siltstones, and sandstones

Red and green shales, gray sandstones

B   (Vertical exaggeration × 300)

Sea level

Nonmarine floodplain

Marine shelf

Marine slope and basin

C

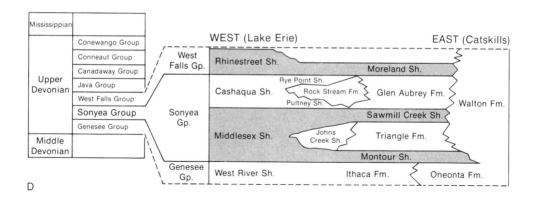

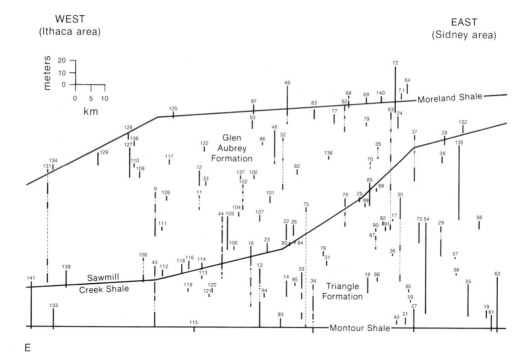

FIGURE 4-15

Paleontologic sampling of the Upper Devonian Series of New York State. A: Map of outcrop belt of the Sonyea Group. B: Generalized stratigraphic cross section of the Sonyea Group. The Montour Shale and Coshaqua Shale are believed to be time-parallel, so that deposition was much more rapid in the east than in the west. C: Reconstructed cross section of environments in which the rock units of B were believed to have been deposited. D: Stratigraphic position of the Sonyea Group within the Upper Devonian of New York State. E: Positions within the Sonyea Group of stratigraphic sections studied by Sutton *et al.* (From Sutton *et al.*, 1970.)

(Waterhouse, 1977). Certainly, some of the species ranges, depicted in Figure 4-16, are abbreviated by stratigraphic gaps or inadequate collecting, but note that even with this bias against longevity, most lineages span more than 3 My and many span more than 5 My. Not only is nothing approaching phyletic transition to a new genus found anywhere in the set of ranges but Waterhouse, like McAlester in the previous example, noted explicitly that species are quite stable, seldom displaying net directional change in morphology that is even noticeable. What is unusual about the brachiopod subfamily Spiriferellinae is not that it is particularly well preserved but that someone has taken the trouble to compile the ranges of all of its species on a worldwide basis. One reason that this viewpoint has not prevailed from the outset in the recent controversy over gradualism is that biostratigraphers have seldom attached absolute time scales to their charts displaying stratigraphic ranges. In general, the late Cenozoic record offers the most favorable opportunities for nearly comprehensive compilations of this type. As an example, the great longevity of Neogene species of Caribbean reef corals is shown in Figure 9-5. All recognized lineages show very slow phyletic evolution or no observable change.

I would suggest that a kind of historical circularity of reasoning has crept into assessment of the completeness of the fossil record. The original hypothesis favored by Darwin and Lyell was that the fossil record must be too poor to support what we now call the gradualistic model of evolution. In time, with the acceptance of Darwinism, this condemnation of the record became well entrenched, and it came to be assumed that fair assessment of the record had provided independent evidence of its inadequacy.

Finally, I have noted elsewhere that to propose a rule that the large majority of genera arise by rapidly divergent speciation rather than by phyletic evolution may contaminate data provided in the future (Stanley, 1978). Some taxonomists may choose to adhere to the rule or violate it. The enormous volume of data already available, however, would seem to permit meaningful, unbiased evaluation without the addition of information from subsequent endeavors.

LIMITATIONS OF PHYLETIC EVOLUTION

Placing the preceding analysis in perspective, it seems reasonable to claim that the fossil record imposes more severe constraints on evolutionary theory than have generally been acknowledged. The primary stricture is that large, well-established species can play little part in large-scale transition, except indirectly, by casting off small populations that undergo rapid quantum shifts.

Presumably species characterized by small population size and restricted geographic occurrence do, on the average, evolve phyletically more rapidly than large species occupying varied habitats (pp. 48–51). Even so, data of the sort evaluated in the previous section indicate that major transitions do not occur even within established lineages whose population sizes are moderately small.

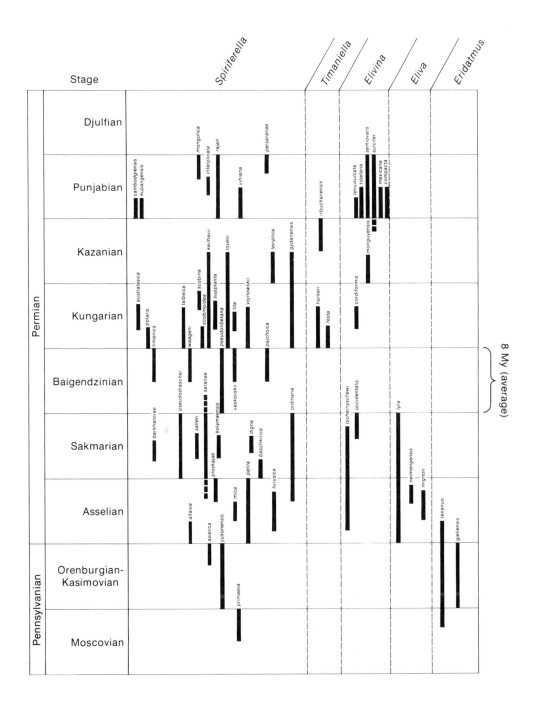

FIGURE 4-16
Worldwide stratigraphic ranges of late Paleozoic brachiopod species of the subfamily Spiriferellinae. Stages are of varying duration but average about 8 My. (Data from Waterhouse, 1977.)

What is particularly striking is that we lack any indication of a distribution of rates of change ranging from a near-static condition to truly substantial values. The comprehensive data summarized in Figure 4-8, 4-14, and 4-16 show *no* lineage evolving rapidly. The abundant data we possess demand such an abrupt discontinuity between the sluggish rates observed and the rapid hypothetical rates demanded by gradualism that the gradualist seems forced to defend his position by retreating to some kind of bimodal distribution of phyletic rates, which seems not only fanciful but absurd. This situation narrows our focus to very small populations—populations so small that they are seldom recognized in the fossil record. A small population of this type, evolving as an independent entity, can form either by splitting off from other populations of a pre-existing species (speciation, as defined here) or by remaining after the severe constriction of a lineage as the sole surviving deme. This second phenomenon, which I will term bottlenecking, might be regarded as a potential mechanism for the acceleration of phyletic evolution. In Chapter 6, however, it will be shown that small populations of the sort that might evolve rapidly must form far less frequently by bottlenecking than by isolation from large species. In fact, dramatic phyletic divergence by bottlenecking must be exceedingly rare.

ACCOMPLISHMENTS OF
PHYLETIC EVOLUTION

Fossil data, of course, reveal little about evolutionary change in variables about which ecologists commonly theorize: reproductive output, growth rate, breeding season, and the like. Thus, the analysis of the present chapter does not necessarily negate the possibility that phyletic evolution accomplishes significant changes in these variables. Even so, the demonstrable weakness of gradual structural transformation, when coupled with theoretical considerations (pp. 48–51), suggests that only minor ecologic adjustments—what I have collectively termed fine tuning—are to be expected.

Structurally, change in body size may be the most conspicuous accomplishment of phyletic evolution. Size changes that are modest enough so that scaling problems require no major changes in bodily proportions can occur, in parallel fashion, among populations of a species that are united by little gene flow. Increase in size is far more common in the evolving form than decrease. For vertebrates, a pervasive agent of size increase may be sexual selection, in which large males father a disproportionately large percent of offspring. In fact, sexual selection, in operating directly upon reproductive output, may also be the primary source of some other kinds of phyletic change. It is conceivable that horns and antlers that serve males in combat or competitive display for the winning of mates have, to some extent, been produced phyletically. Elaboration of such features often accompanies increase in total body size.

Hallam (1975) reported apparent phyletic size increase in a number of fossil marine invertebrates and noted that for these animals this simple kind of change may be more significant than other types of phyletic evolution. It seems reasonable to suggest that a particular kind of sexual selection may be operating here. In most species of marine invertebrates, large numbers of gametes are produced annually. We would expect to find a general correlation in such animals between body weight and gonad weight, which is to say productivity of gametes. This condition might be expected to lead to gradual increase in mean body size for an entire species, up to a point where sexual selection for size increase is balanced by opposing selection pressures.

Previously, I argued that size increase generally prevails in evolution largely because most higher taxa, for reasons of scaling, originate at relatively small body sizes (Stanley, 1973b). The idea here is that increase in size requires morphologic specialization, which reduces ancestral potential. I also suggested that increase in size might usually occur in discrete steps. Løvtrup et al. (1974) have also favored this punctuational idea, based on the observation that body sizes of certain living and extinct species of mammals seem to be discontinuously distributed. It is the relative ease with which phyletic size increase can occur, along with fossil evidence, like that of Hallam mentioned in the preceding paragraph, that has persuaded me that evolutionary size increase is commonly phyletic in origin.

SUMMARY

Most net evolutionary change in phylogeny occurs during adaptive radiation. Certain enormous groups, like the Metazoa and the angiosperms, which once seemed to appear abruptly in the fossil record, can now be shown to have radiated during brief intervals of time. The gradualistic model cannot easily accommodate adaptive radiation because there is no apparent reason why any established species should evolve rapidly (abandon its original niche) simply because adjacent ecospace is available for occupancy. It is much easier to envision such ecospace being invaded through rapidly divergent speciation.

The gradualistic model can be tested more rigorously by examining the proposition that rapid phyletic evolution must yield chronospecies of short duration (such evolution will produce lineages that must be divided taxonomically into small segments). Given fossil data of unusually high quality, a survivorship curve for chronospecies can be derived from the recognized antiquity of living species with fossil records. From such a curve, we can plot a histogram of chronospecies longevities and also a histogram depicting rates of phyletic evolution. Plots of this type for Plio-Pleistocene mammals of Europe reveal very slow phyletic evolution: Very few mammalian chronospecies have spanned less than one-third of a million years, which means that very few lineages have evolved at rates exceeding three chronospecies per million years. The average is only about one chronospecies per million years. At least 15 genera of mammals evolved in Europe during the Pleistocene, which lasted 1.8 My, and only about 100 continuous phyletic pathways were available for the potential phyletic production of these genera. So slow were phyletic rates that hardly any new genera could have formed by phyletic evolution. It seems to be a general rule that genera of animals originate primarily by quantum speciation. Even the well-known phylogeny of the Elephantinae, a group characterized by relatively rapid phyletic evolution, conforms to this generalization. Similarly, human evolution, noted for its overall rapidity, is characterized by demonstrably slow rates of phyletic evolution—perhaps slower even than average rates for other contemporary groups of mammals. Thus, net rates of large-scale evolution in late Cenozoic mammals can only be explained in terms of punctuational evolution.

Although an equally rigorous analysis of invertebrate phylogeny seems beyond our grasp at present, the vast store of data available on the longevity of fossil invertebrate species provides powerful supporting evidence for the generalizations derived from the mammalian record. If the gradualistic model were valid, numerous generic transitions should be documented in the thousands of invertebrate lineages that have been traced through spans of 5 to 10 My, yet such transitions are rarely discerned.

5

Rates of Speciation
and Extinction

*By treating adaptive radiation as a pattern of exponential increase, we can esti-
mate rates of speciation and extinction for many taxa. This procedure is valuable
because major evolutionary changes tend to be associated with adaptive radiation. When we compare different taxa, we find that rate of large-scale evolution
tends to be correlated with rate of speciation (fraction of species added per unit
time), which is a primary determinant of the total number of species formed during a given interval. Rate of speciation varies enormously among taxa. Thus, a
typical radiating clade of mammals, when compared to a similar clade in the
Bivalvia, speciates at a much higher fractional rate, produces many more total
species per unit time, and undergoes much more rapid large-scale morphologic
change. On the other hand, within a particular radiation of any taxon, the most
rapid large-scale evolution usually does not occur late in radiation (when a large
total number of species are being added per unit time) but early, because at the
outset a higher percentage of speciation events are markedly divergent.*

*It is also instructive to consider the macroevolutionary fate of nonradiating
clades. The punctuational view of evolution offers the prediction that most clades
that persist for long intervals at low diversity will exhibit little evolution because
there will be few opportunities for quantum speciation. The paleontologic record
bears out this prediction. Virtually all extant members of such clades are primitive forms designated as "living fossils."*

INTRODUCTION

If speciation is the locus of accelerated evolution and therefore accounts for most large-scale change, it is important to seek ways to assess the rates at which species multiply. The punctuational view implies that rate of macroevolution within and among higher taxa should vary with rate of speciation, but focusing on this inference should not cause us to ignore another variable: degree of divergence per speciation event. Speciation can produce increments of divergence that are almost imperceptibly small, as in the formation of sibling species, but can also contribute sizable morphologic steps. In this chapter, techniques will be presented for estimating rates of speciation and extinction. These techniques will be applied to particular problems that relate to the importance of quantum speciation, the nature of which will be considered in the following chapter. An important conclusion of the present chapter will be that production of a descendant species is a rare event in the lifetime of an average chronospecies, yet only a small fraction of such events can represent quantum speciation. The inference will be that a major evolutionary transition is typically accomplished by a small number of punctuational steps of great magnitude.

ADAPTIVE RADIATION
AS EXPONENTIAL INCREASE

Among the most important contributions of Simpson (1944, 1953) has been his focus upon adaptive radiation as the site of most large-scale morphologic transitions. Accordingly, in order to examine the origin of higher taxa, we will now direct our attention to speciation within adaptive radiation. How rapidly do species multiply in radiation, is our question: How many successive steps along any single phylogenetic pathway are available for possible punctuational transition? Another of Simpson's major points has been that radiation generally follows (1) the emptying of ecospace by extinction, (2) the achievement of an adaptive breakthrough (origin of an evolutionary innovation), or (3) both of these occurrences. A fourth opportunity for radiation, the formation of a new habitat, can be viewed as being equivalent to (1).

Sometimes an adaptive breakthrough is so novel as to permit the invasion of ecospace never before extensively occupied. In the marine realm, the evolution of infaunal modes of life (life within the substratum) represents an example. The basic obstacle to be overcome here is the difficulty of living in a buried position while maintaining communication with the overlying water mass for respiration and, often, feeding. Early in the Paleozoic, most skeletonized marine invertebrates were epifaunal (surface-dwelling). Late in the Paleozoic and continuing into the Mesozoic, there occurred trends toward the origin and refinement of burrowing modes of life within groups of gastropod and bivalve mollusks and echinoids (Stanley, 1968; 1977). Presumably, comparable trends within soft-bodied taxa were initiated earlier, near the start of the Cambrian (Valentine, 1973; Stanley, 1976). In all cases, adaptive innovations are evident. For bivalves and gastropods, the key development was the evolution of siphons by mantle fusion; for echinoids, it was the origin of upward-directed petaloid tube feet; and for soft-bodied, worm-like taxa it seems to have been the evolution late in the Precambrian of a hydrostatic skeleton in the form of the coelom.

Perhaps the most famous example of radiation triggered by extinction is the rapid divergence of early Cenozoic mammals into ecospace vacated by the disappearance of the dinosaurs (Figure 4-5). Obviously, adaptive innovations played a role in the success of particular taxa that participated in the mammalian radiation.

A more recent and restricted radiation, but a dramatic one nonetheless, was the diversification of hydrobioid snails in the Mekong River, analyzed in an exemplary study by Davis (1979). The Mekong River system came into being in the Middle or Late Miocene, with the uplift of the Himalayas, as plate tectonic movements forced what is now peninsular India against the central landmass of Asia. Davis reconstructed an odyssey wherein hydrobioid snails of the southern hemisphere were rafted northward, with the fragmentation of Gondwanaland, as passengers on the small triangular landmass that became part of India. Their arrival automatically coincided with the Himalayan uplift and the formation of the Mekong drainage system approximately 12 My ago. The ensuing monophyletic radiation gave rise to at least 93 extant species of triculine snails in the Mekong River. The rate of radiation is uncertain, but the present level of diversity may have been approached very quickly. Davis attributed the success of the triculines to a lack of competitors in the newly formed river system and also to an adaptively plastic reproductive system that has allowed the triculines to occupy a wide variety of habitats. (Having evolved in southeast Asia, this subfamily of hydrobioids now ranges from India to the Philippines, with its highest diversity in the Mekong River.) Thus, Davis invoked both the presence of free ecospace and the origin of a key adaptation to explain the diversification of triculines. He noted, however, that in radiations of this type (those of newly formed lake or river systems) whether hydrobioid or cerithiacean snails dominate seems to be a matter of which group happens to arrive first.

The cichlid fishes of large African lakes and other bodies of water represent particularly striking examples of more recent radiations triggered by adaptive flexibility. As Liem (1973) has elegantly shown, the key development in cichlid evolution was the origin of a unique pharyngeal jaw that processes food and has thereby permitted the anterior mouth parts to evolve specializations for food gathering. When cichlids invade poorly inhabited bodies of water, they tend to radiate in the manner described for Lakes Victoria and Malawi (page 45; Figure 3-4). Among the feeding specialists that have evolved within Lake Victoria during the past 500,000 to 750,000 years are insectivores, mollusk crushers, crustacean eaters, phytoplankton collectors, croppers of macroscopic algae, and predators upon other fishes. Some of the latter feed upon the embryos and yolk sacs of other cichlids, and one even sustains itself by scraping scales from other fishes! This radiation seems to have been nearly monophyletic (Greenwood, 1974).

Speciation is a multiplicative process, and adaptive radiation, whatever its cause, is fundamentally a phenomenon of geometric or exponential increase. Therefore, the appropriate measure for rate of addition of species is not the absolute number, but the percentage, added per unit time (the fractional rate of increase). A general analogy with populations of individuals is readily apparent. Here the net fractional rate of increase equals per capita birthrate minus per capita death rate. The equivalent for adaptive radiation is speciation rate minus extinction rate. Thus we can borrow from demography the standard exponential equation for population growth:

$$\frac{dN}{dt} = RN,$$

(5.1)

where N is number of species, t is time, and R is fractional increase per unit time.
   Integrating,

$$N = N_0 e^{Rt},$$

(5.2)

where N_0 is the original number of species (unity for monophyly) and e is a constant, the base of natural logarithms.

Clearly, no radiation will follow exponential increase precisely or even approximate it for a long period of time. Presumably, a sigmoid curve is followed (Figure 5-1,A), like the one that typifies population growth in a confined habitat. If, however, we restrict our analysis to taxa in the early stages of adaptive radiation, increase will be approximately exponential. Suppose, therefore, that we consider an extant Cenozoic taxon well represented in the fossil record and currently in the midst of adaptive radiation. The immediate goal will be to calculate R, the net rate of increase. Knowing the approximate geologic time of origin of the taxon, we have a good estimate of t. We also have N, number of living species, to a good approximation. If we assume monophyly, N_0 will equal one. It is important to appreciate that both N and N_0 enter into the calculation of R logarithmically, allowing considerable room for uncertainty in the value of each. In other words, it is no problem that number of living species is often only approximately known, and the assumption of monophyly can be relaxed somewhat without major effect. It will be shown below, and in more detail in Chapter 10, that the characteristic value of R differs systematically among higher taxa, yet is remarkably consistent within a higher taxon (for example, among families of the Mammalia or Bivalvia, as will be shown below).

Equation 5.2 is equivalent to the equation describing radioactive decay,

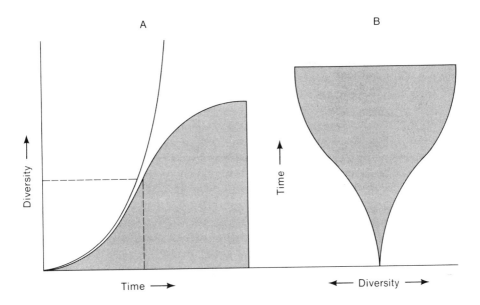

FIGURE 5-1
Graphic portrayal of unbridled and damped exponential diversification. A: Convex upward curve illustrates purely exponential increase. Sigmoid curve bounding the shaded
area represents damped exponential increase, in which the fractional rate of increase
declines as diversity increases. In the early stage of radiation (for example, the stage
represented by the portion of the graph between the origin and dashed lines), there is
little separation between the two curves. B: Shaded portion of A rotated 90° and replotted
symmetrically about a vertical axis to form a balloon diagram of the sort commonly used
to depict taxonomic diversity through time.

except that in the latter the exponent is negative. Just as radioactive decay is
characterized by a half-life, exponential increase entails a doubling time (t_2),
which can be calculated by setting $N/N_0 = 2$ in equation 5.2:

$$(5.3) \qquad\qquad t_2 = \ln 2/R$$

Since first making use of the preceding exponential techniques to compare
rates of radiation of different taxa (Stanley, 1975a; 1977), I have learned that
Yule (1924), a mathematician who developed an interest in evolution, employed
similar ones long ago and took them in directions not followed here. The same
general methods can be applied to the study of higher taxa (Sepkoski, 1978).

Figure 5-2 displays estimates of t, N, and R for currently radiating families of
Mammalia and of Bivalvia. What is particularly notable is that calculated values
of R are relatively consistent within each class and do not overlap between the
two classes. The highest value for mammals (0.35 My^{-1}) represents the Muridae

	t(My)	N(Species)	R(My^{-1})	$\bar{R}$(My^{-1})
Bivalvia				
Mesodesmatidae	47	40	0.078	
Cardiliidae	31	5	0.052	
Tellinidae	122	350	0.048	
Semelidae	47	60	0.087	
Veneridae	122	500	0.051	
Petricolidae	47	30	0.072	0.061
Myidae	61	20	0.049	
Teredinidae	61	66	0.069	
Lyonsiidae	47	20	0.064	
Mactridae	110	150	0.046	
Donacidae	76	50	0.051	
Mammalia				
Bovidae	31	115	0.15	
Cervidae	19	53	0.21	
Hystricidae	14	20	0.21	
Muridae	19	844	0.35	0.22
Cercopithecidae	19	60	0.22	
Cebidae	19	37	0.19	
Cricetidae	35	714	0.19	

A

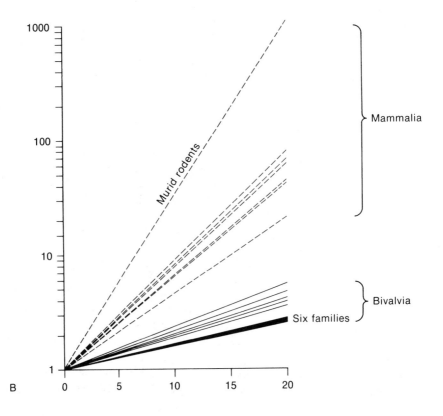

B

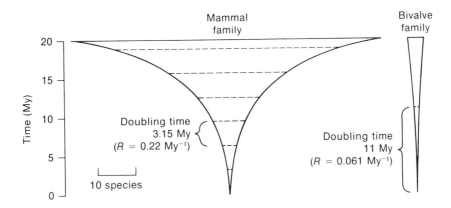

FIGURE 5-3
Patterns of diversification (plotted with an arithmetic scale) for an average mammal family and an average bivalve family during an initial interval of radiation lasting 20 My. The patterns are idealized in that it is assumed that radiation is perfectly exponential.

(Old World rats and mice), which possibly have radiated at an even higher rate than this because many sibling species may remain unrecognized; the lowest value (0.15 My^{-1}) represents the Bovidae (cattle, antelopes, etc.), which have been in existence for a rather long time and may have slowed from their initial rate. Clearly, exponential increase is very much more rapid in families of mammals, as illustrated in Figure 5-3, which depicts net rate of increase in an average family of each class. Doubling time (t_2) is approximately 11 My for the Bivalvia, compared to only 3 My for the Mammalia. The disparity between net rates of radiation in the two classes will be interpreted in Chapter 10.

It was my initial inclination to exclude very young families of bivalves from this kind of evaluation because most contain so few species as to seem aberrant in some way. Examples depicted in Figure 5-2 are the Cardiliidae, Petricolidae, and Lyonsiidae. As it turns out, these families are radiating as rapidly as somewhat older families. This finding serves to illustrate how, because we are accustomed to thinking in arithmetic terms, exponential increase can be misleading.

FIGURE 5-2
Rates of diversification of rapidly radiating extant families of bivalve mollusks and mammals. A: Values of R and data used for their calculation. Mean values ($\bar{R}$) for each class are shown at right. B: Logarithmic plots of paths that monophyletic exponential radiation would follow for intervals of 20 My at net rates calculated in A. (Data for bivalves from Stanley and Newman, 1979. Data for mammals from Romer, 1966, and Walker, 1975.)

The many groups of marine invertebrates that are characterized by very low rates of increase will, if they survive, automatically persist at very low diversities for long intervals of geologic time. Thus, the nature of exponential increase explains the characteristic pinched tail at the base of balloon diagrams depicting the stratigraphic distribution of well-studied taxa that have undergone an initial adaptive radiation (Figure 5-1,B and 5-3).

EXTINCTION AND SPECIATION

We have thus far considered only net rate of increase in number of species. This number (R) equals speciation rate (S) minus rate of termination of lineages (E). (A familiar analog here is the fate of money in a savings account, where net rate of increase equals interest rate minus rate of withdrawal.) Thus, if we can obtain an estimate of E, we can estimate speciation rate as

(5.4) $S = R + E$

There is no way of estimating values of E as accurately as we can estimate values of R. It is, however, possible to employ a method that yields approximate values of E. For mammals and bivalves, these values differ greatly, and in the same direction as values of R for the two classes. These estimates show quite conclusively that rate of speciation in adaptive radiation has been much higher for mammals. (In fact, simple inspection of Figure 5-3 will suggest that the situation could hardly be otherwise.) We will be following rather circuitous pathways in the estimation of rate of termination of lineages, and the procedure may seem tedious, but the results will prove to be quite valuable.

It should be noted that rate of termination of lineages (E) is only one component of total rate of extinction, the other component being rate of extinction by phyletic transition. Total rate of extinction will be labeled E'. We will proceed to estimate values of E' and then convert them to values of E. [*Note:* This usage of the symbols is the reverse of that employed elsewhere (Stanley, 1977).]

First, we will simplify things in order to make an approximation. Let us assume that all lineages of a hypothetical taxon lasted the same length of time, whatever the relative proportions of extinction by termination and extinction by phyletic transition happen to have been. Then, if the lineages are distributed evenly through time, we can easily calculate overall rate of extinction (E') as the reciprocal of this uniform duration. As shown in Figure 5-4, for example, if lineages all last 5 My, then one in five will go extinct every million years, or E will equal 0.2 My^{-1}.

Even if lineages vary in longevity, each one can be viewed as having contributed its own extinction rate to the total. Thus, if one lineage survived for 5 My.

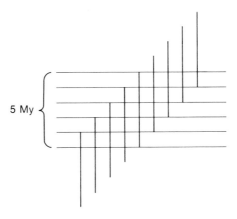

FIGURE 5-4
Species lasting 5 My, spread evenly through
time (vertical bars). During every million-
year interval, five species exist and one goes
extinct. Therefore, the rate of extinction is
0.2 My^{-1}. (From D. M. Raup and S. M.
Stanley, *Principles of Paleontology*, 2nd ed.,
W. H. Freeman and Company, San Francisco. Copyright © 1978.)

it can be viewed as having gone $\frac{1}{5}$ extinct every million years, or its contribution
to the mean extinction rate (E) will be 0.2 My^{-1}. The individual extinction rates
of all lineages must be averaged to give E. This average provides another explanation for E being 0.2 My^{-1} if all species last 5 My.

The lineages for which individual extinction rates are averaged to give mean
extinction rate must be the set of lineages existing at an *instant* in time, because
rate of extinction is a measure of instantaneous change. If instead we tally
longevities of species that existed during an *interval* of time, we will obtain a
different distribution of longevities. As shown in Figure 5-5, a slice of time,
which provides the appropriate data for our calculation, preferentially encounters long-ranging members of the set of species that existed during an interval. In
fact, as the figure illustrates, the sampling bias for the slice of time is directly
proportional to species longevity. This means that to convert a histogram of
lineage durations for an interval of time into a histogram representing an instant
in time, it is necessary to multiply the size of each duration class by the duration
represented (or by a number proportional to this duration).

Figures 5-6,A and 5-7,A display empirical "time interval" histograms of
species longevities for two taxonomic groups that have extraordinarily good
fossil records: European mammals that existed during the interval from the
Eburonian Stage through the Mindel Stage of the Pleistocene and planktonic

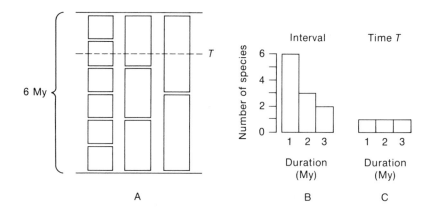

FIGURE 5-5
Conversion of a distribution of species longevities that represents an interval of time to a distribution that represents an instant in time. During the 6 My interval of time depicted in A, a total of 11 species existed, with ranges represented by vertical bars. B: Histogram of species longevities for the interval. C: Histogram of longevities of species existing at an instant in time (T); as shown in A, at time T or any other time, only one species of each duration class is in existence. Time plane T encounters a larger percentage of long-ranging species than short-ranging species. In fact, the fraction (or probability of encounter) for each duration class is directly proportional to the duration. Therefore, the histogram for time T can be derived from that for the interval by altering the fraction of species in each duration class in proportion to the duration that the class represents.

Foraminifera that existed throughout the world during the interval extending from the Late Miocene to the present. The high quality of the record of Pleistocene mammals has already been discussed (page 74). Planktonic forams also have a superb record because they rain down upon the floor of the deep sea, where sedimentary sequences are relatively complete. Some uncertainity is introduced to the data for mammals by the inclusion of species that originated at unknown times long before the start of the interval sampled. In addition, the histograms for both taxa include many species that are alive today. If full durations for these species were known, the histograms would be shifted to the right, or toward longer mean duration. These factors contribute to the lack of uniformity of the distributions, and of course it must be borne in mind that the plotted durations underestimate *lineage* durations because they represent *species*, some of which were "terminated" by pseudoextinction. This problem will be dealt with below.

The two empirical histograms, which represent intervals of time, have been converted to instantaneous distributions by adjusting the fraction of species in each duration class in proportion to the duration of the class, as illustrated in

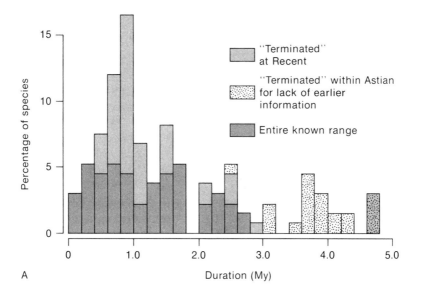

A

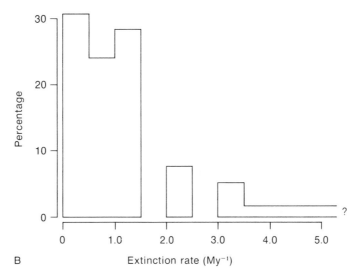

B

FIGURE 5-6

Durations and rates of extinction of mammal species in the Plio-Pleistocene of Europe. A: Histogram of durations. Data plotted are for all species considered to have existed during the interval extending from the base of the Eburonian Stage to the top of the Mindel Stage (Kurtén, 1968). For many species, only partial ranges can be plotted. B: Corresponding histogram of rates of extinction for an instant in time. To obtain this histogram, A has been converted to a histogram of durations for an instant in time by the technique shown in Figure 5-5. Next, the latter has been converted to the histogram depicting rates of extinction (B) by the technique shown in Figure 4-8. If data for full species ranges were available and if pseudoextinction could be excluded, the distribution of rates would be shifted to the left.

112

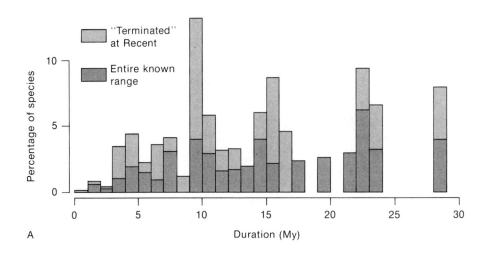

A

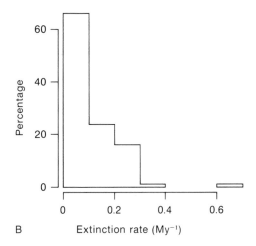

B

FIGURE 5-7
Durations and rates of extinction of planktonic foram species of the late Cenozoic. A: Histogram of durations. Data plotted are for all species considered to have existed during the interval of time extending from the Late Miocene to the Recent (Blow, 1969). B: Corresponding histogram of rates of extinction for an instant in time, derived from A in the same way that Figure 5-6,B was obtained from Figure 5-6,A; this histogram, like Figure 5-6,B, would be shifted to the left if full species ranges were available and if pseudoextinction could be excluded from the tabulation.

Figure 5-5. Mean extinction rates have then been calculated for mammals and planktonic Foraminifera by averaging the individual extinction rates of species (the reciprocals of species durations) for the set of species represented in the instantaneous distributions. The sets of individual extinction rates are displayed as Figure 5-6,B and Figure 5-7,B. The estimate of E' for planktonic forams is about 0.1 My^{-1} and for mammals, about ten times as high. Because the reciprocal of a duration, rather than the duration itself, is averaged, seemingly large errors in estimated longevities have a weak effect on the mean when lying toward the left-hand portion of the histogram, where all individual extinction rates are very low. There is such an enormous disparity between mean species durations and mean extinction rates for the two groups considered that the empirical data, though imperfect, are quite adequate for comparative purposes. For both groups, rate of termination of lineages is overestimated because of the inclusion of incomplete ranges of species and ranges terminated by pseudoextinction, but obviously a difference in the impact of these biases cannot account for the enormous difference in calculated rates of extinction.

It is no accident that mammals and planktonic forams were chosen here, because data of comparable quality to those plotted in Figures 5-6 and 5-7 are lacking for nearly all other taxa. Fortunately, there is another simple procedure that enables us to obtain a rough estimate of E' for a taxon with a reasonably good Cenozoic fossil record. This method employs the simple fact that as we look backward through the Cenozoic history of a taxon, we will find a decreasing percentage of still living species. I will refer to the plot depicting this decrease as a Lyellian Curve, after Charles Lyell (1830), who noted that percentages of extant species in fossil faunas decrease with the geologic antiquity of Cenozoic deposits.

Given past confusion, it is important to clarify the nature of a Lyellian Curve. Each point on such a curve represents faunas of a given age and indicates for these ancient faunas the percentage of species that are alive today. The curve does *not* depict percentages of living species that extend backward to various points in geologic time. A curve of the latter type would be much less accurate because of the incompleteness of the record.

A Lyellian Curve for mammals is shown in Figure 5-8,A. The beginning of the Würm serves as an endpoint here, in place of the Recent, because of the disruption that would otherwise be introduced as a result of the Würm and post-Würm mass extinction of mammals (page 75). At some point on such a curve, 50 percent of the species in an average fossil fauna will be extant. Doubling the age of this fauna will yield the approximate interval of time required for complete faunal turnover. This, then, is a rough estimate of mean species duration (Figure 5-9). The use of a single slice of time (that represented by the 50 percent point) is appropriate here because we wish to estimate rate of extinction, for which we require a sample of the lineages coexisting at an instant in time. Our estimate would be quite accurate if all species were of equal duration. In fact, like most distributions in nature, the histogram representing species dura-

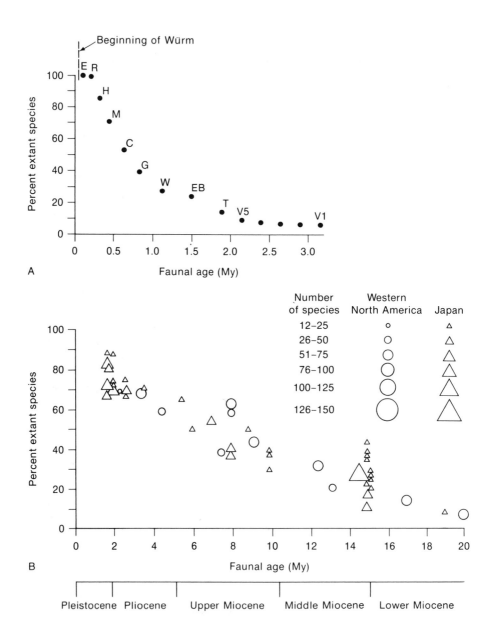

FIGURE 5-8
Lyellian Curves. A: Plot for Plio-Pleistocene mammals of Europe. This plot is terminated at the beginning of the Würm stage 40,000 years ago. Each point represents, for faunas of a given stage, the percentage of species that survived to the beginning of the Würm. Abbreviations used are the same as in Figure 4-7,A. (Data from Kurtén, 1968.) B: Similar plot for temperate and subtropical Bivalvia of the Pacific. This curve is terminated at the Recent. (From Stanley et al., 1979.)

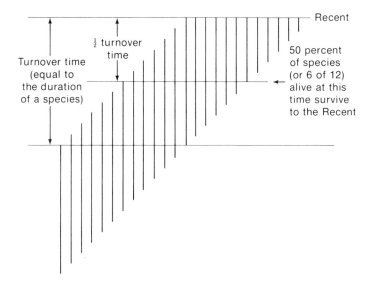

FIGURE 5-9
Diagram showing how, if all species are of equal duration and are evenly distributed through time, faunal turnover time will equal an average species duration, or twice the age of faunas in which 50 percent of all species survive to the Recent. (From D. M. Raup and S. M. Stanley, *Principles of Paleontology*, 2nd ed., W. H. Freeman and Company, San Francisco. Copyright © 1978.)

tions is presumably skewed to the right, meaning that more than half of the species will have ranges shorter than the mean. It should be evident that this condition will cause us to underestimate mean duration slightly when we use the 50 percent point, which relates to the median, in estimating one-half of the mean species duration. It will be shown below, however, that the error introduced here is relatively unimportant.

We can test the method described in the previous paragraph by comparing its results to the direct estimate of mean duration of 1.6 My obtained from the data plotted in Figure 5-6. The latter is manifestly an underestimate, because some species last longer than their recognized records indicate and because some ranges abut backward against the starting point of the interval considered or forward against the Recent time plane. The estimate (also a slight underestimate) obtained by doubling the age of the 50 percent point of a Lyellian Curve for mammals (Figure 5-8,A) is 1.4 My, which is in close agreement. Numbers of species of Late Cenozoic planktonic forams are too low to permit mean species duration to be meaningfully estimated from the 50 percent point of a Lyellian Curve. The general agreement of the two techniques, as applied to mammalian data, however, supports the notion that the indirect "50 percent point" method should provide a reasonably accurate estimate.

If we take mean duration of a mammal species to be 1.4 My, as estimated by the indirect method, total extinction rate (E') will be the reciprocal, or 0.71 My^{-1}. It might be assumed that the calculated value for E' is erroneously high because, as noted, the value for mean species duration obtained from the Lyellian Curve is an underestimate. Operating in the opposite direction, however, will be the error introduced by the use of mean species duration to calculate E'. This can be appreciated by referring back to Figure 4-8,B. Here rate of phyletic evolution is equivalent to rate of pseudoextinction. Recall that this plot is obtained from the plot for durations (Figure 4-7,B) by altering the horizontal scale of the survivorship curve in a nonlinear fashion. The result is that the true rate of pseudoextinction (1.1 My^{-1}) is considerably higher than the rate (<1 My^{-1}) that would have been estimated from Figure 4-7,B as the reciprocal of mean chronospecies longevity (>1 My). In a similar way, the "individual rate" of extinction of a species of average longevity based on data from a Lyellian Curve must represent an underestimate of true rate of extinction. Perhaps the degree of underestimation here approximately offsets the error in the other direction that results from the use of the 50 percent point of the curve. At least there must be a partial cancellation of errors.

For comparison with mammals, let us now return to the Bivalvia, for which we have excellent data for values of R in radiating families (Figure 5-2). The 50 percent point on a Lyellian Curve for temperate or subtropical bivalve faunas falls in the Late Miocene, representing a time about 7.5 My ago (Figure 5-8,B). Doubling this interval and taking the reciprocal of the result gives an estimate for E of 0.07 My^{-1}. This estimate is fully an order of magnitude lower than the one for mammals, reflecting much greater species longevity among bivalves. This disparity should dispel concern over the imperfect estimation of E' discussed above. At least in relative terms, the estimates are quite revealing. We can, then, combine the values of R calculated for these two classes (Figure 5-2), with estimates of E to obtain estimates of rates of speciation. First, a digression is necessary, to deal with the contribution of extinction by phyletic transition (pseudoextinction).

The preceding methods fail to take into account pseudoextinction. What is desired for estimation of S is not total rate of extinction (E'), but only rate of termination of lineages (E). The incidence of pseudoextinction is by no means negligible, as illustrated by the phylogeny of the Elephantinae (Figure 4-11), in which nearly half of the extinctions are thought to have been of this type (and some terminations may be the aberrant products of human activity). Other rather well-studied phylogenies, like that of the scallop *Argopecten* (Waller, 1969), reveal a similarly high incidence of pseudoextinction (Figure 2-2).

In the absence of precise information, we can obtain a maximum value of S by assuming that there is no pseudoextinction. We can then establish boundaries in the other direction by postulating various incidences of pseudoextinction, expressed as values of the ratio E/E'. This procedure is followed in Table 5-1,A, in

TABLE 5-1   Estimates for exponential variables in the Mammalia and Bivalvia, assuming various incidences of pseudoextinction (E/E'). A: Estimates of E and S. B: Estimates of $\hat{N}$ for radiations of 20 My and, for mammals, 10 My.

	A			B	
	E/E'	$E(\mathrm{My}^{-1})$	$S(\mathrm{My}^{-1})$	$\hat{N}(20 \mathrm{\ My})$	$\hat{N}(10 \mathrm{\ My})$
Mammalia	1.00	0.71	0.93	345	38
($R = 0.22 \mathrm{\ My}^{-1}$,	0.80	0.57	0.79	293	32
$E' = 0.71 \mathrm{\ My}^{-1}$)	0.50	0.36	0.58	215	24
	0.30	0.21	0.43	159	18
Bivalvia	1.00	0.090	0.15	8	
($R = 0.062 \mathrm{\ My}^{-1}$,	0.80	0.072	0.14	8	
$E' = 0.09 \mathrm{\ My}^{-1}$)	0.50	0.045	0.11	6	
	0.30	0.027	0.09	5	

which a ratio of unity represents a condition in which there is no pseudoextinction, and decreasing values of the ratio represent an increasing incidence of pseudoextinction (to a level of 70 percent). It is very important to recognize that the degree of flexibility represented here greatly exceeds uncertainties in the value of E' discussed above.

Thus, while I am making use of Lyellian Curves in only a crude manner, by employing only the 50 percent point, I see no way in which they can practically be used more precisely for the calculation of lineage durations and E. There is simply no way of factoring out pseudoextinction. Furthermore, as will now be shown, the estimates of E employed here are accurate enough to be of value.

An estimate of speciation rate (S) is obtained by adding various estimates of E to the mean value of R in Figure 5-2. Note that even if pseudoextinction constituted 70 percent of E in mammalian phylogeny and were nonexistent in bivalve phylogeny (an absurd contrast), estimated speciation rate would still be more than twice as high for mammals (0.21 My^{-1} versus 0.09 My^{-1}). Certainly, the actual disparity in values of S must be much greater, because it is unlikely that incidence of pseudoextinction is appreciably higher for mammals (compare Figures 2-2 and 4-11). If incidence of pseudoextinction is more or less the same in the two classes, the value of S for mammals is estimated to be about five times as high. The importance of the difference in rate of speciation can hardly be overemphasized. Because S appears exponentially in the calculation of N, a five-fold increase in its value has an enormous effect. Recall that less than a three-fold difference in value of R produced the disparity in net diversification depicted in Figure 5-3. (Note that it has been tacitly assumed that the value of E in adaptive radiation, for which S is being calculated here, is approximately equal to its value in phylogeny in general, from which it has been estimated.)

Clearly, rate of extinction and rate of speciation in adaptive radiation are much higher within the Mammalia than within the Bivalvia. While it happens that this contrast accompanies a similar disparity for values of R, it is important to recognize that, arithmetically, this need not be the case. A given value of R can result from high or low values of S and E. It will, nonetheless, be shown in Chapter 9 that the comparison between the Mammalia and Bivalvia is typical: Values of R, E, and S within taxa tend to be intercorrelated throughout the animal world.

RATE OF SPECIATION AND
RATE OF LARGE-SCALE EVOLUTION

In asserting that speciation is the locus of most large-scale change, the punctuational model offers the prediction that taxa characterized by high rates of speciation will also be typified by high rates of large-scale evolution. Taxonomically, we would expect to find that families and orders of such taxa originated in brief intervals of time and, morphologically, we would predict that adaptive innovations should have appeared very rapidly. These predictions seem to be borne out by an abundance of fossil evidence, part of which will be discussed in this and subsequent chapters. It is interesting to compare rates of appearance of families in the phylogenies of the two classes whose rates of speciation have been contrasted: the Mammalia and the Bivalvia (Stanley, 1977). For Cenozoic mammals, peak diversity was attained early in the Oligocene, only about 30 My after the modern radiation began (Figure 5-10,A). Number of families has actually declined slightly since Oligocene time. Bivalves show a much slower build-up of family diversity: There is no evidence that a limit is being approached even after more than 400 My of radiation (Figure 5-10,B).

There is, of course, no reason to believe that a family of bivalves is in some way equivalent to a family of mammals. Judging from the adaptive traits of families in each group, however, it would appear that after only 30 My of radiation, the basic adaptive limits of the mammalian body plan had been approached. Subsequent diversification has been largely a matter of adaptive refinement. In contrast, origins of the basic modes of life of modern bivalves required long spans of geologic time. For example, an interval of more than 200 My was required for the evolution of epifaunal attachment by cementation and also for the evolution of habits that simply entailed an unattached existence on the substratum. The origin of deep-burrowing habits, except perhaps in the unusual Lucinacea, required about 150 My. The ability to burrow rapidly, by modern standards, and to secrete shell ornamentation that aids in burrowing took even longer to appear.

Body size is perhaps the simplest and most directly comparable morphologic feature that serves to contrast rates of diversification for bivalves and mammals.

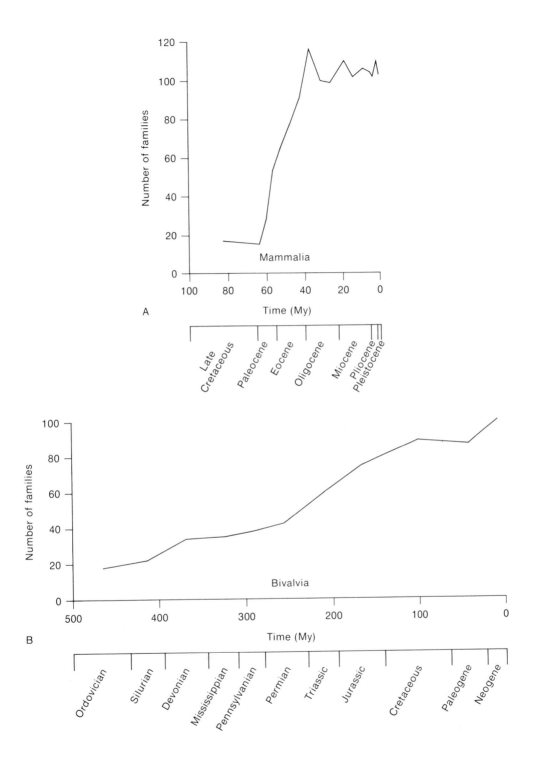

FIGURE 5-10
Diversity of families through time for the Mammalia (A) and Bivalvia (B). (Data for A from Lillegraven, 1972, and for B from Stanley, 1973c.)

Terrestrial mammals that were large by modern standards had evolved from small ancestors by the Late Paleocene and Early Eocene, after only about 10 to 15 My of mammalian radiation. By the Early Oligocene, only about 30 My after the radiation began, the largest land mammal of all time, *Indricotherium*, was in existence (Kurtén, 1971). The mammalian record is replete with examples of rapid phylogenetic size increase. In contrast, the first relatively large free-burrowing clam was *Megalomoidea*. This animal did not appear until after nearly 100 My of radiation. Even more strikingly gradual was the filling out of the modern spectrum of sizes in the epifaunal realm (the adaptive zone of bivalves that rest on the surface of the substratum). Not until the Jurassic and Cretaceous did the epifaunal inoceramids and rudists independently attain huge sizes (sizes comparable to those of modern tridacnids).

The many fundamental adaptations of the bivalves that were slow to originate cut across adaptive zones and taxonomic boundaries, and there is no evidence that general rate of speciation within the class was appreciably lower in the Paleozoic than at later times. Furthermore, comparable adaptive transitions within subtaxa arising in the Mesozoic and Cenozoic were equivalent, in rate, to those of the Paleozoic.

The mammals, then, have speciated at much higher rates than the bivalves and have accomplished major adaptive transitions much more rapidly. These conditions are compatible with the punctuational prediction that rate of large-scale evolution should correlate with rate of speciation. It might, however, be argued that the same correlation would be expected if phyletic evolution prevails in phylogeny. The point would be that number of rapidly evolving lineages, as a stochastically constant fraction, should increase with total number of lineages. A way of choosing between these possibilities is to hold phyletic evolution constant, in effect, by examining what happens when speciation is increased within the Bivalvia themselves. A natural experiment of this type has been provided by Late Cenozoic physiographic changes in the region now occupied by the Caspian Sea.

During the Pliocene Epoch, the Caspian region was occupied by the Pontian Sea, a brackish body (or group of bodies) of water isolated from the Mediterranean. Here, in less than 3 My there arose numerous species of cockles of diverse morphologies (Figure 5-11). The exact number is not certain, but more than thirty endemic genera are recognized (Gillet, 1946; Zenkevitch, 1963; Ebersin, 1965). These are assigned to four subfamilies. Many of the genera are quite unusual in shape and ornamentation pattern when compared to typical cockles. Apparently all of the genera arose from one or a small number of species of *Cerastoderma*, the genus of the familiar living European cockle. The point is that speciation occurring at rates which were extraordinarily high for the Bivalvia was accompanied by unusually rapid rates of morphologic evolution. Meanwhile, in the ocean, *Cerastoderma* speciated at the slow rate that typifies marine bivalves. It contains three living species in England, for example, and one in the United States. This low rate of speciation was accompanied by very little evolu-

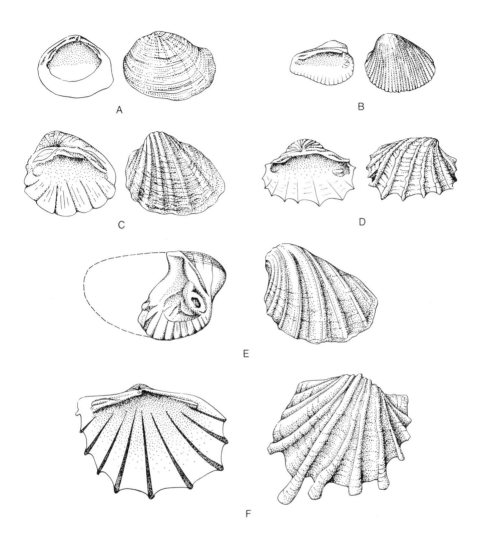

FIGURE 5-11
Some of the cockles of the family Limnocardiidae that evolved in the Pontian Sea. A: *Phyllicardium planum*. B: *Pseudoprosodacna sturi*. C: *Limnocardium squamulosum*. D: *Limnocardium fittoni*. E: *Prosodacna haueri*. F: *Budmania semseyi*. (From Gillet, 1946.)

tionary change. Modern members of the genus look much like those of the Late Oligocene, when *Cerastoderma* first appeared. As noted for cichlids and other taxa that have speciated and evolved rapidly in large African lakes (page 45), we cannot rule out, as an alternative explanation, the phenomenal acceleration of phyletic evolution. Still, this alternative is utterly fanciful: We have neither documentation nor theoretical explanation for any such acceleration. What we do have is direct evidence that speciation was occurring at an extraordinarily high rate. We have, in effect, tested the punctuational model and it has passed, which is to say that it remains intact, if not proven.

I previously suggested that the gradualistic model be put to a complementary test, called the test of generation time (Stanley, 1975a). If rate of speciation should correlate with rate of large-scale evolution in a punctuational scheme, then in a gradualistic scheme, rate of evolution should correlate with the equivalent parameter, birthrate. Net rate of population increase, a parameter generally equivalent to R of this chapter, is closely tied to birthrate, but also to generation time, which is a more widely measured variable. (Figure 7-8, (page 210), displays the inverse correlation.)

It is a self-evident tenet of population genetics that, all else being equal, rate of phyletic evolution should vary inversely with generation time. From this condition, we can predict that if the phyletic component of evolution dominates in the history of life, we should find among higher taxa an inverse correlation between generation time and large-scale evolution. It has, in fact, long been recognized that no general correlation of this type exists (Zeuner, 1931; Simpson, 1949). As we have already noted, marine invertebrates with generation times in the order of one or two years exhibit more sluggish large-scale evolution than mammals that mature after five or ten years. Unfortunately, a problem is now evident for the test of generation time. This is that generation time does not seem to be a major determinant of rate of phyletic evolution even within individual classes. In particular, recall that the most rapid phyletic evolution currently recognized in the Mammalia has occurred within the elephant family (Figure 4-11), whereas the most ancient living mammal species of Europe are animals of very small body size and short generation time (page 79). I would agree with Mayr (1970, p. 344), that the overriding factor here is population size. The stabilizing effects of spatial heterogeneity (pp. 48–51) would be expected to be less effective within small populations than within large ones.

The test of living fossils (Stanley, 1975a), which will now be described, represents a more fruitful avenue of departure from the exponential methodology introduced in the previous section. If should be stressed, at this point, that the analysis of the preceding chapter would seem to refute the gradualistic model. The arguments of the present chapter that relate to the controversy over gradualism are presented as corroboration.

LIVING FOSSILS REINTERPRETED

Rather than examining what happens when species multiply at exceptionally high rates, we can investigate the results of unusually sluggish speciation. A corollary of the punctuational model is that small clades persisting for long intervals of geologic time should exhibit little net evolutionary change. (The most extreme case would be that of a clade consisting of a solitary lineage.) The gradualistic model offers no such prediction, for if phyletic evolution normally proceeds at a moderately rapid pace, lineages belonging to very small clades, like those belonging to large clades, should often exhibit considerable change during long intervals of geologic time.

These deductions permit us to formulate a test of the two models. Do small clades persisting for long periods display very little evolution or varying, and often substantial amounts? The fossil evidence here is quite clear: Such clades seem almost invariably to exhibit morphologic stagnation. They are what are commonly called **living fossils.** This label has been used in a variety of ways, but I would restrict its application to extant clades that fulfill the following qualifications: (1) They must have survived for relatively long intervals of geologic time at low numerical diversity, often as the sole survivors of previously diverse taxa. (2) They must today exhibit primitive morphologic characters, having undergone little evolutionary change since dwindling to low diversity at some time in the past. Thus, it is appropriate to label this test (passed by the punctuational model and failed by the gradualistic model) the test of living fossils (Stanley, 1975a).

The name of this test might seem to suggest circularity, in that if only living fossils were considered, then little evolution *could* be found. But this is not the procedure followed. Rather, the test is conducted by examining *all* clades of low diversity that can be traced by fossil evidence over long intervals. It would be tempting to include here small, extant groups that clearly are living fossils but that are not connected by known fossil evidence to similar, obviously ancestral forms recognized far back in the record. This would be unfair, however, because we would know in advance that all such forms (e.g., *Neopilina, Sphenodon, Tarsier*) have undergone very slow change, yet we would automatically be excluding from our compilation any small, extant clades that might have undergone considerable phyletic evolution without leaving a record to tie them to their otherwise unrecognizable ancestors. We must therefore restrict our analysis to small clades that have fossil records continuous enough to be traced backward for a substantial period of time, regardless of the rate of change recorded. As it turns out, all such clades that I have been able to identify have evolved very little. In other words, their Recent representatives are living fossils—a condition that stands in opposition to the gradualistic model. A compilation of examples appears in Table 5-2. (It should be noted that for a slowly evolving invertebrate group, 100 My is a long interval to survive with little change; for a rapidly evolving group like the Mammalia, 10 to 20 My is a long period.)

A detailed look at examples from Table 5-2 will more fully elucidate the test. The lungfishes represent a particularly well-studied group (Westoll, 1949; Simpson, 1953). They underwent rapid changes in morphology only when speciating rapidly in the mid-Paleozoic, soon after appearing (Figure 5-12). Then, in the late Paleozoic, rate of speciation decreased abruptly and so did rate

TABLE 5-2 Compilation of clades that contain few lineages but have persisted to the present from distant geologic intervals while leaving relatively continuous fossil records. The approximate time at which each clade began existence at low diversity is given in parentheses. Note that none of the clades exhibits appreciable evolutionary change; the extant representatives of each are living fossils. Few invertebrate clades are listed, in part because few that would otherwise qualify display persistent fossil records. Other invertebrate taxa, like the lingulid brachiopods, pleurotomariid gastropods, and pinnid bivalves, are excluded because they contain several species today and may have undergone a rather large total number of speciation events during long intervals of persistence with little evolutionary change.

Echinoneid sea urchins (Late Cretaceous—80 My)

There are only three living species assigned to the Echinoneidae, as now constituted. These are assigned to two genera, one of which is represented by a meager Cenozoic fossil record. Only one other genus, containing a single Upper Cretaceous species, is recognized (Mortensen, 1948). The family has undergone little structural evolution.

Horseshoe crabs (Early Triassic—230 My)

The superfamily Limulacea has a sparse Mesozoic and Cenozoic fossil record. An Atlantic and a Pacific subgroup exist today, the Pacific one bearing the greater resemblance to *Mesolimulus* of the Mesozoic. Daniel C. Fisher informs me that the group displays a larger degree of evolutionary change than has often been assumed.* Still, there are strong similarities among post-Paleozoic members of the Limulacea, considering the long interval during which the superfamily has persisted.

Bairdiid ostracods (Early Triassic—230 My)

The family Bairdiidae has survived since the Paleozoic with little multiplication of species and little morphologic change. *Bairdia*, the likely ancestral genus, exhibits an enormous stratigraphic range (Ordovician to Recent). The genera of the post-Paleozoic portion of the clade, numbering about four, are quite similar to *Bairdia* of the Late Paleozoic (Sylvester-Bradley, 1961).

Galatheid anomuran crabs (Middle Jurassic—170 My)

The family Galatheidae is represented by only two living genera. Its fossil record exhibits much lower taxonomic diversity than is found in other groups of crabs (Glaessner, 1969). During this long period, it has undergone little structural change, remaining as a morphologically primitive decapod.

Notostracan crustaceans (Late Carboniferous—305 My)

The very small clade represented by the order Notostraca exhibits almost no morphologic change whatever. One of the two living genera extends back to the Late Carboniferous and the other, to the Jurassic. No other fossil genus is known. Two Triassic forms are assigned to living species, making them the oldest living animal species on record (Longhurst, 1955).

Bowfin fishes (Albian—105 My)

The family Amiidae has never speciated to speak of and has not been transformed appreciably since first appearing in the record (Figure 5-13). Boreske (1974) has

Table 5-2, continued

documented an excellent Cenozoic fossil record, which never displays more than two contemporaneous species.

Sturgeons (Late Cretaceous—80 My)

Today only two living genera represent the family Acipenseridae. Most species of the family are assigned to the living genus *Acipenser*, which extends back to the Cretaceous with little morphologic change (Romer, 1966).

Garfishes (Late Cretaceous—80 My)

Lepisosteus, one of the two living genera of the family Lepisosteidae, contains only four living species and the other genus, *Altractosteus*, comprises three. The known fossil records of both genera extend back to the Upper Cretaceous, by way of fossil material scattered throughout several Cenozoic stages. The records exhibit little evolution (Wiley, 1976).

Sirens (Late Cretaceous—80 My)

The amphibian suborder Sirenoidea includes only three living species that are assigned to two genera, both of which extend back to the latter part of the Cretaceous without essential change (Estes, 1970).

Snapping turtles (Late Paleocene—57 My)

There are three living species of snapping turtles (family Chelydridae) belonging to two genera (Ernst and Barbour, 1972). There is little morphologic difference between the oldest known species of the late Paleocene and the living representatives (Estes, 1970).

Alligators (Early Oligocene—35 My)

The alligator clade (subfamily Alligatorinae) includes only two living species and apparently has at no time harbored a large number of species (Steel, 1973). It also exhibits little structural evolution.

New World porcupines (Early Oligocene—35 My)

Wood and Patterson (1959) found evidence for very few lineages of the Erethizontidae and very little morphologic change since the Early Oligocene. Of the four living genera, three are monophyletic. The fourth, *Coendon*, contains about 20 named species (Walker, 1975), but the morphologic diversity represented is low, and they may represent taxonomic oversplitting. In any event, *Coendon* is a young genus, leaving a long interval apparently spanned by few lineages.

Aardvarks (Early Miocene—20 My)

There is but a single living species belonging to the order Tubulidentata, and the fossil record reveals only a small number of lineages as far back as the base of the Miocene, where the earliest known representatives are found (Patterson, 1975). Only modest structural changes have occurred since this time. Patterson believes that the distinctive genus *Plesiorycteropus* of Madagascar represents a clade separated from that of other known aardvarks since the Eocene.

TABLE 5-2, continued

Sewellels, or mountain "beavers" (Middle Miocene—15 My)

The family Aplodontiidae contains only one living species and has existed at very low diversity since mid-Miocene times (Romer, 1966). When the group was speciating more rapidly early in the Miocene, it apparently gave rise to the Mylagaulidae, a highly specialized family of burrowers. The subsequent history of the Aplodontiidae accounts for little morphologic change.

Tapirs (Early Miocene—20 My)

The family Tapiridae contains only four living species (Walker, 1975). During the past 15 My, only two genera are known to have existed. Diversity was apparently also low in the Oligocene and Early Miocene, when there were a few additional genera, but "generic distinctions are slight" (Romer, 1966, p. 220). Very little change in form is apparent since the Oligocene.

Pangolins (Early Oligocene—35 My)

There are seven living species of the family Manidae, which has diversified quite slowly, the oldest recognized species being Early Oligocene (Emry, 1970). The living species are similar enough that some workers refer them to one genus. A total of seven extinct species are recognized, apparently documenting low diversity since the Early Oligocene. The overall degree of morphologic change has not been great.

---

* Daniel C. Fisher, University of Rochester, 1977.

of morphologic change. (In the absence of adequate fossil data at the species level, generic diversity is plotted in Figure 5-12 as an index of species diversity.) A better documented example is that of the bowfin fishes (Boreske, 1974). The small extant clade of bowfins can be traced back to the Late Cretaceous. Fossil data, which are available for every epoch of the Cenozoic, reveal the existence of no more than two species at any time (Figure 5-13A, B). During an interval of about 100 My, change of form was restricted largely to phyletic transformation and was also quite trivial (Figure 5-13,C).

Various hypotheses have been advanced to account for the highly problematical nature of living fossils in the context of gradualism, but generally with expressions of uncertainty (see review by Simpson, 1953, pp. 327–335). Darwin (1859, pp. 105–108), who introduced the phrase "living fossils," saw these forms as resulting from a combination of long survival of lineages and remarkably slow phyletic evolution, both of which he attributed primarily to an absence of ecologic competition. Delamare-Deboutteville and Botosaneanu (1970), who reviewed numerous examples of living fossils, saw them as creatures that have "stopped participating in the great adventure of life," being confined by narrowness of adaptation. Simpson (1953, p. 331), on the other hand, considered them to be characterized by broad adaptation. In short, they have defied satisfactory explanation in the framework of gradualism. Why is this?

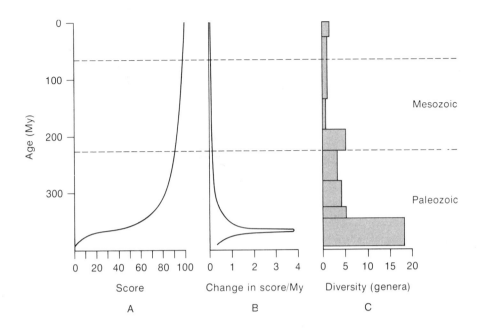

FIGURE 5-12

Rates of evolution in the lungfishes (Dipnoi). A and B: Plots showing changes in skeletal morphology since the group appeared, about 400 My ago. Scores are composite estimates of degree of advancement relative to the most primitive condition. (From Simpson, 1953, based on data from Westoll, 1949.) C: Diversity through time. Note that little morphologic change has occurred since the mid-Paleozoic, when rate of production of species and genera declined to a low level. (From Stanley, 1975a.)

The postulation of aberrantly sluggish rates of phyletic evolution to account for living fossils has persisted in part because of a general misconception about the distribution of rates of phyletic evolution within higher taxa. Many people have believed rapid evolution to be the norm. Simpson (1944; 1953), for example, plotted survivorship curves for extinct genera of carnivorous mammals and of bivalves, and with the tacit assumption that generic durations are inversely related to rates of phyletic evolution, he attempted to convert the data upon which these curves were based into histograms depicting evolutionary rates (Figure 5-14, A–D). Two difficulties must be pointed out here. First, the idea that generic durations are primarily the product of phyletic rates is strongly gradualistic. (Even many gradualists have assumed that most genera go extinct by termination of lineages.) In a punctuational framework, the distribution of generic longevities should be some complex function of (1) the distribution of numbers of lineages within genera and (2) the distribution of lineage *durations* within genera. Second, Simpson unfortunately assumed that, if rate of evolution

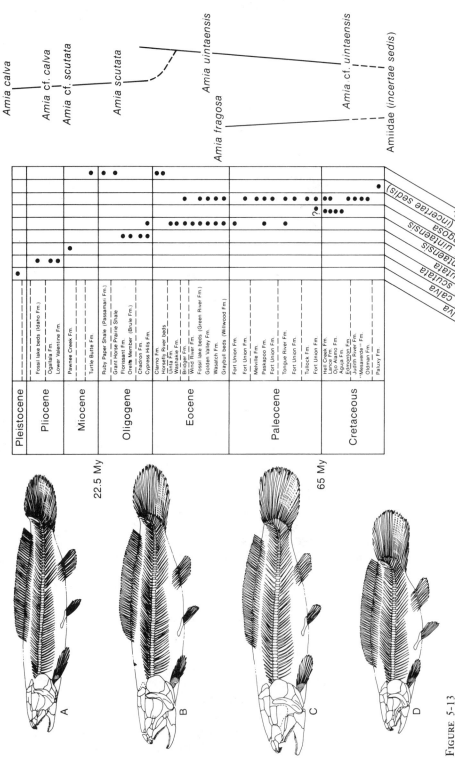

FIGURE 5-13
Low diversity of bowfin fishes (Amiidae) since their origin. Drawings A to D show the modest amount of morphologic change that has occurred since the group's origin, and especially since the beginning of the Cenozoic. Time scale nonlinear. (After Boreske, 1974.)

is the reciprocal of taxonomic longevity, the distribution of rates should be the mirror image of the distribution of longevities. This assumption produces a spurious, negatively skewed shape (Figure 5-14,C and D). As was shown in Chapter 4 (Figure 4-8), to plot a histogram for rates it is necessary not only to reverse the horizontal scale, but to distort it. The distributions derived in this manner from Figure 5-14,A and B are shown as Figures 5-14,E and F, which are positively skewed. These distributions have also been calculated to represent an instant in time (see Figure 5-5), whereas Simpson's histograms (Figures 5-14,C and D) represent an interval.

From Figures 5-14,C and D, Simpson inferred that only a very small fraction of taxa undergo very slow evolution. Given this gradualistic conclusion, living fossils seemed to represent entities that would be expected to exist as the predictably rare products of slow phyletic transformation. (The reason for phyletic stagnation remained a mystery.) Because even the proper histograms for genera primarily represent rate of termination rather than rate of phyletic transition, let us focus upon Figure 4-8, which represents only phyletic evolution. Here the mode lies adjacent to the ordinate. It is rapid, not slow, phyletic evolution that is rare. So low are phyletic rates, in general, that rates for living fossil taxa can easily fall in the vicinity of the mode. A clade that consists of a small number of lineages evolving at the average phyletic rate needs simply to *survive* in order to yield living fossils.

Simpson (1944) coined the term **bradytelic** to describe lineages that seemed to exhibit extraordinarily slow evolution in comparison to rates for related taxa. Based on the idea that that rates of phyletic evolution are negatively skewed within classes (Figure 5-14,C and D), it seemed possible to single out as bradytelic the lineages forming the left-hand tail of an unusually strongly skewed distribution. Given the actual positively skewed shape of distributions, the concept of bradytely loses its utility: The mode of the distribution lies where the bradytelic tail was thought to be.

To be more specific, from Figures 4-8, 4-11, and other data and inferences presented in Chapter 4, it seems reasonable to conclude that phyletic evolution will seldom accomplish a genus-level transition within a mammalian lineage spanning 5 My. Even in 20 My, rather little phyletic change is to be anticipated and, in fact, few mammalian lineages survive as long as this. Still, some lineages would be expected to span 20 My because the distribution of lineage durations is certainly positively skewed. The more youthful representatives of such long-ranging lineages, if alive today, would be predicted to be exactly what they are: living fossils.

It is important to recognize that the punctuational model predicts that some taxa should also persist with little change within clades that undergo extensive branching. This is to be expected, first, because some lineages within such clades will happen to be unusually long-lived and, second, because not all speciation produces marked divergence. In other words, the punctuational model does not predict the absence of primitive taxa from groups that are diverse today.

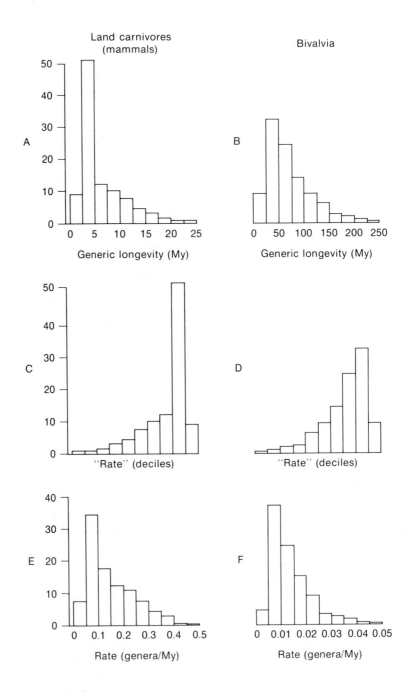

FIGURE 5-14
Inference of the distribution of rates of evolution—or, more properly, rates of extinction—from longevities of genera. A, B: Histograms, for an

Such taxa do, in fact, exist. One example is the moderately diverse nuculoid bivalves, which seem to have changed little since the early Paleozoic. Certain members of the Didelphidae (opossum family) may also represent examples (Simpson, 1953, p. 328). *Dicerorhinus*, the two-horned Malaysian rhino and one of the oldest living genera of mammals, with an antiquity of about 35 My, seems to be another. While surviving with little change, it has apparently given rise to other distinctive rhinos (Kurtén, 1971, p. 91).

It is also important to recognize that an example or two of substantial change within a small clade would not refute the conclusions reached above. As will be shown below, even a single speciation event can produce marked divergence. Although it is improbable that one or two such events will occur within a small clade, it is not impossible.

It might be claimed that in transferring the evaluation of living fossils to a punctuational framework, I am simply switching the problem from the previously alleged difficulty of accounting for unusually slow phyletic evolution to a problem of explaining failure to speciate. Why should a lineage or small clade almost cease to cast off descendants over a long span of geologic time? I find this question in no way perplexing. It will be argued in Chapter 7 that adverse conditions more readily stifle speciation than they cause the extinction of large, established species. Thus, it is not difficult to imagine that a species or small clade, faced with changing biotic conditions, as when more advanced competitors and predators evolve, might be adversely affected to the degree that it would speciate very little, but not to the degree that it would become extinct. While clades of living fossils are geologically long-lived, there is no evidence that the same is true of the individual lineages that form them. If component lineages are characteristically long-lived, however, the resulting co-occurrence of longevity and low rate of speciation might be explained in part by the fact that both traits characterize species that disperse readily (see Chapter 9). Of course, many living fossil forms are not widespread, but occupy geographic refugia. Here the explanation may be that while protection is afforded within such an area, there is little opportunity for speciation because of severely restricted opportunities for isolation and partitioning of the environment.

Finally, it should be noted that the persistence within cichlid faunas in and around the Great African lakes of species of apparently ancestral morphologies (pp. 45–47) represents a kind of microcosmic test of living fossils. Here, in considering the persistence of an ancestral form, we are perhaps evaluating a single

instant in time, of generic longevities for mammalian land carnivores and bivalve mollusks. (Data from Simpson, 1944; 1953.) C, D: Histograms purported to depict rates of phyletic evolution; these rates are the mirror images of those displayed in A and B. (From Simpson, 1944; 1953.) E, F: Correct histograms of rates of evolution (actually rates of extinction) for genera; these are derived from A and B by way of survivorship curves by the method illustrated in Figure 4-7. Like the distributions of durations (A and B), they are strongly skewed to the right.

lineage rather than a clade so that, in effect, rate of speciation is zero. The relatively unchanged species are simply survivors. Their rates of phyletic change, though low, are not necessarily lower than average for cichlids of African lakes. In contrast, we can infer that the ancestry of the divergent forms includes one or more speciation events. The reasoning here is that a large portion of the chichlid radiation of Lake Victoria, for example, seems to be monophyletic. Therefore, only one or a very few of the morphologically diverse living species *could* have descended phyletically from the ancestral species.

NUMBER OF SPECIATIONAL
STEPS IN ADAPTIVE RADIATION

The evidence that phyletic evolution makes only a weak contribution to large-scale transition allows us to focus upon the speciational component. The important question is: Within a particular major taxon, how many speciation events stand between a species that initiated adaptive radiation and an advanced, younger species? If we find that many speciational steps separate the two, then each quantum step may represent modest divergence. If calculation reveals few speciational steps, we will be forced to invoke dramatically divergent speciation as the source of most morphologic change. The first goal will be to estimate total number of speciation events in adaptive radiation. [As before (Stanley, 1977), I acknowledge the assistance of my colleague Owen M. Phillips in providing the following derivation.]

Consider a taxon now in the midst of adaptive radiation and containing N living species. The number of lineages N_T at some previous time T can be obtained by calculating exponential decrease from N, the number of living species, at the rate at which increase actually occurred:

$$N_T = Ne^{Rt},$$

where t is negative, equaling $-T$. The total number of lineages terminated in the next interval of time will have been E times the number of lineages present:

$$ENe^{Rt}.$$

The total number terminated since the start of a monophyletic radiation is the summation for all past intervals of time:

$$EN \int_{-\infty}^{0} e^{Rt}\, dt = \frac{EN}{R}.$$

The total number of lineages that have existed $(\hat{N})$ is this number plus the number (N) still in existence:

$$(5.5) \qquad \hat{N} = \frac{EN}{R} + N = \frac{SN}{R}$$

In general, we would expect the chance of an adaptive breakthrough occurring to be directly proportional to $\hat{N}$ for a segment of adaptive radiation. Thus, calculation of values for $\hat{N}$ indicates quite clearly that a group like the Mammalia would be expected to multiply at the family level much more rapidly than the Bivalvia. In Table 5-1, column B shows that in 20 My of radiation, an average family of mammals will produce between about 159 and 345 species, depending on the degree to which pseudoextinction contributes to total rate of extinction. For bivalves, the comparable range is only 5 to 8 species! Although families are not necessarily perfectly equivalent in the two groups, it is hardly surprising that the Cenozoic radiation of mammals generated about 100 families in only 25 My, whereas the bivalves did not attain comparable diversity until late in the Cenozoic, after nearly 500 My of persistent radiation (Figure 5-10).

The parameter N is also useful for evaluating numbers of successive speciational steps along phylogenetic pathways in adaptive radiation. What we can do, for example, is to estimate numbers of successional steps from which changes could accumulate over some interval of time, to form one family from another. Before embarking on this kind of evaluation, it will be instructive to step back and take a broader view. We can, for example, note that a family undergoing decline that is symmetrical with respect to its earlier diversification (Figure 5-15,A) will experience nearly as many speciation events in its declining phase as in its radiating phase. Thus, in purely numerical terms, it will have nearly as great a chance of budding off a new family in its declining phase, yet abundant fossil evidence suggests that most dramatic breakthroughs are achieved early in adaptive radiation: This is when higher taxa multiply most rapidly (Simpson, 1944; 1953).

Equally striking is the fact that a diverse taxon that simply persists at constant diversity by replacement of extinct lineages will experience more total speciation events per unit time than it did during its earlier, radiating phase, when *fractional* rate of increase (R) was high, but *total* rate of increase (numbers of species added per unit time) was low (Figure 5-15,B).

Note that the preceding observations rest on the assumption that mean duration of lineages does not increase $(E$ does not decrease) markedly during phylogeny. While this assumption may not strictly hold (see Figure 9-7, page 247), it is probably not violated to a degree that the estimates made here are badly distorted. For example, inferred mammalian lineages of the Early Eocene (Gingerich, 1974, 1976; Gingerich and Simons, 1977) seem to have existed for intervals of time resembling those spanned by lineages of the Late Cenozoic.

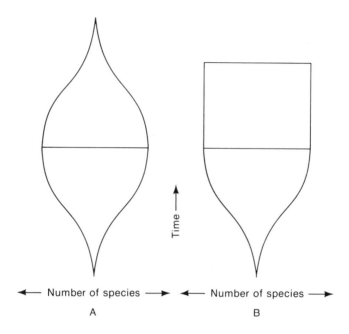

FIGURE 5-15
Relative numbers of speciation events occurring within a higher
taxon during radiation, decline, and persistence without change
in diversity. Balloon diagrams depict species diversity. It is
assumed that species duration is brief with respect to the longevity
of the higher taxon. If the distribution of longevities of lineages
remains roughly constant, approximately the same number of
speciation events will occur during the radiating phase of a
symmetrical phylogeny (A) as during the declining phase. Many
more will occur during a period of stasis at high diversity (B) than
during the preceding radiating phase.

I am sure that I am not alone in previously having assumed that adaptive
radiation accounts for most evolutionary transition partly because radiation is
the site of rampant speciation. This is clearly not the case. The simple observations described above show that the total number of speciation events in radiation is not relatively large. Rather, the critical factor must be the high incidence
of quantum speciation (large fraction of speciation events that are markedly
divergent) at early stages of radiation. The reason that quantum speciation is
distributed in this manner will be considered in the following chapter.

The parameter $\hat{N}$ can be employed to evaluate more rigorously numbers of
speciational steps within a radiating taxon. We will apply this method of estima-

tion to a typical family of Cenozoic mammals. This choice offers excellent opportunities for interpretation of results. Values of R are reasonably consistent among radiating families of mammals (Figure 5-2), so that the average value for R is accurately estimated as 0.22 My^{-1}. A chosen value of t, will yield a particular value of N. In the calculation of $\hat{N}$ for a given initial interval of adaptive radiation (t), the only highly uncertain value will be that of E. The problem is that our estimate (page 108) is for total extinction (E'), and the degree to which the value of E is lower than this value will depend upon the incidence of pseudoextinction. As in the earlier estimation of S (Table 5-1), we can establish boundary conditions by employing maximum and minimum values for this incidence, as shown in Table 5-1,B for $t = 10$ million years. Clearly, there could not be so much pseudoextinction that as few as 10 total speciation events would have occurred, or so little that the number would exceed 40 (see, for example, Figure 4-11, where pseudoextinction represents slightly under 50 percent of total extinction). A liberal estimate would be that 25 speciation events would occur in a typical family of mammals during its initial 10 My of adaptive radiation.

From equation 5.2, it can be calculated that after 10 My of radiation, nine species will be extant. Another class of animals exhibiting a different value of R will, of course, attain a diversity of nine species after a different period of time. Let us ignore the time dimension, however, by focusing on mammals. In this way, we can direct our attention to the geometry of phylogeny, in order to examine numbers of speciational steps along single pathways. Figure 5-16 displays two hypothetical phylogenies representing the adaptive radiation of a family of mammals. For simplicity, each of these is constructed so as to be as nearly homogeneous and symmetrical as possible. In one, the number of speciational steps in each through-going pathway is 3. In the other, it is 13. These differing numbers of speciational steps and the differing durations of lineages in the two phylogenies reflect differing values of E, based on uncertainty about the incidence of pseudoextinction (Table 5-1). The two incidences of pseudoextinction illustrated (70 percent and 20 percent) are taken to bracket the typical value for mammals. Thus, Figure 5-16 indicates that, whatever the normal incidence of pseudoextinction may be, number of speciational steps along any pathway must be quite low.

In all the foregoing discussions, we have been considering homogeneous, symmetrical phylogenies. The only way in which phylogeny can deviate from such a pattern so as to yield a much larger number of steps is to follow a pattern resembling, to some degree, that shown in Figure 5-17. For the multistep pathway on the right of Figure 5-17 to form, a very unlikely set of conditions is required. It is necessary for every lineage of the sequence to have a high probability of speciation soon after forming, and then a low probability after producing one new lineage, even though the descendant lineage immediately has a high probability of speciating. If we make the reasonable assumption that probability of speciation for a full-fledged lineage is not heavily dependent on the age of the lineage, then it is very unlikely that a pattern like that of Figure 5-17 will

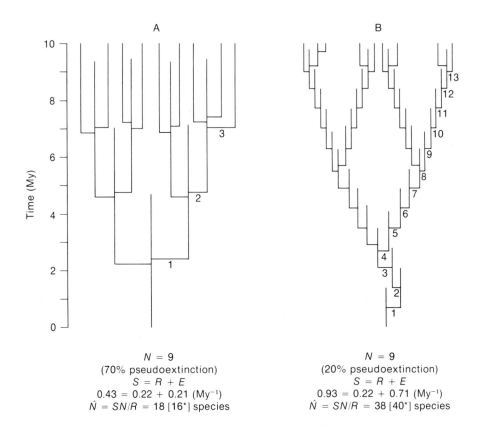

N = 9
(70% pseudoextinction)
S = R + E
0.43 = 0.22 + 0.21 (My⁻¹)
N̂ = SN/R = 18 [16*] species

N = 9
(20% pseudoextinction)
S = R + E
0.93 = 0.22 + 0.71 (My⁻¹)
N̂ = SN/R = 38 [40*] species

FIGURE 5-16
Nearly symmetrical and homogeneous hypothetical phylogenies depicting radiation at rates of speciation and extinction calculated for mammalian families. The two phylogenies display identical net values of R (both yield nine species in 10 My) but differ in rate of extinction according to boundary conditions established by varying the incidence of pseudoextinction (Table 5-1). Average number of speciational steps along a single pathway within actual mammalian families presumably lies somewhere between the numbers estimated here (3 and 13).

develop. Moreover, even the rather ridiculous phylogenetic pathway on the right of Figure 5-17 does not embrace a very large number of speciational steps.

The well-known phylogeny of the Elephantinae (Figure 4-11) represents a concrete example. By the time of its peak diversity, about 0.5 My ago, the subfamily had diversified from its origin, about 7 My ago, at a net rate of 0.33 My⁻¹. This value of R is relatively high even within the rapidly speciating Mammalia (Figure 5-2), yet no species can be shown to have more than four speciation events in its ancestry. Even if more than one speciation event is

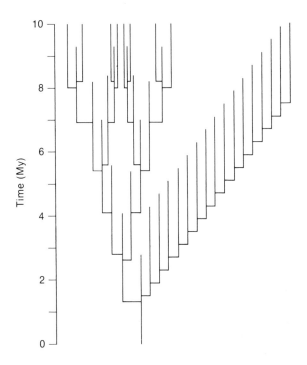

FIGURE 5-17
Hypothetical phylogeny in which a large number of
speciation events form along one pathway. The pattern
differs from Figure 5-16 in its asymmetry and a tendency in the right-hand segment for lineages to speciate
only soon after they appear. Development of such a
pattern is highly improbable.

allowed for the origin of the earliest known species, *Primelephas gomphotheroides*,
no chain of lineages within the subfamily could include more than six or seven.

It seems clear that our estimated values for the parameters R, E, and S, with
support from what is known of elephant phylogeny, impose severe constraints
on the possible number of successional steps in mammalian radiation. We have
used boundary values for E (Table 5-1) and a value of R derived from consistent
empirical values (Figure 5-2). The results have great significance in part because
it is in adaptive radiation that the most rapid sequences of speciation are to be
found. Our conclusion, in effect, is that few speciational steps are available along
a pathway in one radiating family for the origin of another family. The numerical constraint is further tightened by the observation that many speciation events
are known to account for little divergence. These events are, in effect, wasted in
terms of large-scale transformation.

The implication is that evolutionary transitions at the family level entail a small number of quantum speciation events. This conclusion extends to a higher taxonomic level the punctuational claim of the previous chapter that a typical genus forms by one or a small number of markedly divergent speciation steps. We must modify our view of the radiation of Pontian cockles (page 120) accordingly. Presumably here it was not simply number of speciation events but also a high degree of divergence per speciation event that led to the rapid formation of many new genera.

At the beginning of Chapter 4, it was observed that there may be a retrospective factor in the formal recognition of higher taxa (Figure 4-1). We may designate a clade as a discrete family only after it has diversified to a moderate degree. Nevertheless, the preceding paragraphs suggest that forms *deserving* status as families arise very quickly. The monophyletic nature of many transitions is indicated by the distinctive chromosomal patterns of many major taxa of plants and animals. These will be discussed in the following chapter.

Direct evidence is contributed by solitary lineages and clades of low diversity that represent discrete subfamilies and families of rather recent origin. Perhaps the most striking mammalian example is the giant panda, *Ailuropoda*, which will be discussed in the following chapter with reference to the origin of distinctive taxa by means of a small number of genetic changes. *Ailuropoda* evolved from a bear and certainly differs morphologically from living bears at what would normally be considered a family level. Only a desire to emphasize origin from bears rather than from raccoons led Davis (1964) to favor, mildly, status for *Ailuropoda* as a subfamily within the Ursidae. All recognized fossils of the family have been assigned to two species (including the living one) that belong to the extant genus (Pei, 1974; Wang, 1974). None are known from pre-Pleistocene deposits, although the fossil record in the region of the living representatives is admittedly not well known. Still, the absence of evidence of substantial diversification and antiquity, in combination with the simplicity of the genetic transformation underlying the bear-panda transition (page 158), leaves little doubt that very few speciation events were involved in the origin of the family or subfamily of the giant panda. In fact, it seems possible that a single event of quantum speciation accomplished the transition.

Another example, but one entailing only generic transition, is the origin of the white rhino, *Ceratotherium*, from the black rhino, *Diceros*. This event apparently occurred in Africa early in the Pliocene (Hooijer and Patterson, 1972; Hooijer, 1976). It had considerable adaptive significance because the white rhino is a grazer with high-crowned cheek teeth and a square lip, whereas the ancestral black rhino is a browser with low-crowned cheek teeth and a pointed lip (Figure 5-18). Only two species of white rhino are known. The living form, *C. simum*, extends back more than 3 My, and the apparently ancestral form, *C. praecox*, of the earlier Pliocene has a record spanning something like a million years and is intermediate between *Diceros* and *C. simum* in some morphologic characters. The transition from *C. praecox* to *C. simum* took place sometime between 3 and 4

FIGURE 5-18
The African white rhino, *Ceratotherium simum* (above)
and the black rhino, *Diceros bicornis* (below). (From
Hutchinson, 1965.)

My ago, which is brief with respect to the longevity of the descendant species.
Clearly, very few speciation events separate *Ceratotherium* from *Diceros*. The
existing evidence points to just one.

It is interesting to consider whether the preceding calculations for families of
mammals apply to the initial mammalian radiation of the Cenozoic, when many
larger, ordinal transitions were occurring. The radiation of eutherian mammals
that led ultimately to the ecologic replacement of the dinosaurs clearly began late
in the Cretaceous, even though large body sizes did not develop immediately.
This radiation continued into the Cenozoic without interruption at the close of
the Cretaceous (Lillegraven, 1969). It is now apparent that the dominant groups
of early Cenozoic mammals can be traced to a minimum of about three lineages
existing early in the Maestrichtian, about 70 My ago (Figure 5-19). Let us then
postulate a diversity of 1,000 species (about one-quarter of the number alive
today) 15 My later, at the end of the Paleocene. Equation 5.2 yields a value of R
for this radiation of 0.39 My^{-1}. Note that this is only about 75 percent higher

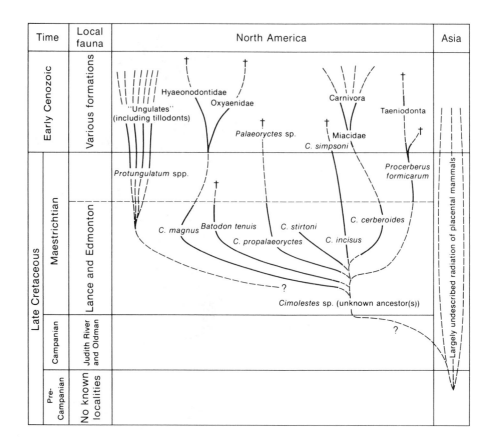

FIGURE 5-19
Interpretation of the early phylogeny of the placental mammals. The dominant modern groups may have shared a common ancestry in the Cretaceous. (From Lillegraven, 1969.)

than the average value for a family of mammals (0.22 My^{-1}), resembling the rate for the late Cenozoic murid rodents (Figure 5-2) and Elephantinae (Figure 4-11). In other words, speciation was perhaps not occurring at an exceptional rate by the standard of a modern subfamily or family, and there could not have been many speciation events in the ancestry of distinctive new orders.

Having calculated that phyletic evolution is generally too slow a process to contribute significantly to evolutionary transitions between higher taxa, we have now looked to speciation and found that few speciational steps are available in adaptive radiation for the piecemeal accomplishment of such transitions. This leaves us no recourse but to adopt a strongly punctuational view. In effect, these lines of evidence uphold the idea of Simpson (1944; 1953) that a higher taxon typically arises rapidly through the occurrence of a sudden adaptive break-through. Ensuing diversification that takes place within the adaptive zone entered via the breakthrough is a matter of elaboration of the ancestral adaptive theme. Occasionally, a simple speciation event or a short succession of events produces such dramatic divergence that the resulting forms will be recognized as a new genus, subfamily, or family. Following some transitions of this type, including the origins of the giant panda and white rhino, adaptive radiation does not immediately ensue. In the latter examples, ecologic specialization or geographic confinement may have reduced opportunities for speciation. In other examples, such as the origin of the Elephantinae, the adaptive transition was followed by persistent radiation (Figure 4-11). In the elephants, the adaptive breakthrough seems to have been the evolution in the Pliocene of adaptations for processing high-fiber food. According to Maglio (1973, p. 87), "once established, the superbly adapted skull and dental complex underwent relatively minor (in a structural sense) alteration in response to selection pressures." Clearly, radiation of major new taxa from an existing taxon should be favored by the opportunity for rapid speciation within the potential ancestral group and by the tendency for speciation events to be markedly divergent. The nature of these optimal conditions will be speculated upon in the chapter that follows.

As a final point, I wish to dispel what might seem to be an inconsistency or paradox. In this section I have concluded that, in the inception and transformation of a clade, it is the incidence of quantum speciation that is most important, *not* the number of speciation events. This is true because the anatomy of a clade is inhomogeneous: Quantum speciation occurs with greatest relative frequency at the outset. On the other hand, early in the chapter, in the comparison of mammals and bivalve mollusks, I related rate of large-scale evolution to rate of speciation in adaptive radiation and, more specifically, to number of speciation events ($\hat{N}$). In such a comparison, however, *totally different* clades are being compared. Early in radiation, when the incidence of quantum speciation is highest, many more speciation events typically occur in a radiation of the Mammalia than in a radiation of the Bivalvia. Thus, in the comparison of taxa characterized by different rates of speciation and extinction, rate of large-scale evolution *does* relate to number of speciation events.

SUMMARY

Adaptive radiation, the site of most evolutionary change, entails geometric, or exponential, increase in number of species. Just as per capita birthrate in a population equals net per capita rate of population growth plus per capita death rate, speciation rate in adaptive radiation (S) equals fractional rate of increase in diversity (R) plus extinction rate (E). The value of R can be estimated for a presently radiating taxon from (1) the number of living species within the taxon and (2) the time interval for the radiation (both of which we commonly know, to a good approximation). Mean extinction rate for a higher taxon can be estimated as being the inverse of mean lineage duration. The latter can be estimated by doubling the geologic age of faunas comprising 50 percent extant species. This is imprecise, in part because some species go extinct by phyletic transition, but boundary conditions for the value of E can be established by specifying various percentages of pseudoextinction. Rate of speciation (S) can be estimated as the sum of R and E. Values of these parameters are much higher in the Mammalia than in the Bivalvia. As would be expected, rates of large-scale evolution are also much higher in the Mammalia. Not surprisingly, exceptionally high rates of speciation of cardiid bivalves in the Pontian Sea yielded extraordinary rates of evolution.

Clades that have existed for long intervals of geologic time at low diversity (with little speciation) invariably exhibit little evolutionary change (are so-called living fossils). This correlation is highly improbable in the framework of the gradualistic model but is predicted by the punctuational model. Because modal rates of phyletic evolution are very slow (as demonstrated in Chapter 4), living fossils are simply the extant members of lineages that happened to have survived for relatively long intervals at normal rates of phyletic evolution. They are not the problematic products of extraordinary phyletic stagnation, as traditionally envisioned.

Calculated rates of speciation and extinction show that only short chains of speciation steps are available in adaptive radiation for the production of new families. This inference is supported by specific examples, such as the phylogeny of the elephants and the origin of the giant panda. No more speciation events are available in the radiating phase of clade development than in the declining phase, and the total rate of speciation in a clade of high, constant diversity is higher than the rate for the radiation leading to this high diversity. We must conclude that major adaptive transitions are typically accomplished by few quantum speciation events and that these tend to occur early in radiation. It can be estimated that fractional rates of speciation early in the Cenozoic diversification of the Mammalia, when many ordinal transitions were occurring, were not greatly higher than fractional rates within families that are radiating today. This estimation suggests that even ordinal transitions are accomplished by few speciational steps. Thus, it is only in the comparison of different kinds of clades (such as families of mammals versus families of bivalves) that rate of speciation explains differences in rates of evolution. Within a clade, it is the uneven distribution of quantum speciation that determines the location of rapid transformation.

6

The Nature of Quantum Speciation

Quantum speciation, by which rapid evolutionary transition is achieved in small populations, is not fully understood, but chromosomal rearrangement and change in gene regulation often play important roles. Many extensive morphologic modifications are achieved rapidly by simple changes in growth gradients or ontogenetic sequences, without wholesale restructuring of the genotype. Although Goldschmidt's "hopeful monster" concept, entailing the instantaneous birth of a fully formed higher taxon, may represent an extreme and fanciful phenomenon, many important evolutionary changes are accomplished within a small number of generations in small populations. The founder effect and genetic drift have significant roles here, but so does natural selection.

Geographically, quantum speciation often occurs during adaptive radiation, by the release of genetic variability within ecologic islands. Rapid evolution of a small population must occur in the process of divergent speciation far more often than by constriction of a lineage (bottlenecking) because, whatever the outcome, lineages are severely constricted much less often than small isolates are cast off.

The punctuational model does not invoke mechanisms of evolution that are unrecognized in the Modern Synthesis, but simply places different degrees of emphasis on various elements of change. The model of quantum speciation advocated here implies, among other things, that evolution has a coarse-grained pattern during episodes of major transition: Variability is particularly pronounced and selection is particularly severe.

INTRODUCTION

The great longevity of chronospecies (Chapter 4) implies that stabilizing selection prevails within most well-established lineages. Clearly, most speciation events do little to liberate populations from conservative ancestral phyletic pathways. On the other hand, adaptive innovations must emerge in the birth of some species, although we are only beginning to understand the modes of formation of these innovations. The preceding chapters show the fossil record to be in accord with the opinion of Mayr (1970, p. 9) that "the origin of new species, signifying the origin of essentially irreversible discontinuities with entirely new potentialities, is the most important single event in evolution." It is therefore unfortunate that the following similie of Fisher (1958, p. 139) also remains valid:

> The close genetic ties which bind species together into single bodies bring into relief the problem of their fission—a problem which involves complexities akin to those that arise in the discussion of the fission of the heavenly bodies, for the attempt to trace the course of events through intermediate states of instability, seems to require in both cases a more detailed knowledge than does the study of stable states.

During the past few years, despite barriers of biologic complexity and limits of generation time, much light has been shed upon the process of speciation and particularly upon the process of quantum speciation. Inasmuch as this is primarily a book about the fossil record, no detailed analysis of speciation will be provided. Rather, I will attempt here to review and interpret evidence that bears on the nature of quantum speciation, the source of major discontinuities in phylogeny. For detailed, though divergent, biologic viewpoints on speciation, the reader is referred to the works of Mayr (1963; 1970), Grant (1971), Bush (1975), Endler (1977), and White (1978).

At present, **quantum speciation** can only be defined in a way that describes its nature superficially. I will define it here, in slightly more detail than earlier in this book, as a form of speciation (branching) that yields marked morphologic divergence, with this divergence taking place (1) within a small population and (2) during an interval that is brief with respect to the longevity of an average, fully established species of the higher taxon to which the new species belongs. There may be more than one genetic mode of quantum speciation. As mentioned earlier, the divergence may precede reproductive isolation (actual speciation), as in the origin of polymorphism; it may more or less coincide with reproductive isolation; or it may follow it, as when character displacement results from secondary contact with the parent species. This possible variation in timing has led me, in preceding chapters, to describe most evolution as occurring "in association with" (not necessarily during) speciation. All that the fossil record can tell us is that the rapid change is concentrated in small populations and, as I will argue later in the chapter, that such populations are likely to be associated with speciation and unlikely to be formed by constriction of an entire lineage.

CHROMOSOMAL ALTERATION

It has already been noted that the traditional gradualistic orientation of geneticists has left the genetic underpinning of speciation little studied. There is,
however, a growing belief that chromosomal transfiguration plays an important
role in the origin of species, as recently reviewed by White (1978). Interestingly,
this viewpoint represents a partial return to earlier ideas that for many years lay
largely dormant.

As was noted earlier, Goldschmidt stood apart from the Modern Synthesis in
maintaining the view that major evolutionary transitions occur within single
generations. Basically, he held that point mutations (changes in single alleles)
and natural selection produce only gradual changes within lineages, failing to
accomplish species-to-species transitions. The other side of his argument was
that new species arise only by chromosomal rearrangements, many of which are
deleterious but some of which instantly yield "hopeful monsters." Only recently
has there occurred a partial turnabout in the evaluation of Goldschmidt's views.

Recently, Bush (1975, p. 357) has written that the concept of a hopeful
monster is "no longer entirely unacceptable." Harking back to Bateson (1894)
and De Vries (1905), Steenis (1969) has laid great importance to this concept in
the macroevolution of higher plants. Hopeful monsters will be considered later
in this chapter. While it may be true that few new higher taxa have arisen during
single generational transitions, in the preceding chapter I offered evidence that
many have arisen by one or a very few speciation events. Evidence is also
mounting that the quantum speciation events themselves may span rather few
generations. No single concept of quantum speciation is emerging. Rather, several different modes seem to operate in nature.

We can infer from evidence discussed in Chapter 4 that Goldschmidt exaggerated the importance of chromosomal alteration, in that taxa recognized as species
do, in fact, arise by gradual phyletic transformation. It follows that new species
also form by the equivalent process of gradual divergence of large segments of
pre-existing species, as would follow the permanent geographic division of a
large, established species into two equal subpopulations, for example. However,
such large-scale geographic splitting cannot be the source of major evolutionary
transformations. Separation simply introduces a dichotomy to the pathway of
evolution, without accelerating rate of change appreciably. Rather, for reasons
that will be made apparent below, it is generally agreed that quantum speciation
takes place within very small populations—some would say populations involving fewer than 10 individuals, although the actual modal number is not known.

It is quite possible that genera and suprageneric taxa normally originate by
episodes of quantum speciation involving chromosomal alteration. With respect
to number of chromosomes, the condition for plants has been summarized by
Lewis (1966, p. 171):

> The significance of basic [chromosome] numbers is twofold: (i) they are very con-
> servative, particularly among woody plants where one frequently finds the same

chromosome number throughout a family, subfamily, or other large taxonomic group. Even among herbaceous groups, basic chromosome number is generally constant throughout a genus or major sections of it, although the number is frequently different from one genus to another in the same family. In general, a difference in basic number is not characteristic of closely related species.... It seems clear that change in basic chromosome number is not a normal consequence of gradual differentiation, but a rare event resulting from unusual circumstances. (ii) In nearly all instances in which a difference in basic chromosome number in plants has been studied in detail, gross differences in chromosome arrangement have been found in addition to those required to effect the change. This suggests that saltation has been involved.

Consistency has also been observed for chromosome numbers within many higher taxa of animals (Figure 6-1). What is particularly noteworthy is that this condition for both plants and animals supports the paleontologic conclusion of the preceding chapter that higher taxa typically arise by a very few critical speciation events.

Wright (1940) established the important point that chromosomal changes are likely to be fixed only within very small populations. The role of chromosomal rearrangement as a reproductive isolating mechanism in speciation has been especially well documented for morabine grasshoppers of Australia by M. J. D. White and his coworkers. In fact, White (1968; 1978) maintains that most closely related species of animals differ in karyotype. He also notes that when, as must often happen, chromosomal change occurs during speciation, it is required that we consider the possibility of the origin of the new species from a single individual.

Apparently an important function of chromosomal reorganization during speciation is in preventing subsequent recombination from dismantling new adaptive complexes of genes that have formed within small transitional populations (Mayr, 1963; 1970). The phenotypic significance of chromosomal rearrangement is poorly understood (White, 1978, p. 50), but it has long been recognized that a correlation can exist between karyotype and ecologic niche (Mayr, 1945). Berry and Baker (1971), for example, concluded that the adaptation of rodents of the genus *Thomomys* to deep, moist soils is related to a small number of acrocentric chromosomes, relative to the incidence of such chromosomes in populations inhabiting shallow, dry soils. In general, however, study of karyotypic adaptation remains in its infancy.

Bush (1975), Wilson *et al.* (1975), and Bush *et al.* (1977) have concluded that there is a general correlation within the Vertebrata between rate of speciation and rate of chromosomal evolution. Such a correlation would support the idea that chromosomal transformation plays a key role in speciation, which these

FIGURE 6-1
Karyotypic relationships within the Artiodactyla. A: Phylogeny of the order. B: Distribution of karyotypic traits within the phylogeny. (From Todd, 1975.)

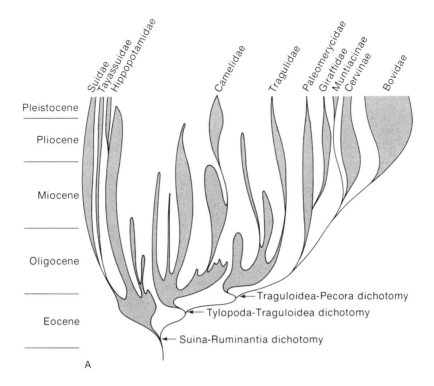

Suidae
Tayassuidae
Hippopotamidae
Camelidae
Tragulidae
Paleomerycidae
Giraffidae
Muntiacinae
Cervinae
Bovidae

Pleistocene

Pliocene

Miocene

Oligocene

Eocene

Traguloidea-Pecora dichotomy
Tylopoda-Traguloidea dichotomy

Suina-Ruminantia dichotomy

A

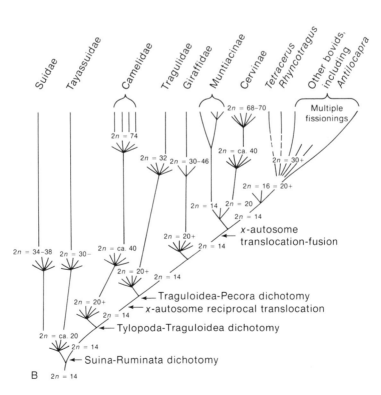

Suidae
Tayassuidae
Camelidae
Tragulidae
Giraffidae
Muntiacinae
Cervinae
Tetracerus
Rhyncotragus
Other bovids,
including
Antilocapra

$2n = 68-70$

Multiple
fissionings

$2n = 74$

$2n = 32$ $2n = 30-46$

$2n = $ ca. 40

$2n = 30+$

$2n = 16 = 20+$

$2n = 14$ $2n = 20$

$2n = 14$

$2n = 20+$

$2n = 14$

x-autosome
translocation-fusion

$2n = 34-38$ $2n = 30-$ $2n = $ ca. 40

$2n = 20+$

$2n = 14$

Traguloidea-Pecora dichotomy

$2n = 20+$

x-autosome reciprocal translocation

$2n = 14$

Tylopoda-Traguloidea dichotomy

$2n = $ ca. 20

$2n = 14$

Suina-Ruminata dichotomy

B $2n = 14$

authors also believe to be the site of most morphologic transformation. They attribute much of the variation in rate of speciation to degree of social structuring, the reason being that severe inbreeding is necessary for a chromosomal mutation to be fixed within a population: Fixation of a new chromosomal arrangement in the homozygous state requires at least two generations, and only in breeding populations of 10 or fewer individuals is it likely to occur. Social bonding, it is claimed, greatly enhances the probability.

The reason that severe chromosomal modifications are likely to be fixed only in very small, inbreeding populations is that karyotypically heterozygous forms (resulting from interbreeding with individuals having the ancestral condition) are generally adaptively inferior.

REGULATORY GENES

Another of Goldschmidt's views that is being resurrected in less extreme form is that change in only a small number of genetic messages can greatly alter the development of the phenotype. Viewing these as being karyotypic in nature, Goldschmidt referred to them as systemic mutations (see, for example, Goldschmidt, 1940, p. 308 ff.). He saw their occurrence as altering the course of development in dramatic ways, not because they transform much of the genome, but because they govern the operation of much of the genome. Goldschmidt related the expression of systemic mutations to **relative growth** (ontogenetic changes in bodily proportions) and to **heterochrony** (changes in the sequence of ontogeny), as elucidated in Goldschmidt's era by Thompson (1917, and later editions), Huxley (1932), de Beer (1940, and later editions), and more recently by Gould (1966, 1977b). In fact, as will be discussed below, a large portion of structural evolution can be fully described in terms of relative growth and heterochrony. As Goldschmidt saw it, even a single, systemic mutation can produce dramatic change in relative growth or development. He suggested that chromosomal arrangement of genes in speciation commonly leads to a change in the pattern of genetic expression and, phenotypically, in development.

In an era when the theory of genetics was focused upon point mutations, Goldschmidt suffered widespread derision for these heterodox ideas. His concept of the systemic mutation seems to have temporarily given way to that of epistatic interactions, or interactions of genes at various loci. These are undoubtedly important, but they have commonly been envisioned as forming complex webs, normally altered in piecemeal fashion by point mutations, to yield gradual phyletic evolution. More recently, many authors have come to lay great importance to something akin to Goldschmidt's systemic mutation. This is change in what are known as **regulatory genes,** or genes that control the operation of **structural genes,** which cause the synthesis of proteins that function in metabolism and biosynthesis. Mayr (1970, p. 183) has written:

Much that is now explained as "epistatic interactions between different loci" might well be due to the activities of regulatory genes. The fact that the macromolecules of most important structural genes have remained so similar, from bacteria to the highest organisms, can be much better understood if we ascribe to the regulatory genes a major role in evolution. ...

The day will come when much of popular genetics will have to be rewritten in terms of the interactions between regulator and structural genes.

Britten and Davidson (1969, 1971) have elucidated the importance of regulatory genes. The particular model that they have proposed amounts to a working hypothesis, but it seems possible that comparable systems exist in the living world. Like Mayr in the passage just quoted, Britten and Davidson suggest that many structural genes may be conservative evolutionary features. Certain differences between species in composition of structural genes may represent useful biochemical clocks for dating geologic times of phylogenetic branching. Some workers relate these conditions to the possibility that most point mutations of structural genes, if not deleterious, are adaptively neutral and become fixed randomly (see Wilson et al., 1977, for a review). In any event, it seems to be control of the operation of structural genes by regulatory genes that is responsible for major changes in grades of biologic organization. Valentine and Campbell (1975) have provided a useful review of structural and regulatory genes that is especially valuable to the uninitiated reader. There are several extremely important implications. For one thing, the regulatory model can account for rapid appearance of evolutionary novelties by alteration of a miniscule portion of the genome. For another, as Goldschmidt anticipated, the complex manner in which genes are held to interact makes position effects important; chromosomal alterations can modify regulatory systems greatly even in the absence of large numbers of point mutations. Valentine and Campbell (1975) have reviewed some of the possible contributions of changes in regulatory systems to large-scale evolution.

Morphogenetic transformation of the sort now viewed in terms of regulatory genetics has long been recognized for vertebrate animals. Kurtén (1953) evaluated "the rule of neighborhoods," whereby within a species or group of related species relative sizes of contiguous morphologic features are more closely correlated than relative sizes of distant features. Kurtén's examples pertain to strong correlations between relative sizes of adjacent molars in mammals. In addition to providing a valuable review of pertinent literature, Van Valen (1970) described numerous gradients in serially homologous structures, especially mammalian teeth. There is much evidence that such structures often have identical or very similar **prepatterns** (genetic programs for development), forming a **developmental field.** The degree of expression of a dental feature, which often forms part of a gradient, is determined by what might now be called gene regulation. A fundamental feature can also be transferred to a new serial position (a prepattern can be altered), and this is known as **homeosis.**

Among the invertebrates, segmented groups like annelids and arthropods should prove amenable to similar kinds of analysis. The same is true for colonial

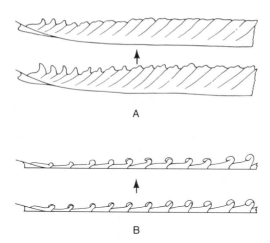

FIGURE 6-2
Evolutionary spreading of a new type of graptolite theca distally (A) and proximally (B) within a
colony. (From Urbanek, 1960.)

animals in which individuals are in some fashion serially arranged. The evolution of the extinct graptolites has, in fact, been interpreted in terms of morphogenetic gradients. Urbanek (1960) suggested that the degree of development
of thecae along a rhabdosome represents the level of phenotypic expression of
underlying genetic coding. As illustrated in Figure 6-2, a new thecal type may
originate proximally or distally. Urbanek saw its evolutionary progression along
the rhabdosome from the site of origin as being under the control of a substance
comparable to a hormone or plant auxin, with any momentary gradient in
manifestation of thecal type reflecting depletion of the substance with distance
from its point of origin. Citing the pioneering regulatory contributions of
Goldschmidt and others, Urbanek (1960, p. 167) proposed that "this substance
would supposedly have the properties of a 'gene controlled substance', i.e. it
would be connected with the presence of genes acting as stimulators or inhibitors
of some other genes (modifying genes)."

Whatever the precise mechanism, there is much evidence that major evolutionary transitions in the graptolites were achieved rapidly in small populations.
For instance, the suborder Monograptina differs from all other orders in shape of
rhabdosome (a meristic trait) and mode of budding. The discontinuities between
these characters and those of other taxa imply that origins were sudden, although
they did not necessarily occur simultaneously or by single steps (Bulman, 1970;
Jaanusson, 1973). Rickards (1973, p. 344) has summarized his experience:

Although in the graptolite lineages that I have examined it is possible to find gradual changes during the time span of a species, such as a gradual shift in the number of thecae in the uniserial portion of *Dimorphograptus erectus* ... or in the thecal spacing of *Monoclimacis flumendosae*, these are the exceptions rather than the rule. More commonly, the new species appear relatively suddenly, and in numbers of cases the reasons are quite clear: thus the change from *Glyptograptus* to *Atavograptus ceryx* is necessarily sudden since the production of a uniserial rhabdosome in this case requires the suppression of dicalycal budding.

It is difficult to envision how numerous other basic steps in evolution could have occurred except through sudden changes in gene regulation or in some comparable agent of morphogenetic control. Torsion in gastropods is a good example. Torsion amounts to a twisting of the body so that the shell is rotated through 180 degrees (Figure 6-3). Its presence defines a snail, and it is conceivable that the Gastropoda came into being as a morphologically discrete class by fixation in a small population of a single mutation that caused differential muscle contraction to yield all or much of what is now seen as full torsion. In the primitive living genus *Acmaea*, torsion is accomplished during embryological development in just two or three minutes (Garstang, 1929). There is much evidence that the origin of torsion led to the great adaptive success of the Gastropoda by affording protection of the body within the shell through the opercular sealing of the aperture when not clamped to a hard substratum. It is nearly impossible for an untorted gastropod-like mollusk to evolve an operculum (Stanley, 1979). Possibly self-fertilization led to the fixation of torsion—simultaneous hermaphroditism is known in snails—but there are alternative ways by which fixation could have occurred. Because fertilization was presumably external, so that morphology and behavior offered no mating barriers, a small genetic difference between the first torted animal and other members of its lineage would not necessarily have blocked interbreeding and the production of fertile heterozygous offspring. Alternatively, one might postulate that a single female underwent a germinal mutation that led to the production of numerous torted siblings. In either case, the new body plan could have been fixed and elaborated upon within a small population by assortative mating or spatial isolation.

As noted above with reference to the work of Goldschmidt, much of the general role of gene regulation in macroevolution can be expressed in terms of the classical concepts of allometry and heterochrony. **Allometry** has been defined in various ways, but in its broadest sense simply denotes a difference in relative growth rates between two morphologic features. Often these rates yield a roughly linear curve for bivariate plots of bodily dimensions using logarithmic coordinates (see the comprehensive review by Gould, 1966). The linearity of such a plot reflects approximate adherence to a power function of the form $y = bx^k$, where x and y represent the bodily dimension and b and k are constants. Presumably, in most cases, this relationship reflects the fact that growth is

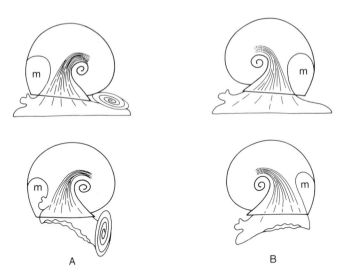

FIGURE 6-3

Protective advantage of torsion in the Gastropoda. A: Diagram of
a 'torted snail. The retractor muscle (striated pattern) curves
backward around the axial columella of the coiled shell. The
mantle cavity (m), which houses the gills, shrinks during retraction. The circular, plug-like operculum is the last structure to be
pulled into the shell. B: Diagram of a hypothetical untorted
gastropod-like mollusk. The shell coils over the head of the animal. The retractor muscle also curves forward, so that, upon
retraction, the head or creeping sole of the foot must enter the
mantle cavity last. These regions of the body cannot bear an
operculum so that the retracted animal is left vulnerable to attack.
(From Stanley, 1979.)

inherently geometric. If rates of geometric increase of x and y differ, then x and y
will be related by a power function in which the value of k differs from unity.
Changes in bodily proportions that are evolutionary rather than ontogenetic are
sometimes referred to as representing *allomorphosis*, rather than allometry. We
might question the need for this term, because its definition includes much that
is simply describable as structural evolution. In any event, the partial synonymy
brings out an important point. Regulatory genes may be largely responsible for
allometry and allomorphosis, and this means that they may play a major role in
structural evolution.

Heterochrony amounts to evolutionary change in the sequence of ontogeny.
Inasmuch as an evolutionary line of descent amounts to a sequence of ontogenies, evaluation of heterochrony also provides a complete morphologic de-

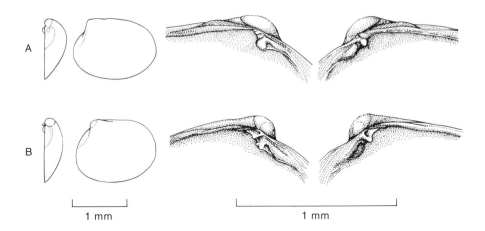

FIGURE 6-4
Paedomorphic features of a leptonacean bivalve. A: Adult shell and dentition of the leptonacean *Turtonia minuta*. B: Similar features in the juvenile stage of the large venerid *Venerupis pullastra*. (From Ockelmann, 1964.)

scription for much structural evolution. Nowhere is the importance of gene regulation more evident than in morphogenesis.

Paedomorphosis is perhaps the most widely recognized form of heterochrony. As defined by Gould (1977b), this process includes the acceleration of sexual maturation relative to the rest of ontogeny (**progenesis**) and the retardation of somatic development with respect to the onset of reproductive activity (**neoteny**). In some instances, the development of only a few morphologic features is retarded. In others, the entire animal is juvenilized. The classic example is that of certain Amphibia in which development is arrested so that the adult retains the form and aquatic habit of the ancestral larval stage. Here the pituitary gland fails to stimulate the thyroid gland sufficiently for ontogeny to run its full course (Frazer, 1973).

Progenesis can yield dramatic evolutionary change. The Leptonacea are a large group of marine bivalves, containing about 300 living species (Boss, 1971). All are minute forms, a fraction of a centimeter in length. Ockelmann (1964) showed that one reputed leptonacean, *Turtonia minuta* (Figure 6-4), is actually a progenetic member of the Veneridae, a family that contains a substantial fraction of the large, burrowing bivalve species in modern seas. This confusion, coupled with the ease by which a postlarval animal can be transformed into an adult through the early onset of reproductive activity (and the termination of subsequent development), suggests that the assemblage formally recognized as the Leptonacea is, in fact, a polyphyletic array of progenetic species (Stanley, 1972).

The origins of many of these species must have represented sudden transitions at the genus or family level. Neoteny can also occur in a punctuational pattern, as suggested by Rickards for the graptolites shown in Figure 6-5. (Note here the temporal overlap of lineages.)

It should be apparent that paedomorphosis and also **hypermorphosis** (evolutionary lengthening of ontogeny, usually by the retardation of sexual maturation) are concepts that bear a close relationship to evolutionary changes in the phenotypic expression of prepatterns, as discussed above. For example, whether a tooth forms part of a developmental gradient or not, a change in the degree to which the tooth passes through its full ontogenetic program can be viewed as a facet of heterochrony. Independent evolutionary change of various portions of the phenotype—a phenomenon that has long been recognized—represents what is now commonly termed **mosaic evolution.** I offer three examples:

1. Mosaic evolution is strikingly evident in the phylogeny of the modern Hominidae (see, for example, Buettner-Janusch, 1973, p. 245). Here the early evolution of structures for bipedal locomotion (especially modification of the pelvic girdle) was accompanied by little change in the form of the skull or in brain size. Both of the latter subsequently evolved quite rapidly to the modern human condition.

2. In the phylogeny of the elephants (Maglio, 1973) *Mammuthus*, the genus of mammoths, and *Elephas*, the genus of Indian elephants, underwent rapid early molar modification but little foreshortening of the forehead. *Loxodonta*, the genus of African elephants, underwent parallel changes, but forehead foreshortening took place very early, while modification of the molars lagged behind. Maglio related the rapid evolution of *Mammuthus* and *Elephas* molars to a shift of habitat.

3. Modern opisthobranch snails include some forms, like *Acteon* and *Architectonica*, that seem frozen at grades of evolution essentially intermediate between the ancestral prosobranch state and the derivative opisthobranch condition (Robertson, 1974). Opisthobranch gills and simultaneous hermaphroditism are established within these two genera, but held over from the prosobranch grade of evolution are the anteriorly directed mantle cavity, the operculum, and the streptoneurous nervous system. It is generally believed that the prosobranch-opisthobranch transition occurred at small body size, which permitted the total loss of the prosobranch gill. Some minute living species seem to represent this grade of evolution. The opisthobranch gill would then have developed as body size increased again. Thus, perhaps the mosaic pattern of transition resulted in part from a kind of scaling bottleneck that affected the respiratory system in particular.

The occurrence of mosaic evolution is hardly surprising. If, within phylogeny, many structural changes occur rapidly and if regulatory systems operate with partial autonomy, then we would expect the independent fixation of some new features in phylogeny, as long as organisms retain sufficient coadaptation of morphologic features to function successfully.

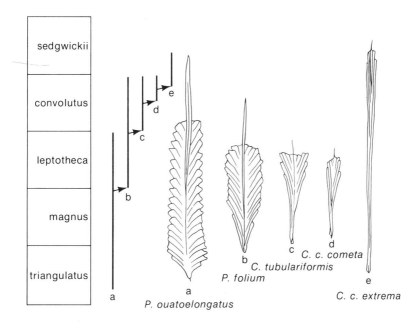

FIGURE 6-5
Stratigraphic ranges and inferred paedomorphic origins of graptolite species
of the genera *Petalograptus* and *Cephalograptus*. (From Rickards, 1977.)

To the extent that regulatory systems lack autonomy and affect enormous
suites of morphogenetic features, mosaic evolution will be opposed because the
fates of morphological features will be linked. **Pleiotropy,** or control of two or
more features of the phenotype by single genetic components, is a well-
established phenomenon. It is also readily accommodated within the Britten-
Davidson Model. It may, indeed, be a source of important new morphologic
adaptations. Darwin laid great stress on **preadaptations,** or structures that
evolved in association with one function but later, with relatively minor evolutionary modification, came to serve another. Perhaps the most famous example is
the evolution of a fish's primitive lung into a swim bladder. This kind of event
forms much of what Simpson has referred to as the **opportunism** of evolution
(page 206). Pleiotropy may have introduced some important features that had
little initial adaptive value. They would represent byproducts of adaptive
evolution—harmless experiments that formed the raw material of subsequent
change. Just how significant has been this mechanism for the generation of
evolutionary novelties? No assessment is possible at present, and a basic problem
obstructs any effort to gain relevant information: We can never be certain that a
structure, even if thought to be a pleiotropic byproduct, evolved without adaptive significance. We can never know that we have exhausted all possibilities in

156

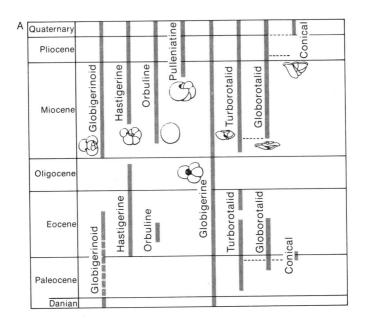

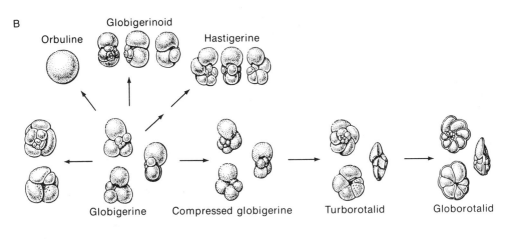

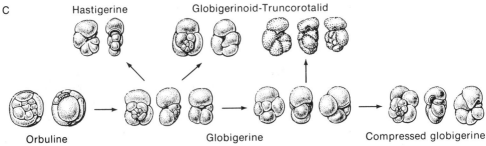

the search for a function. My guess, however, is that pleiotropy may have introduced much important raw material for evolution in the form of initially nonadaptive features. The evolution of the giant panda, to be discussed below, offers support for this notion. Here we may find a way around the traditional problem of the adaptive value of incipient features—a problem with which Darwin and others have grappled unsuccessfully in the context of gradualism.

A common possession of certain regulatory systems must account for many examples of convergence, parallelism, and iterative evolution. These patterns remained problematic in the era when only structural genes were contemplated and particularly when epistatic interactions among such genes were ignored. It would now seem that, given opportunity for diversification, two or more similar taxa may tend to follow certain evolutionary pathways that are readily accessible because of the pre-eminent morphogenetic role of certain regulatory genes. Similarity of regulatory systems would seem to account for much of the marked convergence between marsupial and placental mammals. A remarkable case of iterative evolution, reported by Cifelli (1969), is that of the Globigerinidae (planktonic forams), in which two radiations were separated by a mass extinction at the end of the Eocene. Each radiation issued from ancestors of the *Globigerina* type (Figure 6-6). Continued presence in these ancestral forms of a few labile regulatory systems of great importance may account for the iterative pattern here. Thus, the retention of genetic coding that remains phenotypically unexpressed following regulatory change can permit lost features to reappear in phylogeny. There is, of course, little likelihood of perfect iteration because the "switched off" portion of the genome will inevitably have experienced some mutations before its phenotypic re-emergence and because it will inevitably be expressed in a new morphogenetic context.

The nature of pleiotropy, growth fields, convergence, and other conditions and processes associated with gene regulation are illustrated in a remarkable, but seldom-cited, pioneering study by Davis (1964) of the giant panda, *Ailuropoda*. The conclusions of Davis can be placed in perspective by considering the history of the taxonomic placement of the giant panda mentioned earlier. For many years, it was debated whether this animal is a descendant of the bears or is closely related to the lesser panda and should be united with it in the family Procyonidae (raccoons). Davis offered convincing evidence that the giant panda evolved from bears, but the earlier taxonomic indecision serves to illustrate the marked degree of its divergence from familiar bears. Davis concluded that the giant panda deserves status as either a discrete subfamily of the Ursidae or a

FIGURE 6-6
Iterative evolution of planktonic forams. A: Stratigraphic ranges of the various morphologic varieties. Only globigerine forms pass across the upper Oligocene boundary, from the Paleogene into the Neogene. B: Neogene adaptive radiation from globigerine ancestors. C: Similar Paleogene radiation, also from globigerine ancestors. (From Cifelli, 1969.)

separate monogeneric family and only favored the former as an indication of ancestry.

Davis concluded that the giant panda is essentially an aberrant bear specialized for feeding upon coarse vegetable matter (primarily bamboo). In the origin of *Ailuropoda*, the masticatory apparatus was hypertrophied and specialized: The cheek teeth and masticatory musculature were enlarged, the skull was strengthened, and the mandibular articulation was shifted to a position above the tooth row. Incipient masticatory characters of the sort that became pronounced in the giant panda are present in the partly herbivorous bears. In contrast to the evolutionary changes of the head region leading to *Ailuropoda*, the postcranial region was disturbed and weakened in various ways, yielding the clumsy "teddy bear" appearance of the modern animal. Lacking natural enemies, *Ailuropoda* is seldom required to run or fight.

Davis interpreted the evolution of the *Ailuropoda* in terms of pleiotropic effects and growth gradients. Selection pressures for hypertrophy of the head region, being unopposed by pressures for efficient locomotion, operated at the expense of mechanical efficiency in the postcranial anatomy, and especially in the lumbosacral region. Among other changes, the vertebral column was shortened.

The critical condition in the ancestors of *Ailuropoda*, probably not uncommon in the animal world, was that morphogenetic fields were not coincident with functional units of anatomy. The result was the pleiotropic evolution of nonadaptive features in the postcranial region. Davis pointed to examples of evolution in other taxa that, to varying degrees, parallel the origin of the giant panda. The lesser panda and the spotted hyaena resemble the giant panda not only in their hypertrophied masticatory apparatus, but also in their abnormal genitalia. The bulldog, developed by domestic breeding, is the product of changes resembling those that have led to the giant panda and has been the subject of similar morphogenetic analyses.

Davis avoided any evaluation of the rate at which the giant panda evolved. The genus is known only from the Pleistocene, though a slightly earlier origin cannot at present be ruled out (page 138). What does seem evident is that there has never been a high diversity of giant pandas. Very few speciation events could stand between the living form and its ursid ancestors. What Davis did conclude was that the basic adaptive transition from *Ursus* to *Ailuropoda* required the changing of very few genetic messages—possibly fewer than 5 or 6. Sarich (1973) further found that the transferrin immunological distance between the giant panda and a bear is very small, resembling that between such similar species as a dog and a fox. These genetic similarities suggest that the basic shift could easily have been achieved by a quantum speciation event. I find it difficult to imagine that the drastic structural and ecologic changes could have come about by slow, sequential fixation of the few genetic changes or that an entire species occupying a large geographic area could have made such a remarkable phenotypic transition. Far more likely would have been origin by way of a very small population occupying a local bamboo forest.

Certain other sudden transitions would also seem to have been so drastic as to demand interpretation as localized occurrences. Examples already mentioned are juvenilization of Amphibia, which has accomplished changes of adult life habits, and the origin of torsion in snails. Thus, it seems an inescapable conclusion that rapid, drastic changes governed by regulatory genes have most often occurred in association with speciation. Just how dramatic such quantum shifts have been remains uncertain.

HOPEFUL MONSTERS

Early in this chapter, it was noted that there has recently been renewed expression of support for the importance in macroevolution of what Goldschmidt (1940) termed the hopeful monster (Frazzetta, 1970; Gould, 1977c; Bush, 1975). Goldschmidt's monster was a single animal that served as the progenitor of a new higher taxon. At least in principle, Goldschmidt accepted Schindewolf's extreme example of the first bird hatching from a reptile egg.

The problem with Goldschmidt's radical concept is the low probability that a totally monstrous form will find a mate and produce fertile offspring. For this reason, it may be that the hopeful monster *sensu stricto* will find no place in the modern punctuational view. In fact, some of the authors who have purported to resurrect the hopeful monster seem actually to be softening Goldschmidt's concept somewhat. The degree of freakishness that is permissible in generational transitions can only be considered in terms of probability. At present, we have no means of rigorously assessing the matter in morphologic terms, but two guidelines seem appropriate: (1) If strange new phenotypic traits emerge in a single individual, they cannot be fixed, even within a small population, if the individual is so bizarre that it cannot find a mate and produce fertile offspring (only after the first generation is inbreeding of bizarre individuals possible). (2) Somewhat more aberrant traits may be fixed, however, if they result from the germinal mutation of a female, but only if her offspring interbreed, a qualification that reduces the probability.

All of these considerations raise the question whether self-fertilization is a trait promoting quantum speciation. We do, in fact, have evidence that some clades have been founded by single individuals. On the island of Hawaii, which is only about 700,000 years old, Carson (1970) has found 23 species of picture-winged *Drosophila*. These form 11 chromosomal groupings, and it seems likely that each grouping was founded by a single gravid female from another island. Unfortunately, *Drosophila* here and elsewhere is noted neither for chromosomal lability nor for divergent speciation.

Beginning with the third edition (1861), Darwin opened the second chapter of *On the Origin of Species* with arguments against the importance of rapid transformation in nature:

> It may perhaps be doubted whether monstrosities, or such sudden and great devia-
> tions of structure as we occasionally see in our domestic productions, more espe-
> cially with plants, are ever permanently propagated in a state of nature. ... They
> would, also, during the first and succeeding generations cross with the ordinary
> form, and thus their abnormal character would almost inevitably be lost.

Obviously, Darwin's opposition to rapid transformation was in part erroneously founded on the idea of blending inheritance. He seems also not to have
appreciated the role of isolation. Furthermore, in assessing the probability of
dramatic divergence, such as that associated with chromosomal alteration, we
must not forget that our temporal yardstick is the geologic time scale. Consider,
for example, a family having an average diversity of 100 species over a 10 My
interval. Suppose that an average lineage has a mean abundance of 10^6 females,
each of which produces 10 offspring per year. Then, during the 10 My interval,
there will be something like 10^{16} births within the higher taxon. Even lacking the
additional information necessary for rigorous analysis, we can see that there is an
astronomical number of opportunities for isolation of a small, strange, inbreeding population of the sort that could found a new family. In considering such
numbers and the developmental simplicity of torsion in gastropods (page 151), I
cannot agree with Ghiselin (1966), who opposed the idea of a sudden origin for
torsion because such a phenomenon seemed to represent the hopeful monster
concept and, thus, to violate the tenets of the Modern Synthesis. More generally,
I cannot accept Darwin's negative assessment of the possibilities for sudden
introduction of unusual morphologic features.

That freakish organisms like the dodo have sometimes been discovered on
island refuges is not necessarily evidence of long-term phyletic transformation.
Consider the case of the Stephen Island wren (Oliver, 1955). This unusual,
flightless species (Figure 6-7) was discovered by the lighthouse keeper's cat,
which systematically captured the total population of less than a score of individuals and thereby caused the wren's extinction. There is a strong possibility
that the discovery and demise of the wren occurred almost at the moment (in
geologic time) of its evolutionary origin by way of a handful of individuals.

One of the strongest stands taken in favor of hopeful monsters in recent years
is that of Steenis (1969), whose arguments were restricted to higher plants.
Steenis cited evidence for the saltational appearance in the plant world of bizarre
somatic structures that have turned out to have adaptive value. Without addressing the question of whether hopeful monsters in the sense of Goldschmidt really
do survive in the plant world, we can note that it is to be expected that bizarre
morphologic features will be fixed more frequently within plants than within
animals. Plants, in general, are not constrained by strongly determinate growth
or complex behavioral patterns. Perhaps it was no accident that the prominent
macromutationists De Vries (1905) and Willis (1940) were botanists.

Certainly, in the history of evolutionary neontology the most persuasive circumstantial evidence for sudden transitions has been the mere existence of mor-

FIGURE 6-7
The Stephen Island wren, *Xenicus lyelli*.
(From Carlquist, 1965.)

phologic features that cannot be imagined to have evolved by small steps. (Darwin's difficulty in explaining the adaptive origin of incipient features has already been noted.) Following this line of reasoning, Frazzetta (1970) proposed that the bolyerine snakes originated from the Boidae as hopeful monsters. Although the alleged monstrosity of the earliest forms was not extreme, it represented sufficient divergence that these animals constituted a new family. The most remarkable feature of the bolyerines is unique among living vertebrates—it is a joint dividing the upper jaw into two segments (Figure 6-8). Because such a feature must be either present or absent, Frazzetta reasonably concluded that the divi-

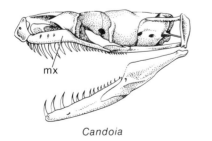
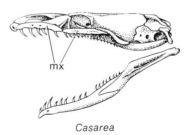

Candoia   *Casarea*

FIGURE 6-8
Cranial features of the bolyerine snake *Casarea*, in which the maxillary bone (mx) is jointed, compared to those of the similar boine snake *Candoia*, in which the maxillary bone is unjointed. (From Frazzetta, 1970.)

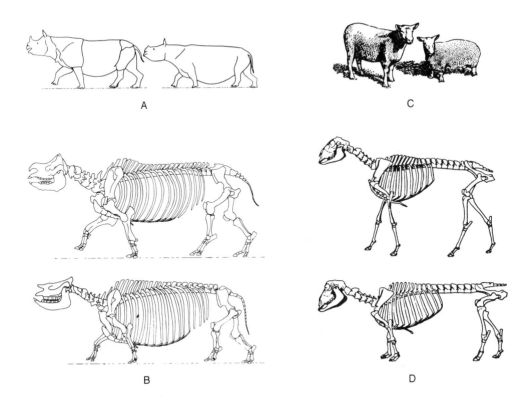

FIGURE 6-9
Effects of achondroplasia. A: The living Indian rhino, *Rhinoceros unicorni*, followed by a reconstructed version of the Late Miocene achondroplastic dwarf rhino, *Teleoceras fossiger*. B: Skeletons of
the species represented in A. C: Normal domestic sheep and member of the achondroplastic Ancon
strain. D: Skeletons of the normal and Ancon varieties. (A and B drawn by Gregory S. Paul; C and
D from Grüneberg, 1963.)

sion must have originated suddenly. He suggested origin by means of developmental alterations that are quite compatible with the Britten-Davidson Model of
gene regulation. The bolyerine snakes evolved on Mauritius, or a neighboring
island. This insular location is the kind of site in which many striking evolutionary transitions have occurred, and this relationship forms one of the most
important subjects of the remainder of this chapter.

Also strongly suggestive of rapid evolutionary transition is the origin by single
mutations in domestic animals of certain distinctive morphologic features—
namely features that closely resemble traits of species that appear suddenly in
the fossil record as the earliest members of discrete higher taxa. Modern sheep
representing achondroplasia, or dwarfing that results from improper development of cartilage at the ends of bones, are compared in Figure 6-9 to the fossil

rhino genus *Teleoceras*. At least twice during the last 200 years, achondroplastic sheep have arisen by single mutations under domestic conditions, and dwarf populations have been maintained by artificial inbreeding. Similar dwarfs have developed in many other varieties of domestic animals (Grüneberg, 1963). It is much easier to imagine that *Teleoceras* evolved by the rapid fixation of achondroplasia in a small, inbreeding population than by the dwarfing of an entire species. Despite the strange appearance of *Teleoceras* and the evident vulnerability of the animal's nether anatomy to substratal abrasion—Matthew (1931) thought it adapted to smooth grassy plains—the genus spread throughout vast areas of the Old and New Worlds, becoming one of the most widely distributed Miocene rhinos. The origin of another rhino taxon seems to have represented a family-level transition of the same kind. The family in question, the Amynodonitidae, appeared suddenly in the Eocene and survived well into the Oligocene (Wood, 1941).

COEVOLUTION

It was noted in Chapter 2 that in many minds coevolution has come to signify a race of phyletic evolution between two ecologically interlocking species (see, for example, Janzen, 1966; Southwood, 1973). What is not widely recognized is that many seminal contributions to the study of coevolution portrayed the phenomenon as being strongly punctuational—a "stepwise pattern," in the terminology of Ehrlich and Raven (1964).

So complex are many coevolved relationships that it is difficult to imagine how, once formed, they can change significantly. The simple fact that pollen vectors represent potential isolating mechanisms suggests that their relationships with plant species may be set up rapidly. In some instances, there is additional evidence. While not excluding the possibility of gradual origins for certain coevolutionary relationships between flowering plants and their pollinators, Baker recounted P. H. Allen's description of two complex interactions involving a single bee and two species of tropical American orchids. The following description refers to the second species (Baker, 1963, p. 879):

> Another orchid, *Coryanthes speciosa*, operates an equally complex but entirely different mechanism. ... Here again, male bees are attracted by the fragrance. As it alights on the mesochile, the bee maintains its position by the vigorous use of its wings. Ultimately, these hit against a drop of fluid (secreted by the glands at the base of the column) which had been hanging over the head of the bee. This dislodged drop then carries the bee with it into the liquid-filled bucket formed by the epichile. The struggle of the bee to free itself from this prison is likely to be prolonged and is successful only when it pushes its way up past the anther and emerges through the lateral opening of the lip, with two pollinia attached to its back.

Baker concluded that the involvement of a single species of bees, *Euglossa cordata*, in two elaborate pollination mechanisms represents strong evidence that at least one of the interactions "is the product of recent, sudden evolution." He noted also that certain floral features are under simple genetic control. The presence of the nectariferous spur of the columbine genus *Aquilegia*, for example, is determined by a single gene. Switching of pollinators to accommodate newly evolved flowers has been documented in a number of cases (review by Baker and Hurd, 1968).

Certainly, it seems imperative that coevolutionary relationships be considered in light of quantum speciation. While some may remain problematic, it must be expected that others will join the examples cited above as likely products of rapid evolution.

THE GEOGRAPHY OF SPECIATION

As mentioned previously, divergence of a large population, isolated by a geographic barrier from the remainder of the species to which it originally belonged, is the general equivalent of phyletic evolution within an entire species. In particular, we can envision the permanent geographic division of a species of very large population size into two equal subpopulations. Because of the random nature of mutation and the impossibility that two geographic regions will be environmentally identical, it is inevitable that the daughter populations will diverge through subsequent evolution, yet their large size should preclude rapid change.

It has been Mayr's firm position, as was noted in Chapter 2, that rapid evolution tends to occur only within small populations (see Mayr, 1970, p. 345 ff.). Historically, this view traces in part to Simpson (1944), whose viewpoint was influenced by the absence of conspicuous fossil evidence for rapid evolutionary transitions. Simpson's **quantum evolution,** however, embraced both speciation and accelerated phyletic evolution within existing lineages, and the latter phenomenon was emphasized in his subsequent work (Simpson, 1953). Mayr, in contrast, has focused on speciation. The idea that small populations are the locus of rapid evolution has been bolstered by the more recent developments in chromosomal and regulatory genetics pointing to speciation via small populations.

The notion that quantum speciation occurring allopatrically must generally be restricted to small populations represents no geographic problem. On purely geometric grounds, we would expect small, geographically marginal populations to be segregated from the remainder of a large gene pool far more easily than we would expect the large gene pool to be divided into two large segments. Steep genetic gradients, which, as will be discussed below, have also been claimed to engender quantum speciation, would also be expected to form primarily within narrow geographic belts.

The low probability that a new genetic feature will spread rapidly throughout an entire widespread species was assessed in Chapter 3. Strong and consistent selection pressures are much more likely to obtain within small populations. Here too the **founder principle** may frequently operate: A minute, localized sample of individuals is likely to embrace only a small fraction of the total genetic variation of the parent species (Mayr, 1942, p. 237). This biased ancestry increases the likelihood that the small population will diverge rapidly. For one thing, new homozygotic combinations are likely to arise by inbreeding. In addition, genetic features, whether new or old, are most likely to spread throughout. Such events are recorded by varying populations of the bank vole *Clethrionomys glareolus*, in Britain. This widespread species displays marked regional variation in morphology, much of which seems likely to be maintained by selection (Corbet, 1964). A particularly variable structure is the M^3 molar, which for two populations at Loch Tay, Perthshire, changed markedly even within 15 years (Figure 6-10). For one of the populations the change may have been produced by immigration from the adjacent population (Corbet, 1975), but in populations apparently introduced accidentally by humans to islands like Raasay, Jersey, and Skomer, extreme degrees of development of the posteromedial ridge of the M^3 molar are apparently maintained by selection.

The many isolates of *Clethrionomys glareolus* illustrate the potential of small populations for divergent speciation. Notably, the species itself may be one of the most ancient of the mammalian fauna of Europe, with a possible age of more than 3 My (Kurtén, 1968). The recent origin of the small distinctive populations underscores the sluggish nature of morphologic change recorded for apparent phyletic evolution within entire, established species of mammals, as in the studies of Gingerich (Figure 4-12).

Mayr, in particular, has championed allopatric speciation, and there is little question that this mechanism is important in nature. One form of indirect evidence here is the occurrence of superspecies (Mayr, 1963, p. 499). A superspecies is a monophyletic group of distinct species that are allopatric in distribution. Mayr has noted that the frequency of superspecies varies according to the frequency of physiographic barriers, as we would predict if allopatric speciation was the rule. More direct evidence is provided by the observation of distinctive small populations at the periphery of the larger gene pools. Mayr (1947; 1963, p. 573) noted that populations of a species that are ecologically most aberrant tend to occur along the periphery of the species' range. This observation, and the belief that isolates should form most rapidly here, led to the conclusion that most speciation occurs peripherally. The common occurrence of these in the plant genus *Clarkia*, as described in the quotation from Lewis on page 174, is a prime example. A fundamental difficulty of sympatric speciation is that, in requiring that a group of individuals become reproductively isolated

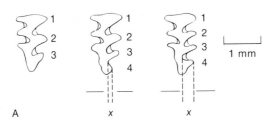

A

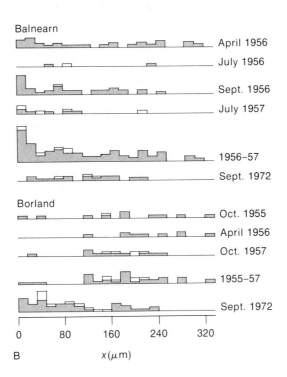

Balnearn

April 1956

July 1956

Sept. 1956

July 1957

1956–57

Sept. 1972

Borland

Oct. 1955

April 1956

Oct. 1957

1955–57

Sept. 1972

0   80   160   240   320

B $x(\mu m)$

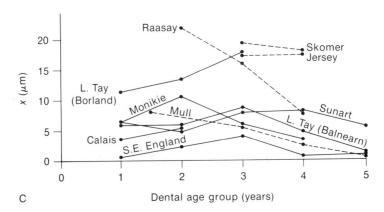

C   Dental age group (years)

without geographic separation, it calls for genetic features that confer reproductive isolation somehow to spread by way of interbreeding. However, Thoday and Gibson (1962) and Maynard Smith (1966), among others, have proposed mechanisms by which they believe that disruptive selection and stable polymorphism might lead to sympatric speciation.

A major source of misunderstanding relates to the meaning of allopatric, or geographic, isolation. The problem is one of scale. Some authors have viewed speciation by local isolation as being sympatric. In fact, entities that are known as geographic regions intergrade with the entities that we call habitats. No clear distinction is possible. Furthermore, what might be viewed as a homogeneous habitat for large or widely dispersing animals might be considered as a patchy geographic setting for small or weakly dispersing forms. For example, the knowledge that 27 endemic species of common ancestry are found on the tiny South Pacific island of Rapa, which measures only about 2½ by 3 kilometers, might be regarded as demonstrating sympatric speciation. The revelation that these species are small weevils of the subfamily Cryptorhynchinae (Zimmerman, 1938) negates this conclusion. In comparison to larger animals, these creatures experience the island in a coarse-grained way. Rapa is a sizable landmass, and there is no *a priori* basis for excluding the weevils' origin by a scaled-down form of allopatric speciation. A similar situation obtains for many parasites and parasitoids. Bush (1975), who specializes in the study of these groups, regards them as including the vast majority of sympatrically formed species. Here shifts of host seem to account for most speciation. It must be appreciated, however, that such shifts entail spatial isolation and do not differ in principle from standard allopatric speciation (White, 1968). Again the contrast is a matter of scale.

Certain more particularized models of speciation seem to hold promise for widespread application. Before looking at two of these, it will be useful to consider briefly some general principles. Chromosomal evidence that quantum speciation occurs by way of a very few individuals, sometimes only one, opens the way for two processes that have received much attention over the years. The first, the **founder principle,** was referred to earlier. The second is **genetic drift,** or the random fluctuation of genetic composition. It is possible that significant genomic shifts are sometimes achieved by chance in isolates that are the propagules of new species. This matter was first treated theoretically in some

---

FIGURE 6-10
Geographic variation in dentition of the bank vole *Clethrionomys glareolus*. A: Right M^3 molar, showing minimum (left), intermediate (center), and maximum (right) development of postero-medial ridge (dimension x is plotted in graphs B and C). B: Comparison of changes in x between the mid-1950's and 1972 for two populations at Lock Tay, Perthshire. The Borland population, in which the ridge was unusually large in the mid-1950's, reverted to the more normal condition of the Balnearn population during this period. C: Variation in degree of development of the ridge (x) among island populations (data for Borland and Balnearn populations gathered during 1955 and 1957). (After Corbet, 1964; 1975.)

depth by Wright (1940 and elsewhere). Genetic drift is of undetermined importance in speciation, but has been invoked by numerous authors.

Both the founder principle and genetic drift may play a role in the **population flush,** a mechanism that Carson (1968) proposed to account for the rapid formation of distinctive new Hawaiian species of *Drosophila*. A population flush is a release of bizarre variability within a small founder population invading new territory that is largely devoid of competitors and predators. Following the flush, according to Carson, comes a crash, with only a few of the variants surviving after a few generations. These represent adaptive peaks. As Carson sees it, the genetic consequences of the flush and crash are essentially stochastic. Surviving propagules may consist of only a single founder individual, which gives rise to a population that is at first subject to genetic drift. Later, as the population grows, adaptive evolution prevails.

The concept of **catastrophic selection** was formulated from a study of the plant genus *Clarkia* in California (Lewis, 1962; see pp. 174–175). At least half of the 28 species of *Clarkia* seem to have had sudden origins (Lewis, 1966). The best studied example is of *C. lingulata*, which is known from only two colonies at the periphery of the range of *C. biloba*, the obvious parent species. The descendant species ($n = 9$) has one more chromosomes than the parent form. The two species also differ karyotypically by a large translocation and at least two paracentric inversions. They differ externally in petal shape, and their hybrids are sterile. The parent species shows little chromosomal variability, and, in fact, all species of the genus are karyotypically stable. The clear implication is that the two small colonies of *C. lingulata* arose recently by a speciation process involving chromosomal alteration. This species, and other similar derivatives, are more xeric (adapted to drier conditions) than their parent species of *Clarkia*. They seem to have evolved from one or a small number of individuals that remained after a severe drought, as local remnants of populations of the parent species. This process, whereby a population is decimated, leaving one or more localized, genotypically distinctive populations, is what Lewis has termed catastrophic selection.

In the past few years, support has grown for a form of nonallopatric speciation that can be termed **parapatric.** This process entails the divergence and reproductive isolation of a population that is never fully out of spatial contact with one or more populations of the parent species. Ford (1975, p. 357) has gathered field data suggesting that distinct races can form in the face of considerable gene flow, and Murray (1972) has claimed that such divergence can be carried to the point of full speciation. Endler (1977) has provided an extensive theoretical treatment that, he concludes, renders speciation by way of steep clines a tenable proposition. He sees the process as occurring quite rapidly in small populations, in a manner explicitly compatible with the punctuational model (Endler, 1977, p. 174). He has also noted the difficulty of distinguishing the results of this process from the results of rapid allopatric speciation. According to Clarke and Murray (1969), steepening of clines to the point of speciation may have occurred within

Partula, a genus of land snails on the Pacific Island of Moorea. Here movement of animals is so slow that genetic differences may be observed between populations only 10 meters apart. Clarke and Murray have suggested the possibility that some of the 10 recognized species of the island may have originated when steepening of genetic gradients in ancestral species led to assortative mating and eventual reproductive isolation.

White *et al.* (1967) have claimed what can be viewed as parapatric origins for certain morabine grasshoppers in Australia. A number of morabine species in South Australia have originated by way of single chromosomal rearrangements, some of which serve as reproductive isolating mechanisms. These, it is contended, may arise within the range of the parent species and, if adaptive, expand as new species. Because chromosomal rearrangement lies at the heart of this model, the possibility of rapid divergence of the sort discussed above is implicit. When individuals with a new karyotype interbreed with members of the ancestral population, resulting heterozygous individuals are normally adaptively inferior. Because of this condition, Mayr opposes the idea that chromosomal rearrangement can lead to speciation within the parent species' range.

THE ECOINSULAR MODEL
OF ADAPTIVE RADIATION

Islands have been the site of remarkable adaptive radiations, the most famous of which is the Galápagos radiation of Darwin's finches (see Lack, 1947). Lakes are, in effect, aquatic islands on the land, and a few of the many lacustrine radiations comparable to those on true islands were discussed in Chapter 3. We can define an ecologic island as any isolated region in which biotic pressures, such as competition and predation, are unusually weak. It seems apt to label adaptive radiations within ecologic islands of any sort **ecoinsular radiations.**

The rates of speciation that characterize ecoinsular radiations are truly phenomenal. The remarkable radiation of cockles in the Pontian Sea was described in the previous chapter. The ecoinsular radiations for which mechanisms of speciation have been most intensively debated are those of the Great African lakes. The nature of this speciation will be considered below. A more immediate question relates to the importance of ecoinsular radiations in the general history of life. Frequently, these episodes have been regarded as evolutionary aberrations. Briggs (1974, pp. 442–445) regards them as phylogenetic cul-de-sacs, which play a generally insignificant role. The fact that all products of ecoinsular radiation visible today are in marginal geographic positions has led Briggs to conclude that such products are doomed to an ephemeral existence. My position, like that of Udvardy (1969), is quite the opposite. It is that while some ecoinsular radiations may be snuffed out, a significant fraction are not. The dramatic radiations for which we find evidence in the living world are well displayed simply because their products have not escaped the original insular boundaries. There is an observational filter working here. Ecoinsular radiations whose products have spread far and wide will tend automatically to remain undetected today. I have already expressed agreement with the idea that most evolution occurs in adaptive radiation. Not only is phyletic evolution generally very sluggish (Chapter 4) but background rates of speciation (rates in the absence of radiation) are also quite slow (Chapter 5). To the view that speciation in adaptive radiation is the site of most large-scale change can be added the idea that most radiation conforms to the ecoinsular model: Most radiation occurs in particular geographic regions where, to use Simpson's phrase, there exists the ecologic opportunity for diversification.

In Chapter 9, it will be shown that terrestrial taxa tend to speciate more rapidly than marine taxa. The insects, for example, radiate at phenomenally high rates. Limited mobility in a heterogeneous environment favors more frequent speciation in the terrestrial realm. Because of the relative ease with which effective barriers arise here, we would expect ecoinsular adaptive radiations to occur with relative frequency. Many have probably occurred in the late Cenozoic, but the terrestrial fossil record is too patchy for reconstruction of most such episodes, and leakage of taxonomic products with the shifting of barriers has obscured past geographic patterns of radiation. In fact, there is no reason to rule out entirely the existence of similar patterns for the marine realm. Evidence here is not altogether lacking. Around the shores of Cocos Island, which lies off Costa Rica near the Galápagos, there have been found about nine endemic marine and brackish-water species of invertebrates (Hertlein, 1963). This volcanic island, which has an area of only about 46 square kilometers, is believed to have formed as recently as the Pleistocene. I would suggest that throughout Phanerozoic time, radiation of shallow-water marine invertebrates in the shallow waters of island archipelagos may have occurred far more often than has generally been realized.

It is especially important to appreciate that mass extinction creates vacant ecologic islands. At the end of the Mesozoic Era, numerous continental areas stood as terrestrial islands available for occupancy, and these were indeed subsequently colonized by way of evolutionary episodes that can be viewed as enormous ecoinsular adaptive radiations. The proliferation of marsupials in Australia from ancestors that arrived from South America, perhaps by way of Antarctica, unquestionably exemplifies ecoinsular radiation. Persistence of the continental boundaries of Australia leaves no question here, yet the radiation on this true island did not differ in principle from simultaneous radiations on other landmasses where geographic changes and resulting biologic migration happen to have blurred our view of history. The radiation of birds in Australia, though presumably stemming from a larger number of ancestors, is well understood, having been analyzed in rather detailed biogeographic terms by Keast (1961). A smaller terrestrial radiation not obscured by geographic vicissitudes is that of the carabid beetle *Trechus*, 26 species of which seem to have arisen quite recently on Mount Elgon, a forest-covered central African volcano that is surrounded by desert (Jeannel, 1961). The genus is otherwise Palaearctic in distribution.

Thus, in some way, an enormous amount of morphologic variability erupts in certain kinds of adaptive radiation. In the previous chapter, evidence was presented that only a few speciation events typically accomplish the adaptive breakthroughs that take place early in adaptive radiation. In the punctuational scheme, it is perhaps fruitful to view adaptive radiation as representing a release of genetic variability—some of it previously present but not phenotypically expressed—under ecoinsular conditions. The idea would be that stabilizing selection pressure, which typifies established species, is released in the absence of a normal complement of predators and competitors. One possibility here is that Carson's concept of a population flush (page 168) is of general application. There is some evidence for the marked relaxation of stabilizing selection within founder populations that occupy vacant habitats. For example, introduction of 67 individuals of the fish species *Bairdiella icistius* to the Salton Sea of California yielded an initial cohort of offspring that retained to adulthood an extraordinary percentage of abnormal individuals (Walker, 1961). In 1953, of the surviving members of the 1952 year class, 6 to 15 percent were blind, about 2 percent had abnormal maxillaries or premaxillaries, 2 percent had malformed lower jaws, 1 percent were snub-nosed, and 3 percent had twisted vertebral columns. The incidence of abnormality declined through time, as competition set in, and the frequency of initial survival of abnormal individuals was also lower for subsequent year classes, but this example serves to demonstrate the enormous store of morphological variability normally held in check by stabilizing selection. In a larger, more complex geographic setting, like a small continent, it might be expected that such extreme polymorphism would become fixed within some small populations.

The manner in which variability issuing from a population flush may become embodied in discrete new species is open to speculation. Carson (1968, 1975) has favored the idea that a few adaptive peaks may survive a population crash following the flush. A basic question is how much of the ultimate variability that emerges in adaptive radiation arises as polymorphism within the original species, perhaps later to be fixed by the interposition of breeding barriers. The possibility that many radiations have begun with a population flush is by no means incompatible with a prevalence of allopatric or parapatric speciation. Owing to environmental heterogeneties and to assortative mating, polymorphism may have a patchy geographic expression. The emergence of geographic or microgeographic barriers or steep genetic gradients could then lead to division of the distinctive populations into discrete species. Rapid divergence in this scheme, though preceding reproductive isolation, would still occur almost simultaneously with it on a geologic scale of time, making the described mechanism a form of quantum speciation. Thus, I am in agreement with other authors (Eldredge and Gould, 1972; White, 1978, p. 9) that the quantum morphologic step may commonly predate a population's attainment of formal status as a discrete species. Jaanusson (1973) has, in fact, cited evidence that many sudden transitions in the evolution of the graptolites originated through polymorphism. Working with the remarkable cichlid fishes, Sage and Selander (1975) have presented evidence that an extraordinary degree of polymorphism may characterize single species. Within the endemic cichlid fauna of the Cuatro Cienegas basin in Mexico, single broods include individuals that differ strikingly from one another in type of pharyngeal teeth. Fishes tending to feed on snails possess heavy molar teeth, while those specializing in algae have few if any molars but a larger total number of teeth. Certainly speciation is of importance not simply in yielding new morphologic features, but also in fixing within entire species features that arise as polymorphic variation.

Even radiation resulting from an adaptive breakthough can be viewed in ecoinsular terms. A single event of quantum speciation that moves a population into a new adaptive zone, in effect, turns the surrounding region into ecospace accessible to radiation. The region was previously available for the *kind* of occupancy in question, but the parent taxon lacked what Simpson (1953, p. 207) has termed "ecological access." One interesting example of recent occurrence cited by Stebbins (1974a) is that of the South American cricetid rodents of the *Ichthyomys* group. These rat-like forms have invaded aquatic habitats, where their diet includes fish. At least 16 living species have formed in a radiation that followed evolutionary entry into what would seem to represent a previously unoccupied mode of life.

VARIATIONS AMONG TAXA

Given our present uncertainties about the relative merits of suggested modes of quantum speciation, there is no reason to assume that any particular mechanism plays the same role in any two taxa. As noted earlier, Bush, Wilson, and co-workers have suggested that rapid isolation and divergence resulting from chromosomal alteration are most likely to occur within taxa characterized by social interactions, which can facilitate inbreeding among divergent individuals. Bush (1975) noted that endemic *Drosophila* of Hawaii have undergone relatively minor chromosomal restructuring in the course of their rampant speciation of the recent past. Whether fundamental phenotypic traits, such as mode of dispersal or reproductive behavior, or genotypic features are responsible for karyotypic conservatism, it is interesting that *Drosophila*, in general, is also notable for its morphologic conservatism (Dobzhansky, 1956).

For birds, the geographic clustering of superspecies in Africa (Moreau, 1966), Australia (Keast, 1961), and South America (Haffer, 1974) has been viewed as evidence that geographic isolation of small populations in the dominant mode of speciation. Indeed, Mayr's widely accepted concept of allopatric speciation has grown largely from a variety of avian evidence, although Endler (1977) argues that parapatric models are equally applicable.

I myself have suggested that quantum speciation may be particularly frequent in taxa characterized by strong competitive interactions because of a high incidence of character displacement (Stanley, 1973c). The tendency of cichlid fishes to undergo quantum speciation at extraordinary rates was discussed earlier. As noted above, the conclusion of Steenis (1969) that the sudden origins of freakish structures represent important macroevolutionary steps in the phylogeny of the angiosperms would seem clearly to rest on the fact that plants have indeterminate growth and lack complex, animal-like behavior. Both of these traits confer considerable morphogenetic flexibility.

THE BOTTLENECK EFFECT

Paleontologic data evaluated in the two previous chapters (especially Chapter 4) led to the conclusion that major evolutionary transitions are generally restricted to populations too small to form recognizable lineages in the fossil record. The idea that quantum speciation normally involves small populations, when combined with the concept of catastrophic selection (Lewis, 1962), raises an important question. Does it frequently happen that an *entire* species declines almost to extinction to be "reborn" as a new species? Although resembling speciation, such a process would not be true speciation because it would not be multiplicative. It would represent a special kind of phyletic evolution, what I have termed the **bottleneck effect** (Stanley, 1977). If repeated within a lineage, it might produce significant evolutionary trends.

In fact, the great longevity of chronospecies (Chapter 4) rules out bottlenecking as the predominant source of macroevolutionary change. As illustrated in Figure 6-11, the implication is that bottlenecking events within lineages cannot have come in rapid succession. Furthermore, a general arithmetic argument can be marshaled against the idea that bottlenecking occurs with frequency in phylogeny. If quantum speciation entails transition by way of very small populations, then saltational bottlenecking should also occur only by the extreme constriction of a lineage. Mayr (1970, p. 294) estimated that of all isolates having the potential to evolve into new species, only one in 50 or 100 or 500 actually do so. The rest become extinct. The critical point is that a bottleneck population small enough to be equivalent to such an isolate should have the same probability of extinction as the isolate. Because every bottleneck population is, by definition, the sole remnant of an entire species, every termination of such a population represents the extinction of a species. Rates of extinction are, however, so low as to imply that the bottleneck effect cannot be a generally significant mode

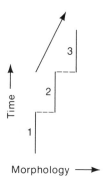

FIGURE 6-11
Slow net rate of evolution produced by bottlenecking, in which each
of a succession of species is suddenly transformed into the next.
Diagonal arrow represents rate of evolution. Compare this pattern to
that of Figure 4-6,B, whereby divergent speciation produces trends
very rapidly.

of quantum speciation. Consider, for example, a taxon in which an average
lineage, of the set existing at any moment, lasts 5 My. Then the extinction rate
will be about 0.2 My^{-1}. If we accept Mayr's conservative conclusion that something like 1 percent of isolates having the potential to form new species actually
do so, then we should expect the same success rate to apply to bottleneck
populations having the same population size and potential for divergence. Thus,
if extinction rate is 0.2 My^{-1}, rate of rebirth by bottlenecking will be 0.002
My^{-1}. Even if the taxon being considered contains a standing diversity of 1,000
species, bottlenecking will yield an average total of only two new species every
million years.

The inferred low incidence of bottlenecking stands in sharp contrast to the
high frequency with which marginal or intrapopulational isolates form. It would
appear that for many taxa, the rate of successful speciation from isolates is even
lower than Mayr has estimated. Pertinent here are the observations of Lewis
(1962, p. 262) for the plant genus *Clarkia:*

> Deviant marginal populations are so frequent in *Clarkia* that one often stumbles
> onto them without making a deliberate or systematic search. They may differ from
> the parental species in any one or more ways, including morphology, genetic
> compatibility, breeding structure, and chromosome arrangement. They all have,
> however, two features in common; they are ecologically marginal, and the discon-
> tinuity with the parental populations is abrupt.
>
> One may assume that most deviant marginal populations are ephemeral and
> destined to become extinct in situ. Few warrant taxonomic recognition, but some
> have persisted and flourished as distinct species, and occasionally, we may suppose,
> a deviant may set the stage for a successful new phylad of species. ...

The fact that *Clarkia* is a genus of annual plants offers evidence of the frequency with which its species engender isolates. If an average lineage of *Clarkia* survives for 5 My and yields only 10 isolates per annual generation, then before dying out it will generate a total of 50 million isolates! If standing diversity within the genus remains constant during the existence of the lineage, which must be a rough approximation of the true condition, the odds are that only one of the isolates will become a fully established new species.

Some unknown percentage of the myriads of isolates issuing from any species must technically attain the status of a distinct species without actually being recognized as such because they are snuffed out before expanding to become fully established. These ephemeral entities, which occur with unknown frequency, by definition play no role in macroevolution. I have labeled them **aborted species** (Stanley, 1978). It is reasonable to assume that they, like isolates in general, form far more often than fully established species. If they could be represented on a histogram like Figures 5-6,A and 5-7,A, they would form an extremely narrow mode along the vertical axis. Earlier analysis of the completeness of the data for established species upon which these histograms are based rules out any possibility that the valley between the mode for aborted species and the mode for established species would disappear with the plotting of complete data. For example, the excellent fossil data for Pleistocene mammals of Europe (page 75) have yielded hardly any species recognized in only one stage. This fact takes on great significance when we recognize that although degradation of a perfect fossil record might preferentially eliminate "single-stage" species because of their limited geographic ranges, it would also fragment the ranges of species actually spanning two or more stages, to yield numerous artificial single-stage species. The great rarity of apparent single-stage species indicates, first, that here the fossil record of fully established species is of high quality, and, second, that few species that became fully established lasted for less than two or three stages: There is a true discontinuity between these entities and aborted species.

It seems evident that speciation is a "boom or bust" phenomenon (Stanley, 1977). If a species of mammals expands beyond the local isolate stage, it is likely to last at least one-third of a million years. Similarly, a species of planktonic forams, once established, is likely to last at least 2 or 3 My. This interesting pattern is not difficult to assess: Whether reproductively isolated or not, few isolates survive. When one does, it is in response to the presence of generally favorable conditions. Thus, a newborn species faces an incredibly high risk of "infant mortality" during an extremely brief interval of geologic time. Local conditions that permit the brief emergence of species are common but short-lived. In contrast, propitious conditions so widespread as to permit a new species to expand and become fully established have a low probability of being quickly reversed to a degree that the species will go extinct.

What frequently happens in nature must be what Lewis observed for *Clarkia*. Namely, reproductive isolates representing varying degrees of divergence must constantly be tested, first against their immediate surroundings and subsequently against more general conditions. To an extreme degree, many are called but few are chosen. Numerically, all that is required is that one population in perhaps several million pass beyond the "small isolate" stage. Given the small number of genetic changes required, it is not at all difficult to understand how, occasionally, an adaptive innovation is developed that constitutes, or represents a large step toward, a new genus or family. In particular, it is within a weakly populated ecologic island, whether large or small, that the chances of immediate expansion and long term survival of aberrant populations are greatly increased.

THE CONTINUITY OF EVOLUTION

Having argued that unusual new features are normally fixed rapidly in association with phylogenetic branching, I feel compelled to address the question of continuity of both rate and process in evolution. To some workers, punctuational schemes seem inherently to imply a belief in discontinuity of rate and process, threatening the very foundations of the Modern Synthesis. Bock (1970, p. 705) has seen as the "fundamental cornerstone" of the Modern Synthesis what he has called the **synthetic assumption,** "that macroevolutionary changes can be explained completely by known microevolutionary mechanisms and that no additional or special macroevolutionary mechanisms exist. Microevolution and macroevolution constitute a continuum of change." Is this continuum challenged by the punctuational scheme advocated here?

The question of continuity is strongly linked to scale of observation. In the distant perspective of geologic time, phylogeny in the punctuational model takes on a disjointed appearance. This picture is misleading, however, in that as we magnify our view, almost any mode of transition, except special creation or the extreme hopeful monster concept, entails a continuum of descent—a graded series of generations from ancestral taxon to descendant taxon. For any rapid change of the sort proposed in this chapter, a minimum of three generations is required: A transition from parent to aberrant offspring to a generation in which the new feature is fixed by inbreeding. Usually, more than three generations and more than one genetic change will be required for quantum speciation.

Thus, I see the punctuational model as differing from the gradualistic Modern Synthesis largely in *emphasis.* It is convenient to divide this difference into two components: (1) that relating to the production of phenotypic variability, and (2) that relating to the factors that guide evolution by acting upon this variability.

In the production of variability, point and chromosomal mutations may be quite discrete features, yet point mutations must play a role in quantum speciation. Especially important here must be mutations of regulatory genes, but certainly also, to some degree, mutations of structural genes. If we look in the opposite direction, synthetic theorists were forced, by the very evidence of its occurrence, to accommodate chromosomal transformation, though they did so with little emphasis. (Among other things, it has long been accepted that new species of plants often arise in the form of single, polyploid individuals.) Thus, both chromosomal and point mutations are accepted in each model of evolution. The punctuational model simply stresses "high-amplitude" sources of variability—ones, like changes in gene regulation, by which relatively pronounced morphologic modification issues from few genetic alterations. The Modern Synthesis laid greater importance to "low-amplitude" point mutations and sexual recombination, from which major transformations were though to have been wrought gradually, over many generations, by the selective accumulation of infinitesimal steps.

In the guidance of evolution, the Modern Synthesis focused, of course, on natural selection. It also acknowledged the founder effect and genetic drift and held them to be most effective in small populations, but in playing down the overall evolutionary role of small populations, automatically attached little importance to these processes. It is quite likely that in the punctuational model natural selection will continue to be viewed as the dominant guiding process of generation-by-generation change. It will, nonetheless, be seen as accomplishing major transitions mainly by operating on more pronounced and discontinuous phenotypic variation than was envisioned in the Modern Synthesis. For example, selection presumably often weeds out karyotypically heterozygous individuals. Furthermore, the founder effect and genetic drift must be accorded a larger role in the punctuational model.

Intergradation of the modes of evolution within small and large populations implies that there is no definite maximum population size for what can be called quantum speciation, nor any clear separation, on any grounds, between this form of change and more gradual speciation or phyletic evolution. Still, there is likely to be a rather narrow band of overlap. What I would envision is a continuum, but one characterized by rather sharp changes in the importance of various elements of evolution. For example, if we could plot mean degree of involvement of chromosomal change in evolution versus population size, the resulting curve would, for the most part, lie near the horizontal axis, approaching it asymptotically in the direction of large population size but bending sharply upward near the origin (Figure 6-12).

As mentioned earlier, synthetic theorists saw epistatic interactions as accomplishing the kinds of change that are now being credited to regulatory genes, but believed the phenotype to be restructured gradually. The new concern with regulatory genetics in no way violates Bock's synthetic assumption, that macroevolutionary change can be explained by microevolutionary mechanisms. Additional components of the Modern Synthesis may give way to new genetic concepts, but it seems likely that most new elements will be viewed in terms of conventional processes at the population level. Løvtrup (1976) has challenged the

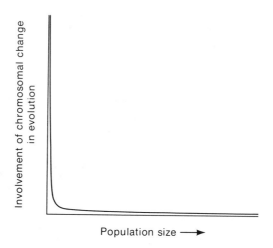

FIGURE 6-12
Hypothetical plot of the relationship between population size and mean degree of karyotypic involvement in evolution. Though not always involving karyotypic change, quantum speciation is restricted to very small populations.

adequacy of the Modern Synthesis and stressed the importance of rapid transitions in a manner that seems generally consonant with the thesis of this book. He has proposed the name "comprehensive theory" for the resulting paradigm, which subsumes neo-Darwinism while emphasizing that many evolutionary steps are achieved rapidly, by few genetic changes.

Thus, the punctuational model (or comprehensive theory) represents a less radical departure from many conventional notions of evolution than an initial assessment might suggest. Even so, this model yields important corollaries that run counter to certain inferences of the Synthetic Theory. A number of these corollaries will be explored in the chapters that follow.

SUMMARY

There is evidence that much quantum speciation is associated with chromosomal alteration. The implication is that such speciation occurs by way of very small populations within which inbreeding is likely. Inbreeding serves to reduce the problem of heterozygotic inferiority once a potentially advantageous new chromosomal arrangement is introduced. Chromosomal change may function to preserve useful gene combinations, but it also seems likely, through position effects, to introduce major phenotypic transformations, especially through the agency of regulatory genes. Gene regulation can account for pleiotropic effects and for major morphologic changes in the ontogenetic expression of structural genes. Profound transformations commonly described in terms of heterochrony and allometry can result from changes in small numbers of regulatory genes, as in the origin of the giant panda from a species of bear. Although it is difficult to assess probabilities, the vastness of geologic time leaves no basis for the argument that such rapid transitions are unlikely. It is possible that the class Gastropoda, as universally defined, evolved by fixation of a single genetic change that induced torsion of the body. Rapid morphologic transitions may occur with greater frequency among plants than among animals because of the flexibility allowed by indeterminant growth and an absence of behavioral constraints. Although coevolutionary relationships, such as those between flowering plants and their pollinators, have commonly been viewed in a gradualistic context, they must in at least some cases have developed quite suddenly.

Although it seems evident that quantum speciation occurs within very small populations, the degree to which total geographic isolation is required is currently a matter of debate. The founder effect and genetic drift are invoked as factors in many models of quantum speciation, but the nature of such speciation may vary among higher taxa. Ecoinsular adaptive radiations occurring in all kinds of ecologic "islands," where there are weak biotic constraints (competition and predation), may, in particular, foster quantum speciation. One likely mechanism is the population flush, which releases adaptive variability through the removal of stabilizing selection. Catastrophic selection, or the decimation of a population, leaving one or more discrete new propagules, may represent another important source of quantum speciation, but the infrequency of total extinction rules out a dominant evolutionary role for the bottleneck effect, whereby an entire lineage is drastically constricted, to be reborn as a new species.

Speciation is a "boom or bust" phenomenon. In other words, small populations technically attaining the status of species appear with great frequency in nature, but once a species is fully established (abundant and geographically widespread), it is likely to survive for a considerable interval of geologic time.

The punctuational model, even with its focus upon quantum speciation, perpetuates many elements of population genetics of the Modern Synthesis. It simply lays great emphasis on small populations and the kinds of genetic change that characterize them.

7

Large-Scale Trends and Species Selection

It has long been recognized that most macroevolutionary trends represent net changes within branched phylogenies, rather than evolution within single lineages. Still, the idea has erroneously persisted that such trends result primarily from phyletic evolution. In a punctuational framework, macroevolutionary trends become analogous to microevolutionary trends: Species take the place of individuals, speciation and extinction substitute for birth and death, and speciation generates variability in the way that mutation and recombination do within a population. Apart from the relatively weak contribution of phyletic evolution, there are three sources of macroevolutionary trends: (1) phylogenetic drift (comparable to genetic drift), (2) directed speciation (comparable to mutation pressure), and (3) species selection (comparable to natural selection among individuals). Except within small clades, species selection is the dominant mode of change. Its significance stems partly from the fact that the direction taken by speciation has a strong random element (the evolutionary direction of the next speciation event from a given set of lineages has low predictability). The nonrandom (directive) components of species selection, analogous to the components of natural selection among individuals, are (1) differential rates of speciation among lineages and (2) differential rates of extinction (differential longevities) of lineages. The agents of species selection are the familiar limiting factors of ecology: competition, predation, habitat alteration, and random fluctuations in population size. These agents function within both components of species selection, causing differential extinction among established species and selectively suppressing the multiplication of species. The prevalence of species selection in the framework of punctuational phylogeny underscores the opportunistic nature of evolution, with speciation events representing the experimental probing of uninhabited ecospace.

INTRODUCTION

Traditionally, it has been assumed that many large-scale evolutionary trends, like those embodied in the transition from *Hyracotherium* to a modern horse, were phyletic in nature, representing persistent selection within long-ranging lineages. This gradualistic view is dealt a major blow by the previously cited calculations (Lande, 1976) that long-term trends documented in the Mammalia, if gradual, would have represented such slow evolution that the selection coefficients involved would have been miniscule—so small that their maintenance cannot reasonably be envisioned (pp. 56–57). Even the idea that phyletic evolution of a more episodic type typically produces major trends is contradicted by the many arguments against gradualism presented in the preceding chapters. Darwin viewed large-scale evolution as an enormous race of phyletic transition, but it now seems evident that established species do not respond to environmental change to the degree that Darwin and most contributors to the Modern Synthesis believed.

If most change is concentrated in speciation, then a minimum of two lineages have formed part of any change comparable to the one leading from *Hyracotherium* to *Equus*. Most such transitions, of course, have involved more than two lineages. Thus, the punctuational view is that the observed change represents a *net* trend in phylogeny (Eldredge, 1971), or a phylogenetic trend. Simpler phyletic trends produce only minor adjustments of adaptation, as discussed in Chapters 3 and 4.

Actually, the view that single lineages have accomplished most large-scale trends began to crumble long ago, as a byproduct of the neo-Darwinian assault on orthogenesis. Simpson (1944, Chapter 5) observed that evolution does not proceed with perfect rectilinearity, and also that the pathway from an ancestral species to a much younger descendant species often includes segments of many lineages, some of which represent evolution in a direction opposite to that of the overall trend (see also Romer, 1949). Phylogeny has a bush-like configuration, with many short branches, so that in comparing two species separated by a substantial interval of time, we are seldom examining fragments of a single lineage. In particular, Simpson showed that a complex phylogenetic pattern stands between *Hyracotherium* and the modern horse. Such a pattern need not be punctuational, of course, if the primary accomplishment of speciation is simply to produce new directions of phyletic evolution. Also, as is true of the horse phylogeny, there may be few reversals of certain evolutionary changes. Nevertheless, while the gradualistic model merely accommodates the notion that most large-scale trends are phylogenetic rather than phyletic, the punctuational model *demands* this interpretation.

On the other hand, the punctuational model does not require *a priori* that large-scale trends occur within a complex phylogeny. There is the theoretical possibility that oscillations in the direction of speciation will result in a statistically static condition analogous to the Hardy-Weinberg equilibrium within a

TABLE 7-1   Mechanisms that produce
macroevolutionary trends, in the punctuational
scheme, and analogous mechanisms that produce
microevolutionary trends.

Mechanisms	
*Macro*evolution	*Micro*evolution
1. Phylogenetic drift	Genetic drift
2. Directed speciation	Mutation pressure
3. Species selection	Natural selection

population. Why phylogeny departs from this condition—as shown by major
adaptive transitions in the history of life—will be the subject of this chapter.

Comparison of phylogenetic stagnation with the Hardy-Weinberg condition
represents a continuation of the analogy with population biology introduced in
Chapter 5, where speciation and extinction were treated in a way parallel to the
demographic evaluation of birth and death. Similarly, the mechanisms that
underlie phylogenetic trends can be analogized with those that yield phyletic
trends. Species take the place of individuals, and, in a general way, speciation
becomes equivalent to mutation and recombination.

First, we will consider **phylogenetic drift.** Here the direction of speciation is
random, but stochastic fluctuations yield significant net change. This mechanism is equivalent to genetic drift within a population (Table 7-1). Second, we
will consider **directed speciation,** which represents a tendency for speciation to
move in one adaptive direction. This process bears a resemblance to mutation
pressure within a population. Finally, we will consider **species selection,**
which is equivalent to natural selection at the level of the individual, with the
character of phylogeny shifting according to variation in longevity and rate of
speciation among lineages. Here there is a more precise demographic analogy,
namely with selection (differential survival and fecundity) among individuals.

PHYLOGENETIC DRIFT

Raup and Gould (1974) made a valuable contribution in showing, by computer
simulation, that random walk can produce notable trends within small clades.
As would be expected, however, their simulations showed rate of random divergence from the ancestral character state to decrease with an increase in diversity. Raup and Gould expressed uncertainty as to the importance of random

effects in nature, but stressed that for any observed trend, random factors cannot be ruled out *a priori*.

It seems likely that random factors have had an important influence on phylogenetic trends within small clades, just as genetic drift seems likely to have played a significant role in phyletic evolution within some very small populations (Wright, 1931 and elsewhere). On the other hand, there is a wealth of evidence that species selection has predominated within sizable clades and even within small clades in which adaptive breakthroughs have occurred. I would concur with the opinion of Raup and Gould (1974) that the surest evidence that trends are adaptive lies in the functional analysis of adaptations rather than in the purely numerical evaluation of changes in diversity, which always entail some probability of random causation.

In this light, I find it inconceivable that a typical Paleocene family of adaptively primitive, diurnally carnivorous mammals could flourish for long at high diversity if introduced into the modern world, and I cannot believe that the absence today of such forms is largely a matter of chance. As another example, the fossil record traces a reduction in diversity of the trilobites from the Middle Cambrian onward. Combining the observation of this decline with the evidence that trilobites lacked powerful jaws, claws, and armor like that of a modern crab, as well as the ability to swim as efficiently as fishes, there is little doubt that trilobites, too, became adaptively outmoded. Can we reasonably entertain the idea that the few types of trilobites that survived to the middle and late Paleozoic were not selected by changes in the biotic environment? I suspect that many kinds of trilobites were easy targets for jawed fishes and also for some cephalopods. Some disagreement will be found for any single example proposed, but paleontologists have almost universally accepted the idea that certain body plans have been rendered obsolete during modernization of the world ecosystem. From evidence of adaptive morphology alone, the progressive nature of evolution has long been recognized (Rensch, 1959, [1954] Chapter 7). This kind of progressivism relates to temporal trends in ecologic adaptation (largely, success in biotic interactions). It does not necessarily entail increase in morphologic complexity or advancement toward the human condition—concepts that have aroused much controvery since the time of Darwin. Traditional gradualistic schemes can in theory accommodate complete randomness in the location of speciation and extinction within phylogeny. To explain the progressive nature of phylogeny, the gradualist has recourse in an appeal to directional phyletic evolution. Acceptance of the punctuational model, on the other hand, leaves one hard-pressed to uphold a dominant role for phylogenetic drift.

DIRECTED SPECIATION

A series of speciation events can produce a phylogenetic trend most simply through a bias in the direction of divergent speciation. In the simplest case, a

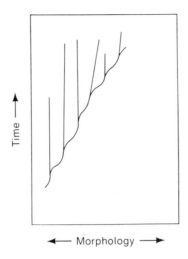

Time

←— Morphology —→

FIGURE 7-1
Macroevolution of the type envisioned by Grant (1963), with
successive speciation events moving in a particular direction.
(Axes reversed relative to Grant's
original diagram.)

trend will form if every step of a single chain of speciation events moves in the same adaptive direction. While, as noted above, there is a crude analogy here with mutation pressure, the apparently weak role of the latter in microevolution is no reflection on the possible significance of directed speciation in macroevolution.

Grant has proposed directed speciation as a mechanism by which large scale trends commonly occur. His illustration is reproduced here as Figure 7-1. One example cited by Grant (1963, pp. 416 f., 561 ff.) is the apparent history of the herbaceous perennial plant *Polemonium* in northern California. A geographic transect shows *P. carneum* now to occupy the cool moist region of the coast; *P. caeruleum*, the coniferous forest; *P. californicum*, the upper part of the coniferous forest and subalpine zone; *P. pulanerrimum*, the alpine zone above treeline; and *P. eximum*, high mountain peaks. The genus is believed to have a tropical or sub-tropical ancestry and to have speciated in stepwise fashion along an ecologic gradient toward the cold conditions of high elevations.

Environmental gradients are ubiquitous, and it seems clear that speciation will often proceed along them, as seems to have happened in the case of *Polemonium*. On this scale, the direction of speciation may be highly nonrandom. In a broader sense, however, I believe speciation to have a strongly random aspect. This view implies that some mechanism other than directed speciation often guides major trends—trends within phylogenies containing tens or hundreds of species. The view will be spelled out in the following section because it represents one of the premises of species selection.

SPECIES SELECTION

Within a given higher taxon, while one chain of speciation events may proceed from warm to cold or from wet to dry habitats, other successions of speciation are likely to be proceeding in the opposite direction. If we consider the example of *Polemonium* cited in the preceding section, Stebbins (1974b, pp. 175–181) has presented evidence in opposition to the commonly held botanical view that xeric plants (those adapted to dry conditions) often evolve from mesic plants (those adapted to moderately moist conditions) but that evolution in the opposite direction is uncommon. Stebbins shows that trends in each direction have been common and reversals in direction have often occurred in phylogeny.

The direction taken by any particular speciation event or succession of events must be determined largely by historical accident—in particular, by the nature of the habitát in which a potential ancestral species happens to exist, by the availability of niche space for potential occupancy, and by the nature of geographic changes that happen to effect divergence. Environmental gradients and inhomogeneities are ubiquitous in nature, and the species of a large, higher taxon will exist in varied settings. Inevitably, during an interval of time in which several species arise, speciation will move small populations in various directions. If we consider a single variable of adaptation that might be expressed by position along a single axis of niche hyperspace, some speciation events should move in one direction and some in the other. At any time during the existence of a higher taxon, the geographic location and direction of the next speciation event will be unpredictable. In this sense, speciation will have a strong random element. (If perfectly random, speciation would have a 50 percent probability of moving in either direction along any axis representing some aspect of niche space or morphology.) Rensch (1959 [1954], Chapter 4), Wright (1956, 1967), and Mayr (1963) long ago recognized this point, and Mayr (1963, p. 621; 1970, p. 9) has likened species to mutations, viewing them as biologic experiments, the origins of which represent irreversible discontinuities that offer new possibilities for large-scale evolution. Thus, in adopting Mayr's punctuational scheme, Eldredge and Gould (1972) accepted Wright's view that large-scale evolutionary trends have a stochastic element.

Of course, while speciation is irreversible in detail, its effects can be turned backward in a general way. An example from personal research that led me to accept the presence of randomness in speciation is the back-and-forth oscillation of life-habit transitions within families of bivalve mollusks (Stanley, 1972). Members of all major primitive clades of bivalves seem to have been burrowing animals (Pojeta, 1971). The byssus, an array of threads employed for attachment to the substratum, apparently evolved as an organ of post-larval fixation and was passed polyphyletically to adult animals by paedomorphosis (Yonge, 1962). A sizable foot remains in the juvenile stages of nearly all species that are byssally attached as adults; the byssus develops preferentially during ontogeny and the foot declines in relative size. But species that burrow and lack a byssus as adults have reoriginated from many adult-byssate stocks by secondary hypertrophy of the juvenile foot. Both a foot and a byssus are, in effect, stored within the juvenile so that whether the adult habit is one of free burrowing or byssal fixation, speciation can produce a switch, simply by bringing forth to the adult stage the required organ. Thus, back-and-forth life habit transitions have been commonplace in the evolution of groups like the Arcacea (Figure 7-2). There has been no strong tendency for speciation to move in one direction or the other. The direction of particular speciation events seems to have been determined by local environmental conditions and accidents of speciation that have varied haphazardly in space and time.

A heuristic analogy—though one representing a system that is more perfectly random—would be a coin-tossing exercise in which some coins are unweighted, but of the rest, half are weighted in favor of heads and the other half are similarly weighted in favor of tails. If the identity of the next coin to be flipped is determined by chance, heads and tails have the same probability of appearing.

Speciation events are obviously not simply oscillations that provide nothing substantially new. Periodically, adaptive breakthroughs occur. As Mayr (1963, p. 621) has emphasized, a divergent new species may happen to represent an evolutionary cul-de-sac, or it may mark the opening of a major adaptive zone. In other words, the emergence of the species in one small area carries no guarantee that the traits of that species will confer success in a broader environmental context.

I have suggested that Mayr's punctuational reasoning be carried a step farther (Stanley, 1975a). If rapidly divergent speciation interposes discontinuities between rather stable entities (lineages) and if there is a strong random element in the origin of these discontinuities (in speciation), then phylogenetic trends are essentially *decoupled* from phyletic trends within lineages. Macroevolution is decoupled from microevolution. Then what determines the course of most phylogenetic trends must be a selection process in which species are the units.

Natural selection, in which individuals are the units, has two components. The kinds of individuals that are favored are ones that produce an unusually large number of offspring, either (1) by reproducing at unusually high rates or (2) by surviving for unusually long periods of time. The comparable traits favoring certain kinds of new species will be (1) tendency to yield descendant species at high rates and (2) tendency to survive for relatively long intervals of geologic time.

The two levels of selection are compared in Table 7-2. It is important to understand that species selection is actually analogous to natural selection within

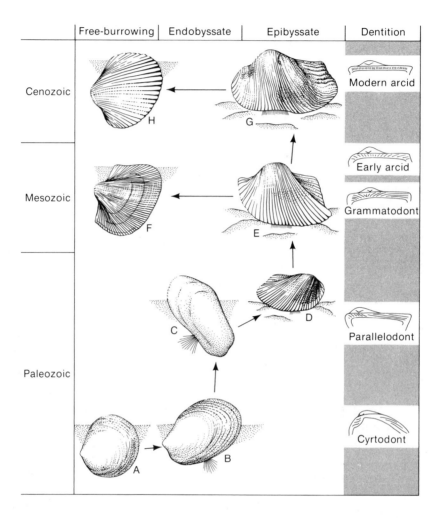

FIGURE 7-2
Life-habit changes in the phylogeny of the Arcoida, showing an initial evolutionary transition to an epifaunal habit of byssal attachment (epibyssate habit) by way of the endobyssate habit, which entails partial burial and attachment within the substratum. This was followed by evolutionary reversions to the ancestral burrowing habit. A: *Vanuxemia*. B: *Cyrtodonta*. C: *Parallelodon* (*Cosmetodon*). D: *Parallelodon*. E: *Grammatodon*. F: *Cucullaea*. G: *Arca*. H: *Anadara*. (From Stanley, 1972.)

asexual, rather than sexual, taxa. Within sexually reproducing species, the units of selection (individuals) interbreed, whereas, by the very definition of species, the units of species selection do not exchange genetic components from "generation" to "generation." Thus, the proper comparison is between an asexual clone of individuals and a clade of species.

TABLE 7-2   Components of individual selection and species selection.

Process:	*Micro*evolution	*Macro*evolution
Unit of selection:	Individual	Species
Source of variability:	Mutation/recombination	Speciation
Type of selection:	Natural selection	Species selection
	A. Survival against death	A. Survival against extinction
	B. Rate of reproduction	B. Rate of speciation

The nature of species selection is further elucidated in Figure 7-3. In general, toward the right-hand side of the hypothetical phylogeny depicted, there is an increase in species duration and an increase in speciation rate per unit time. Note that rather subtle gradients in the values of these two variables produce a marked phylogenetic trend. The gradients result in an increase in *total* rate of speciation toward the right.

It should be understood that the gradient in total rate of speciation does not represent nonrandom speciation, as randomness has been considered here. Randomness refers to the *direction* of speciation events, not to their *position* in phylogeny. To emphasize the allegedly strong random element in speciation within real phylogenies, speciation in Figure 7-3 is shown as occurring with equal frequency in each direction. Speciational steps moving to the left also contribute the same total amount of morphologic change as speciational steps moving to the right.

For phylogenetic drift to prevail, not simply the direction of speciation but also the phylogenetic position of speciation must be largely random. Gradients in rate of speciation and rate of extinction must be quite weak. It was, however, shown in Chapters 4 and 5 that quantum speciation supplies an enormous amount of phylogenetic variability upon which selection can operate. Particularly in large clades, drift is unlikely to play a major role. Williams (1975, Chapter 13) explicitly considered selection among taxa; but thinking in gradualistic terms, he asserted that such selection is not a creative agency, only a weeding out process that can accomplish little. Williams regarded even the sorting process (extinction) as being random, but, as already noted, such a notion can be entertained only in a framework of gradualism, where phyletic evolution accounts for obviously adaptive large-scale trends. Belief that the disappearance of species is nonselective seems to reflect a misconception that extinction usually strikes suddenly. In fact, as will be discussed below, long-term environmental trends—especially changes in the biotic environment—often work persistently against particular kinds of species, sometimes actually causing extinction and sometimes reducing population sizes greatly, to increase the likelihood of demise by other agents, including chance factors.

To help to alleviate certain past misunderstandings about randomness in speciation, I stress that *it is quite possible for directions of speciation within a clade to be randomly determined even if evolution is guided solely by natural selection.* Undoubtedly, mechanisms other than natural selection also operate during speciation,

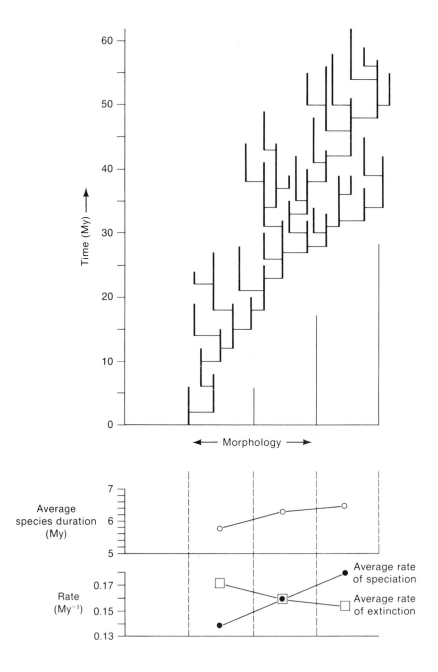

FIGURE 7-3
Hypothetical phylogenetic trend produced by species selection. Average rate of speciation increases toward the right, and average rate of extinction, which is inversely proportional to average species duration, decreases in the same direction. Speciation events moving to the left are equal in number to those moving to the right and contribute the same total amount of morphologic change, so that direction of speciation plays no role in the formation of the trend.

but their importance has yet to be determined. To an unknown degree, these factors add to the random nature of speciation. One of them, genetic drift (page 167), is a phenomenon that Wright (1931, 1940) showed can produce substantial evolution within small populations such as the ones from which many divergent species evolve. Another is the condition described by the founder principle (page 165), which holds that before divergence a local isolate may embody a small, nonrepresentative sample of the gene pool of the parent species. Still another kind of randomness is that contributed by sexual recombination, which exposes to selection only a few of many possible associations of genes, and by mutation, which is widely regarded as being largely random. Mutation and recombination are, of course, stochastic factors within evolution in general, but the concentration of evolution within speciation implies that the randomness of these processes will emerge in another way, as stochastic elements in the direction of speciation.

Another source of misunderstanding has been the identification of species selection with the variety of group selection propounded by Wynne-Edwards (1962), whose controversial concept is actually much narrower than the one originally advanced by Wright (1945) as "intergroup selection." Wynne-Edwards considered specifically interdeme selection within species. His argument, following the suggestions of Wright, was that in certain taxa selection for traits valuable to individuals is overwhelmed by selection for traits valuable to semi-isolated populations. Wright considered selection at all levels above that of the individual to be important and made no distinction between interdeme selection and selection at higher levels. Williams (1966; 1975) has opposed group selection regardless of level, also without attributing different characteristics to different levels.

I wish to establish a fundamental difference between interdeme selection, as espoused by Wynne-Edwards, and selection between taxa, or more particularly, between species. This difference has been widely overlooked. In the process of species selection advocated here, no assertion is made that adaptations become *fixed* within species because they are of value to entire species, rather than to individuals. It seems evident that fixation results primarily from selection at the level of the individual. It is the fate of adaptations, once established, that is determined by species selection. Implicit here is the view of Mayr (1963, 1970) that the species is the temporary holding vessel for genetic material and is the fundamental unit of large-scale evolution. Thus, species selection is quite distinct from Wynne-Edwards' concept of intraspecific group selection, in which the interests of the individual are, in effect, sacrificed to the interests of the group in the very process of phyletic evolution. By definition, selection among species is selection among reproductively separate entities. At this level, individual and group selection are not pitted so directly against one another. Still, some conflict is to be expected. My colleague, Robert T. Bakker, has pointed out to me that it is most likely to occur where sexual selection—an unusually effective source of phyletic evolution—has produced large bodily dimensions or bizarre features of sexual display that are of no adaptive value with respect to the environment. If,

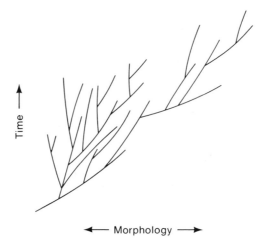

Time →

← Morphology →

Figure 7-4
Hypothetical phylogenetic trend in which
species selection tends to move the mean morphologic condition in a left-hand direction (speciation rate increases and extinction rate decreases
toward the left), but the effect of this condition is
negated by the prevalence of phyletic evolution
moving lineages toward the right. (Note that
while the lines representing lineages in the
right-hand portion of the diagram are relatively
long, the time represented [the vertical com-
ponent] is relatively short.)

as must often happen, such features increase probability of extinction, their
preservation and proliferation within phylogeny will be opposed by species
selection.

Selection among species is by no means excluded from gradualistic schemes of
evolution. In fact, Darwin clearly recognized it, and Wright (1945; 1956) has
discussed it. In a gradualistic context, however, the effect of selection at the
species level is greatly diluted by the alleged prevalence of phyletic trends. This
point is illustrated in Figure 7-4, which represents a hypothetical phylogeny
within which the phyletic component of evolution greatly predominates over the
speciational component. This phylogeny is devised so that the two components
oppose each other, but the speciational component is so weak that selection at the
species level is of little consequence. The phylogenetic trend here is determined
mainly by the orientation of strong phyletic trends.

The enormously important implication of the punctuational model is that
species selection must be elevated in status. It is not just one among many levels
of selection, but the chief source of major trends within higher taxa. I tend to
agree with those who have viewed natural selection as a tautology rather than a

true theory (see review by Peters, 1976). It is essentially a description of what has happened, with only weak powers of prediction, in that the kinds of individuals that are favored can often be recognized only in retrospect. The doctrine of natural selection states that the fittest succeed, but we define the fittest as those that succeed. This circularity in no way impugns the heuristic value of natural selection as a generation-by-generation description of evolutionary change.

The concept of species selection is equally instructive, but for a higher level of evolutionary analysis. Species selection deserves status as a distinct, though analogous, description because evolution above the species level is largely decoupled from evolution within lineages. Selection among individuals provides an inadequate description of large-scale evolution because unpredictable factors extrinsic to natural selection play a major role in the origin and survival of divergent new species. To a considerable degree, these factors randomize the average evolutionary direction taken by a set of speciation events in any interval of time. The decoupling of large-scale evolution from small-scale evolution led me to make the antireductionist statement that molecular genetics alone cannot explain large-scale evolution (Stanley, 1975a, p. 650).

As already noted, Eldredge and Gould (1972) adopted the notion that a stochastic element exists in speciation and in the development of evolutionary trends. These authors accept the concept of species selection (Gould and Eldredge, 1977), but in their initial paper on the punctuational model (Eldredge and Gould, 1972) did not spell out such a process, making no mention of the basic components of species selection: gradients in rates of speciation and extinction. In fact, the diagram illustrating their concept of a macroevolutionary trend (Figure 7-5) does not depict species selection, but directed speciation of the sort advocated by Grant (Figure 7-1). In the following section, it will be shown that much directed speciation can itself be viewed as a kind of intergroup selection, but selection among isolates (small populations) rather than among established species.

Gould and Eldredge (1977) have accepted my phrase "species selection," although Gould (1977) earlier expressed the view that this process should properly be considered a subdivision of natural selection, not deserving the separate label that I have advocated. In order to forestall possible objections from other quarters, I offer the following justification.

My opinion is perhaps most effectively supported by historical considerations. For propounding his concepts of macromutation and hopeful monsters, Goldschmidt was ostracized by adherents of the Modern Synthesis. His ideas were considered utterly non-Darwinian. It must be appreciated, however, that, if correct, his views today would easily be accommodated within the punctuational model, albeit in its most extreme form. The point is that Goldschmidt's ideas actually imply that transpecific trends should be guided by species selection. What does "hopeful monster" describe if not a new form that might succeed or fail as an ancestral species? Both a random direction of speciation and a severe form of species selection are tacitly asserted here. Goldschmidt's scheme

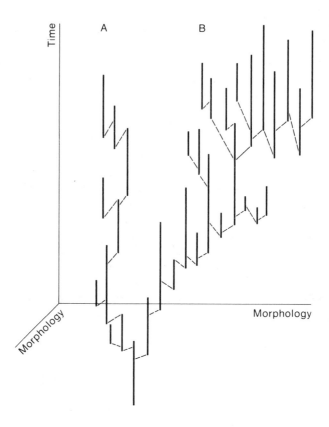

FIGURE 7-5
Diagram presented by Eldredge and Gould (1972) to illustrate a macroevolutionary trend. Here speciation is shown as a stochastic phenomenon only to the degree that some speciation events move in each direction. The trend, which moves toward the clade labeled "B," results from the fact that the first 11 speciation events move in a right-hand direction. The rate of speciation in the direction of the trend is actually lower than the rate in the opposite direction. Although average longevity of species increases slightly toward the right, this feature plays no role in the formation of the trend because nearly all speciation events proceed from newly formed species.

amounted to a clear message that the species is the unit of macroevolution and that macroevolution is decoupled from microevolution. All this was perhaps not appreciated by Goldschmidt or his contemporaries, but the universal recognition of Goldschmidt as a nonselectionist serves to illustrate the historical disassociation of selection at the level of the species from selection at the level of the individual. Willis (1940), a macromutationist, made a clear distinction, entitling his book *The Course of Evolution by Differentiation or Divergent Mutation Rather than by Selection.* De Vries (1905, pp. 799–800), an earlier macromutationist, explicitly considered a restricted form of species selection, recognizing the importance of differential extinction, yet he too was an outcast in the eyes of neo-Darwinians. On these points I will rest my case.

Parenthetically, it is worth noting that a directional bias in speciation could conceivably result from a tendency for evolution to move in a particular direction whenever stabilizing selection within established species is relaxed. The idea here would be that the unity of the genotype might resist certain widespread selection pressures except during genetic revolutions. Given the wide variety of local selection pressures to be expected among speciation events, it seems unlikely that such a phenomenon could be of general importance.

Certainly, the parallel and iterative patterns of evolution commonly seen in the fossil record may attest to a bias in the direction of speciation. To a degree, these patterns may reflect the fact that certain morphogenetic directions are inherently likely to be hit upon because they happen to involve simple genetic changes (page 157). It must be remembered, however, that parallel or iterative directions of change are ones that we single out as curious, while, in fact, adjacent phylogenetic groups may have taken quite different directions of evolution. Thus, some apparent bias in the direction of evolutionary steps may reflect an observational prejudice reflecting our interest in parallelism.

Finally, it should be mentioned that because species selection actually entails the branching and termination of lineages, some of which include two or more chronospecies, it would perhaps be more precise to employ the phrase **lineage selection** for the process in question. Because the phrase "species selection" seems to convey the desired meaning rather readily and has gained some familiarity, I hesitate to recommend a change of terminology, but the question deserves some consideration.

ISOLATE SELECTION: DIRECTED SPECIATION RECONSIDERED

For convenience in discussing more fully the source of directed speciation, I will define an **isolate** as a small population that may be a spatial isolate, with the potential to become a new species, or may represent an incipient species that could not interbreed with its parent population even in the absence of physical barriers. Isolates formed sympatrically or parapatrically are not excluded by this definition. Isolates, including the aborted species discussed on page 175, will almost never be recognized in the fossil record. Species selection, as defined here, represents selection not among isolates, but among fully established species.

Although isolates clearly intergrade with established species, a substantial discontinuity between the two entities is demonstrated by distributions of geologic

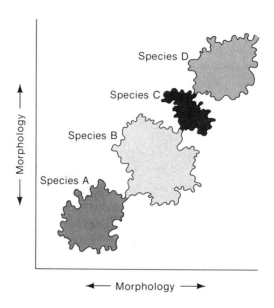

FIGURE 7-6
Phylogenetic trend resulting from isolate selection,
in which new species tend to arise from subpopulations that have relatively pronounced morphologic
features of a particular type.

longevities of established species within higher taxa. Recall the analysis of
aborted species, showing that speciation is a "boom or bust" phenomenon (page
175). The addition to Figure 5-6 or Figure 5-7 of data for isolates and aborted
species would form a single discrete mode lying along the vertical axis. This
bimodality seems sufficient justification for distinguishing two evolutionary
units and for recognizing both species selection and isolate selection. Even in the
living world, it is seldom known whether an isolate has evolved reproductive
isolation. In the present context, there seems to be no possibility or practical
need to distinguish between incipient species and simple spatial isolates.

    While selection at the species level is a matter of both differential extinction
and differential multiplication, selection at the isolate level must be primarily a
matter of differential extinction or failure to diverge and expand. As discussed in
the previous chapter, catastrophic fragmentation of established species can yield
two or more descendant species more or less simultaneously, but it seems doubt-
ful that many isolates themselves generate descendant species before expanding
into established species. Mere survival is so rare for any kind of isolate (page 174)
that multiplication must be an exceedingly rare event.

    Selection is envisioned as occurring among isolates in the following way (see
Figure 7-6). While simple isolation may take place in many parts of a species'

range, ecologic opportunities for establishment of a new species should vary from place to place. Whether an isolate stands to become established as a full-fledged species instead of being exterminated or reunited with its parent population (if not reproductively isolated) will represent selection at the isolate level. The result will be a bias in the direction of speciation.

Another important point, which can perhaps be understood most easily by simultaneous consideration of Figures 7-3 and 7-6, is that isolate selection takes place among the set of isolates derived from a single parent species. To appreciate this point, the reader should recall that the significant random element of species selection is in the *direction* of speciation at specific positions within phylogeny, not in the *position* of speciation within the entire phylogeny. The latter represents one of the two components of species selection: differential rate of speciation (as opposed to differential species longevity). Thus, isolate selection is quite distinct from this aspect of species selection.

THE AGENTS OF SPECIES SELECTION

Historically, ecologic competition has received more credit than it deserves as the biotic driving force of evolution. Time and again, *On the Origin of Species* describes competitive interaction as determining the success or failure of particular kinds of organisms. Darwin's struggle for existence, which was envisioned for both individual and species, perhaps set the tone for the emphasis on competition that pervades modern evaluations of large-scale evolution (Simpson, 1953; Fisher, 1958, p. 50; Wright, 1967). Until studies like those of Connell (1961a) and Paine (1966) popularized the notion that predation or other forms of disturbance commonly prevent competitive exclusion, the prominence of Gause's experimental work and the popularity of studies seeming to document niche partitioning in birds (e.g., Lack, 1933, 1947, 1954; MacArthur, 1958) helped to perpetuate the early bias.

Species selection involves more than competition. Taking a more general view, it seems evident that the agents of species selection must, to a considerable degree, be the agents of extinction. MacArthur (1972) listed the basic agents of extinction as (1) competition (2) predation (including disease), (3) alteration of habitat, and (4) random fluctuations in population size. These are, in fact, the basic factors that limit population densities in nature. Extinction is simply limitation of population density carried to the extreme.

It is self-evident that the agents of extinction account for one aspect of species selection (differential extinction). Given the fact that the same agents limit population size, it becomes clear that these must also contribute to differential speciation. The reason is that failure to speciate represents a failure of small populations to expand into new species. When conditions favoring a successful taxon deteriorate, we must expect not only that its lineages will go extinct at a higher rate, but also that remaining lineages will be less likely to cast off isolates that blossom into new species. We might, in fact, predict that adversity would manifest itself most readily in suppressing speciation, rather than in accelerating the extinction of established species. The small populations involved in the speciation process should be more susceptible to extirpation or to restraint from expansion than a full-fledged species should be vulnerable to total extinction. The analogy here would be with vulnerable planktonic larvae in the marine realm (Thorson, 1950), plant seedlings on the land (Janzen, 1970), or other fragile infant or juvenile stages.

Extinction of a species is often a complex business. In treating competition, predation, and habitat alteration separately below, I do not mean to imply that they operate independently of one another or even that they are entirely separate categories. It seems likely that two or more factors often conspire to bring about the demise of an established species or an isolate. In this light, it seems appropriate to treat random fluctuation in population size at the outset.

Clearly, sudden chance extinction is most likely to befall a small, localized population. This probability implies that random fluctuations should play a larger role in isolate selection than in species selection and may play a particularly important role in the final disappearance of an established species, after other agents have reduced its population size to a dangerously low level. Consider, for example, the plight of a hypothetical species surviving in sparse numbers during or following an extended catastrophic food shortage that nearly causes extinction. It might happen that nearly all of the few remaining females, though well nourished again, by chance produce a relatively small number of offspring within their normal range of fecundity. Customary death rates the following winter might then reduce population size so drastically that the following spring no male and female encountered each other for mating. Here, chance reproductive factors would have finished the job of extinction begun by food shortage (or, in MacArthur's classification, by habitat alteration).

While competition has been granted an undeservedly large role as a biotic agent of evolution at the expense of predation, I would also argue that both of these kinds of biotic chance received too little emphasis relative to vicissitudes of the physical environment. A paper by Bock (1972) represented a welcome reversal of this tendency.

It seems evident that different higher taxa should be characterized by different primary agents of species and isolate selection. From the fact that extinction represents the imposition of a limiting factor to an extreme degree, it follows that primary agents of extinction should tend to be distributed among taxa in the same manner as primary limiting factors (Stanley, 1975a).

The idea that a single biotic agent of selection may tend to dominate for a particular taxon stems from the relationship between competition and predation. When severe in their effects, these two agents tend to be mutually exclusive: heavy predation prevents competitive exclusion. We might reason, conversely, that for a particular taxon the dominant limiting factor should not be the dominant agent of extinction, because the taxon should be especially well adapted to withstand the effects of precisely this limiting factor. The counterargument, which seems to be more reasonable, is that in a taxon dominated by predation or competition, deleterious conditions are most likely to arise within the existing system: one game should have a single set of rules. In the punctuational model, coevolution is constantly taking place in the context of species selection. In a heavily preyed upon higher taxon, for example, species should periodically be eliminated by predation as a normal aspect of species selection, representing failure in the coevolutionary race. Extinction of a particular prey species is likely, of course, only if the predator has alternative food sources. Otherwise, an endless oscillation of predator and prey abundances is predicted by the Lotka-Volterra equation (see for example Krebs, 1972, pp. 245–250). Rate of extinction of the prey should be highest following an adaptive breakthrough within the predatory taxon. An occasional change in rules of the game—for example, the entry of a devastating new kind of predator into a system previously dominated by competition—might be expected to cause mass extinction, but, as will be discussed below, it would represent a disruptive event, differing from the form of selection normally in effect.

One would expect competition to be especially severe in behaviorally advanced groups, like birds and mammals. Advanced behavior is frequently accompanied by territoriality, even when resources are not in short supply. Numerous examples have been reviewed by Miller (1967). It appears that large territories, though unnecessary when food and space are plentiful, take on value at times when one or the other resource is in short supply.

One would also expect competition to increase with position in the food web. Top carnivores, whose populations are not checked by predation, must normally be limited by competitive interactions (often interference competition). The mere presence of trophic pyramids of species numbers within ecosystems must reflect this condition. Of course, in reality, the proximate factor limiting diversity may be space, but territoriality in carnivores serves as a guarantee of food availability at times of shortage, as noted above.

It is possible that harsh competition, as among top carnivores, may tend to force selection from the level of the species down to the level of the isolate. The reasoning here would be that where competition is severe, speciation will tend to be constrained to particular adaptive pathways. In a small clade of top carnivores occupying a restricted geographic region and feeding in a generalized manner, it is possible that very little "experimentation" in speciation is tolerated.

Certainly, competition must also exert a powerful influence in the macroevolution of many taxa occupying positions lower in the food web. For example, today the two dominant groups of acorn barnacles are the Balanoidea, which evolved in the Eocene and include about 273 living species, and the Chthamaloidea, which evolved earlier, in the Cretaceous, and now include only about 53 species (Newman and Ross, 1976). The classic study of competition by

Connell (1961b) showed that *Semibalanus balanoides* overgrows, undercuts, and crushes *Chthamalus stellatus* by virtue of more rapid growth, restricting *Chthamalus* to the upper portion of the intertidal zone. It seems evident that the advanced Balanoidea, in general, are competitively displacing the more ancient Chthamaloidea, causing a decline in their diversity (Stanley and Newman, 1979). The present ecologic distribution of the chthamaloids supports this idea: Except for a few localized forms that inhabit offshore areas, all species occur in the upper intertidal zone, which seems to represent a refuge from physiologically less tolerant balanoids. If barnacles that compete as effectively as the balanoids eventually break through, physiologically, into the upper intertidal zone, we can predict that the chthamaloids will become virtually extinct. According to Newman and Ross (1976, p. 24), the principal advance in the evolution of the Balanidae (the largest balanoid family) was the development of a tubiferous wall structure. The evolution of a porous skeleton that permits rapid growth seems to represent the key feature that has given the Balanoidea competitive supremacy over the Chthamaloidea. Such a skeletal breakthrough would parallel that of the reef-building scleractinian corals, which are believed to have displaced densely calcified sclerosponges through the evolution of a porous skeleton that permits rapid growth (Jackson *et al.*, 1971). On tropical coral reefs, competition for space is known to be particularly severe.

In order to balance the traditional emphasis upon competition as the dominant biotic agent of large-scale evolution, I will devote the following paragraphs to apparent examples of species selection guided by predation. Personal experience leads me to concentrate upon the marine realm. Let us first consider some ecologic features of the bivalve mollusks (Stanley, 1973c, 1977). Lacking even a head, this group of animals, in general, exhibits weak perception of biotic neighbors. It also possesses limited mobility. Food may be limiting in some unusually dense populations, but, especially at juvenile stages, bivalves are subjected to such intensive predation that it is difficult to imagine competitive interactions pervading the range of a species. Passive larval dispersal, relative immobility, and local disturbance yield disjunct and fluctuating populations. Sympatry of congeneric species is much more common than among mammals or birds, for example, in which advanced behavior and conscious dispersal foster territoriality.

The bivalves represent one of several groups of bottom-dwelling marine animals that seem to have absorbed the impact of a major advance in the sophistication of predation late in the Mesozoic Era (Stanley, 1974, 1977; Vermeij, 1977). Marine crabs, teleost fishes, and carnivorous snails all radiated at this time, and together these groups account for most predation upon modern bivalves. I have singled out the evolution of predators as a likely cause for the dwindling of endobyssate bivalves since the Ordovician, when they represented the dominant life-habit group of the marine Bivalvia (Stanley, 1977). These are forms that live mostly or entirely buried in soft sediment, attached by a byssus. Endobyssate bivalves in subtidal areas, being largely immobile and often partly exposed, are

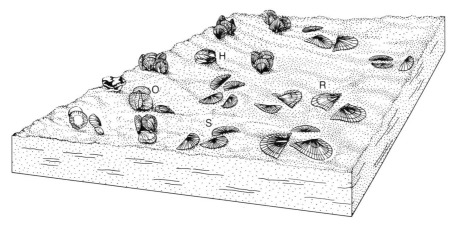

Sowerbyella-Onniella community

FIGURE 7-7
Reconstruction of the *Sowerbyella-Onniella* community, a typical example of a Paleozoic soft-substratum marine assemblage dominated by epifaunal brachiopods. R: *Rafinesquina*, S: *Sowerbyella*, O: *Onniella*, H: *Hebertella*, or a similar genus. (From Bretsky *et al.*, 1969.)

subject to heavy predation: It is no accident that by far the most abundant endobyssate species in shallow-water habitats of eastern North America is *Modiolus demissus*, which lives so high in intertidal marshes as to be well protected from most marine predators. Today, in sharp contrast to the condition in Ordovician seas, endobyssate species form only a small fraction of the world fauna. The fossil record shows that the decline of endobyssate taxa began in the Paleozoic (Stanley, 1972). What seems to have occurred is what was described earlier as a diffuse process, whereby endobyssate taxa have suffered significantly higher rates of extinction and lower rates of speciation than many free-living, fully infaunal groups.

The same pulse of predation seems also to account for the Late Mesozoic and Cenozoic decline of the articulate brachiopods, which today are mostly small forms living under rocks and coral colonies (Stanley, 1974). During the Paleozoic Era, articulate brachiopods reclined on the seafloor in great abundance and diversity in both temperate and tropical latitudes (Figure 7-7). Today the peak diversity of the group is in temperate seas, rather than in the tropics, where predation is particularly heavy. Articulate brachiopods had reradiated in the Mesozoic, following drastic decline at the end of the Paleozoic, but then dwindled slowly to their present, relatively inconspicuous state. For the most part, living representatives are small, thin-shelled creatures that grow slowly and therefore resemble juvenile bivalve mollusks, lacking an escape from predation by way of ontogenetic size increase. Among modern bivalve mollusks occupying hard substrata of the marine realm, predation is reduced by growth of spiney or thick shells, by occupation of crevices in or beneath rocks or coral colonies, or by colonization of the intertidal zone. The only sizable living group that lies free on subtidal seafloors in the manner that typified many Paleozoic articulate brachiopods is the Pectinidae (scallops). Can it be pure accident that this singularly successful free-living group has the remarkable ability to swim away from predators, as is often demonstrated by squeezing a little juice from a starfish into an aquarium full of scallops? I cannot believe that today any immobile, nontoxic group of organisms could recline openly on the seafloor in abundances and diversities comparable to those of the Paleozoic brachiopods.

The stalked crinoids may have suffered a similar evolutionary fate, being restricted to offshore areas of the modern marine realm, where predation is less intense than in reef environments. (Meyer and Macurda, 1977; Stanley, 1977). Paralleling, to a degree, the example of the scallops, the unattached comatulid crinoids still flourish on reefs, but also possess the ability to swim and to take refuge in crevices. These habits seem to have given them enough of an edge that, of all crinoids, they alone have been able to survive in shallow-water habitats. It is noteworthy that the only remaining group of stalked crinoids, the Flexibilia, have a reduced calyx with little exposed tissue.

Vermeij (1976) has studied what he has termed the "arms race" between crushing crabs and some of the snail species that they victimize, in both the Caribbean and the Indo-West Pacific regions. Crabs tend to be more powerful in the Indo-Pacific, but the antipredator skeletal armor of co-occurring snails is also more highly developed. Experiments show Caribbean snails to be more easily crushed by Pacific crabs. It seems evident that the coevolutionary race between predators and prey has been carried farther in the Indo-Pacific. On a broader scale, during the Late Mesozoic and Cenozoic throughout the world, there has been a proliferation of predator-resistant snails and a decline of taxa having vulnerable shell morphologies (Vermeij, 1977). Features that seem to have rendered shells liable to destruction were planispiral geometry, open coiling, and the presence of an umbilicus. On the other hand, strong external ornamentation, a narrow aperture, and apertural dentition are predator-resisting features that have proliferated primarily in Late Mesozoic and Cenozoic time. This pattern of change seems to have been brought about by species selection during the great radiation of benthic marine predators discussed in the preceding paragraphs.

Like the dwindling of endobyssate bivalves, the decline of the articulate brachiopods, crinoids (perhaps along with other pelmatozoan echinoderms), and weakly armored snails has been a gradual event, not a sudden extirpation. Species selection here has been a weak but persistent process, in which differential predation has resulted in slightly higher than average rates of extinction and lower than average rates of speciation. It is possible that the process has been somewhat more intense in other examples, although precise patterns of decline have yet to be assessed.

Possible additional examples that are difficult to assess because of their

Paleozoic timing are the decline of the trilobites and of echinoderm taxa having flexible skeletons. The dominant large arthropods of modern seas are decapod crustaceans. They survive predation by two divergent modes of adaptation. (1) Brachyuran crabs, on the one hand, possess a rigid carapace, formed in part by the folding of the abdomen under the thorax. Hermit crabs have hit upon a different solution, housing the abdomen in a snail shell. In addition to armor, crabs, of course, possess weapons in the form of chelae. (2) Shrimps also employ folding of the abdomen, but folding of a temporary sort, which permits rapid propulsion by flapping of the telson. Lobsters represent a compromise, being chelate and armored, but able to swim at moderate speeds in the fashion of a shrimp. It seems reasonable to suggest that the problem for trilobites was that they were basically dorso-ventrally flattened, with laterally splayed legs, so that their bodies hugged the substratum. The preservation of many trilobites reveals a habit of temporary enrollment under adverse circumstances, but this defensive posture left them otherwise defenseless and immobile. It seems likely that the decline of the trilobites, which was certainly underway by mid-Ordovician time, resulted from the evolution in the Ordovician of large, predaceous nautiloid cephalopods and, in the Devonian, of jawed fishes. Trilobites lacked the defensive weaponry of crabs and lobsters, and probably never developed the swimming capability of modern shrimps. A key feature for the evolution of decapod crustaceans was the presence of long, ventrally directed legs, which allowed flexure of the abdomen both for protection without incapacitating enrollment and for rapid locomotion.

It seems likely that the disappearance of Paleozoic sea urchins with flexible tests (Kier, 1965) and the success of those with rigid tests, may also, in part, reflect predator-driven species selection, as may the Cenozoic decline of sea urchins in general.

Finally, it is interesting to note that among full-time swimmers (most cephalopods and fishes), species selection has favored speed at the expense of protection. Animals that are fundamentally adapted for swimming have apparently had greatest success if particularly adept at this activity. Perhaps because early locomotory systems were incapable of producing rapid movement, both early cephalopods (nautiloids) and early fishes (ostracoderms and placoderms) were heavily armored. Following adaptive breakthroughs, such as shell coiling in cephalopods and the development of advanced fins and swim bladders in fishes, mobility has ultimately prevailed in the evolution of both groups. What combination of biotic pressures has caused this parallelism is uncertain (see, for example, Packard, 1972), but vulnerability to predation, and not simply competitive interaction, deserves consideration.

Disease may have caused the extinction of many species. Unfortunately, this special form of predation is seldom detectable in the fossil record.

It seems evident that a taxon can, by adaptive breakthrough, depart from a dominance of predator-limitation. This kind of escape may have been achieved by the rudist bivalves of the Late Mesozoic. These aberrant forms, which descended from burrowing Megalodontidae, took on a conical shape and according to Kauffman and Sohl (1974) seem to have excluded previously well-established scleractinian corals from reef habitats. Functional and preservational considerations led Kauffman and Sohl to suggest that rudists possessed a biochemical defense system against many types of epibionts and predators. While feeding and respiring, many rudists seem to have left tissue exposed to potential predators without ill effects. It was only with the rudists' mass extinction at the end of the Cretaceous that corals again came to dominate tropical reef habitats. The rudists may have differed from most other bivalve taxa in being limited by competition for space (Stanley, 1973c), as are the modern scleractinian corals, which undergo intense competitive interaction by overgrowth and extragastric digestion of neighbors (Lang, 1971).

The notion that predation can cause extinction by no means contradicts the idea that it can also promote diversification by relieving resource limitation (a notion popularized especially by Paine, 1966). I have suggested, for example, that until the first animal-like creatures able to consume algae evolved, Precambrian habitats may have been nearly saturated with algae (Stanley, 1973a). In other words, algal populations may have grown to the limits established by competition for environmental resources. A catch-22 may then have delayed the transition to animal-like creatures: The latter had to evolve from algae, but algae could only evolve (speciate) very slowly in the absence of these very creatures. When croppers finally did evolve, algae should have begun to speciate much more rapidly. The distinction between this kind of phenomenon and what has been discussed in this chapter is a matter of degree: Moderate predation permits diversification, while unusually intense predation, such as might result from the evolution of an efficient new carnivore, may lead to widespread extinction and a decline in diversity.

Habitat alteration, as an agent of species selection, entails not simply change in conditions of the physical environment but also change in aspects of the biotic environment not classified as competition or predation. In particular, disappearance of a food resource, including the host species of a parasite, may cause extinction. This, of course, represents the other side of the predation coin. Other biologic sources of extinction are the disappearance of pollinators or of species that form habitats for others or engage in obligate symbiotic relationships.

Examples of alteration of the physical habitat are changes in climate and topography or, for the marine realm, in equivalent conditions like salinity, temperature, and water depth. Several observations are worthy of note here. One is that major long-term phylogenetic trends—those spread over intervals in the order of 100 My or more—are not likely to be guided by habitat alteration. Gross climatic changes do sometimes span tens of millions of years but nonetheless can hardly be expected in such long intervals to specifically favor adaptive changes other than those related directly to climate. Furthermore, as we have seen for the Pleistocene Epoch, latitudinal migration often averts extinction. Biologic interactions provide more persistent long-term pressures, in that they themselves are guided by phylogenetic trends in what amounts to a complex system of feedbacks, as in the punctuational coevolution of predators and prey. Evolutionary trends are inherently more directional over tens of million years than are changes in the physical environment.

It is also important to note that the punctuational model poses a paradox with regard to the guidance of trends by climate. This paradox is that climatic trends must often ultimately produce phylogenetic trends in directions that are opposite to those that might be expected. For example, at first consideration, it might seem that a severe shrinkage of tropical forest area and complementary spread of arid territory would favor phylogenetic trends toward xeric adaptations. During the Pleistocene in South America, however, the result of such changes has apparently been quite the reverse. Haffer (review, 1974), studying birds, and Vanzolini (review, 1973), studying lizards, both uncovered evidence that tropical forest species have multiplied in pulses during brief intervals of the Pleistocene when forest habitats deteriorated. Catastrophic selection seems to have occurred, with new species evolving from small populations isolated in residual islands of forest. The effect has by no means been trivial. Haffer (1974) concluded that many new bird species have arisen by this process. Vanzolini estimated that at least three episodes of forest deterioration have occurred during the last 100,000 years, with the most recent two taking place only about 25,000 and 11,000 years ago. The composite result has been a considerable enrichment of the avian and lizard faunas of tropical forests. Whether there has been complementary enrichment of xeric forms during forest expansion is unknown, but there would not have been if xeric habitats have never been as severely decimated as forests. This is a likely possibility here, which illustrates the principle that asymmetric deterioration of habitats with oscillations of geographic conditions can yield a selective trend toward the kinds of species whose habitat is most adversely affected.

This kind of paradox is of great consequence because of the dominant role of divergent speciation. If phyletic evolution were to prevail, large-scale evolutionary trends would simply track environmental change. There would be no paradox. At the very least, the punctuational events of the South American forest faunas illustrate that large-scale transitions in the physical environment cannot be expected to produce simple phylogenetic trends. It is conceivable, of course, that decimation of populations by a predator or competitor could have similar effects, but in most cases we would not expect the geographic impact to be highly patchy or the surviving populations to be easily expanded.

Given the macroevolutionary premium on the survival and splitting of lineages, it seems evident that certain biologic properties of species that relate only indirectly to limiting factors may be favored in phylogeny. An example would be a behavioral or morphogenetic feature that promotes the divergence or reproductive isolation of small populations. The presence of such a trait can be viewed as something analogous to the life history "strategy" of a single species. This kind of demographic analogy will be considered in Chapter 10.

THE OPPORTUNISM OF
PUNCTUATIONAL EVOLUTION

A reversal of viewpoint causes another gradualistic notion to melt away in the context of the punctuational model. Instead of observing habitat alteration and predicting its evolutionary consequences, our Recent vantage point usually places us in the position of observing trends in the fossil record and postulating environmental causes. Here, gradualistic thinking has at times led to an excessive tendency to look for external controls of large-scale change. What *drove* fish out of the water to become the earliest amphibians, or thallophytes out of the water to become the earliest land plants? What drove small dinosaurs into the air to become birds? What drove hominoids out of trees into the savanna? These questions, which are frequently heard in evolutionary circles, sound absurd in a punctuational context. The point is that in the punctuational scheme an entire, established lineage does not make such a transition—only a small population. Ancestral lineages do not abandon their modes of life, nor do they necessarily die out immediately. Thus, what Simpson (1949) has termed the **opportunistic** nature of evolution (see also Rensch, 1959 [1954], Chapter 4) becomes particularly evident in the context of the punctuational model, as manifested in Mayr's simile likening speciation to experimentation. Normal fluctuations of the environment promote continual speciation. Adaptive breakthroughs are to be expected as standard events within ecosystems. As Simpson (1944, 1953) has pointed out, when vacant ecospace exists, all that is needed for invasion by potential ancestors is adaptive and geographic access.

In the punctuational scheme, evolution abhors a vacuum. Speciation, a heavily accidental process, is a constant source of experimental forays into ecospace contiguous to that already occupied. The belief that major evolutionary breakthroughs must be guided by external controls prevailed in the early part of this century, when, for example, the geologically sudden appearance of metazoan fossils in the Cambrian produced much speculation that the chemistry of the world oceans changed abruptly, to permit calcification. Calcification was initiated near the start of Cambrian time, but we now recognize, first, that it was accompanied by the origin of biomineralization of many other types, and, second, that the polyphyletic origins of skeletonization were spread through an interval in the order of 100 My. It now seems evident that skeletonization was simply one aspect of the initial adaptive radiation of the kinds of Metazoa that came to predominate in the Paleozoic (Stanley, 1976). In part, skeletonization served to thwart predators, but it also served to support muscle systems that triggered the invasion of vacant ecospace. "Experimentation" with mineralization, occurring as soon as was possible (soon after metazoan groups evolved), led to varied modes of evolutionary success.

If torsion in gastropods occurred suddenly, as outlined on page 151, it represents another example, as does any instance of extreme progenesis, in which during a single generation a species becomes juvenilized. Gould (1977b, p. 285) has suggested that progenesis must be selected for by particular environmental conditions. It seems quite reasonable to me, however, to view a sudden event, such as torsion or extreme progenesis resulting from simple genetic change, in the way that a mutation has traditionally been viewed—as an evolutionary experiment. A new form of this type might be the product of interbreeding among siblings whose parent has experienced a germinal mutation. If the nature of the new form quickly removes it from the gene pool of the parent species, as must often happen, then the fitness of the new form will not be tested against the fitness of the parental phenotype. The new creature may survive and itself speciate even if it inhabits the kind of environment occupied by its parent lineage. If, as must often be the case, limiting factors are such that competitive interactions are unimportant, the parent and descendant forms will, in fact, have little to do with each other. In short, the new adaptive complex will be tested primarily in the long run, by species selection. This is very much the view that I hold with regard to the origins of forms like the minute leptonacean bivalves (page 153), which have, in effect, been frozen at a post-larval stage of ontogeny. The apparently polyphyletic origins of the leptonaceans have certainly been sudden, and I find it difficult to believe that they have entailed the juvenilization of sizable populations. Rather, it seems likely that in most cases a single female (probably in most cases a venerid) founded a particular juvenilized species.

Even if major transitions have occurred by way of many generations, it is unnecessary to invoke wholesale environmental change. A *deus ex machina* quality places many alleged evolutionary controls—especially those postulated in past decades—in the realm of expedient, farfetched speculation. Thus, I suggest that before invoking general environmental change, we should consider major transitions in the context of punctuational opportunism within heterogeneous environments.

MASS EXTINCTION AND SPECIES SELECTION

Historically, it has been widely believed that most mass extinctions result from alterations of habitat. This conclusion seems inescapable, although secondary agents involving biologic interactions must come into play as well. Of particular interest here is the fact that mass extinction may be only weakly selective. (Like mass mortality, it must represent a rather uniform imposition of lethal conditions.) Therefore, to the degree that mass extinction is the site of species elimination in the history of life, the importance of differential extinction as a component of species selection may be diminished. Only if mass extinction is both incomplete and nonrandom can it represent species selection. Bakker (1977) has documented in detail and with modifications the general assertion of Cope that terrestrial tetrapods surviving mass extinctions have tended to be of small body size, and it is from these that renewed radiation has proceeded. The phrase

catastrophic species selection seems appropriate here, paralleling catastrophic selection at the level of the population. In the final chapter, I will assess the relative importance of mass extinction among various taxa. The general point is that the significance of differential extinction as an aspect of species selection may be low if average longevity of lineages is great relative to the average interval of time between mass extinctions. Among planktonic forams, for example, an average species has lasted something like 20 My (Figure 5-7), yet the interval between mass extinctions has been only about 30 My. Here mass extinction has accounted for a relatively high percentage of extinction events. It will also be shown, however, that variability among taxa in the ratio [lineage longevity/interval between mass extinction] is damped because of linkage between the two. For mathematical reasons, taxa with high rates of turnover (rates of speciation and extinction) are inherently susceptible to mass extinction and therefore suffer it with relative frequency. In this sense, mass extinction selects *among higher taxa.*

Mass extinction, which amounts to rapid exponential decline, can occur because rate of extinction suddenly increases, or because rate of speciation suddenly decreases, or because changes of both kinds occur. We might expect a combination of the two effects commonly to occur, because one represents the termination of established species (large populations) and the other, in effect, the termination of isolates (small populations). As noted earlier, the latter must be especially vulnerable, which suggests that a severe decline in rate of speciation may be the dominant factor in mass extinction. Empirical evidence favoring this idea will be presented in Chapter 10, where mass extinction will be considered more fully. At this point, however, it is apt to note that the punctuational view of evolution leaves higher taxa more vulnerable to mass extinction than they seemed in a traditional gradualistic context. If it is true that a braking of speciation contributes heavily to mass extinction, then there is no way in which evolution can effectively respond to environmental deterioration. The dominant source of variability—speciation—is effectively shut down. In the conventional gradualistic scheme, there was a much greater possibility for survival through escape by phyletic evolution. At times of mass extinction, variability would still exist within populations and selection, if anything, should be unusually intense. As noted earlier, the potential for phyletic response to environmental change has been greatly exaggerated in the Modern Synthesis.

RATES OF LARGE-SCALE EVOLUTION:
FUNDAMENTAL CONTROLS

Continuing to pursue the parallelism between natural selection and species selection, it is illuminating to consider what factors govern rates of large-scale evolution that is guided by species selection. R. A. Fisher's Fundamental Theorem of Natural Selection states that rate of evolution at the level of the individual increases with genotypic variance. Rate of evolution automatically increases with a decrease in generation time, for each generation amounts to a single selection event. It also, of course, increases with selection pressure, but this amounts to a tautology because selection pressure is measured by the selection coefficient, which is also a measure of rate of evolution. By analogy, rate of change via species selection should increase with variability within a clade. (Variability will amount to variety of species.) In a tautological sense, rate of change should also increase with intensity of selection.

It is interesting that Fisher, a gradualist, overtly considered selection at the level of the species, but rejected it as being relatively unimportant (Fisher, 1958, p. 50). His explanation placed an overly heavy emphasis on competition. More importantly, he ignored the possibility of quantum speciation, suggesting that selection would occur primarily among competing species, which, he concluded, should be adaptively similar to each other and therefore few in number. In other words, Fisher viewed variability among species as being too low for effective selection. The much larger number of individuals providing raw material for natural selection seemed to offer greater variability and therefore to provide for much more effective selection. Fisher concluded that the great longevities of species relative to the lifetimes of individuals also weighed strongly in favor of the dominance of selection at the level of the individual. Here, in effect, he was contrasting "generation" times.

Fisher's impeccable logic entirely justifies the gradualistic tradition of playing down the importance of selection at the species level. In a gradualistic scheme, variability on a taxonomic scale is generated primarily by slow phyletic divergence. The problem, then, lies in the assumption of gradualism. Evidence for the punctuational model becomes a double-edged sword, undercutting two of Fisher's premises. In the first place, it presents a strong empirical case that natural selection within established species is somehow stifled, forcing us to look beyond phyletic evolution to account for large-scale transition. In the second place, it shows that quantum speciation is a real phenomenon and a source of great variability, thus opening the way, on a theoretical plane, for rapid macroevolution via species selection.

Generation time has no precise counterpart in species selection, but the parameter R is equivalent to net per capita growth rate of a population, and the latter is inversely related to generation time (Figure 7-8). In general, taxa exhibiting high rates of speciation and extinction (which normally correspond to high values of R) are equivalent to populations consisting of individuals with short generation times. High rates of turnover of lineages clearly account for most differences in rate of evolution between groups like the Bivalvia and Mammalia.

Within a particular taxon, however, turnover rate is less important than the variability upon which species selection operates. In Chapter 5, it was shown

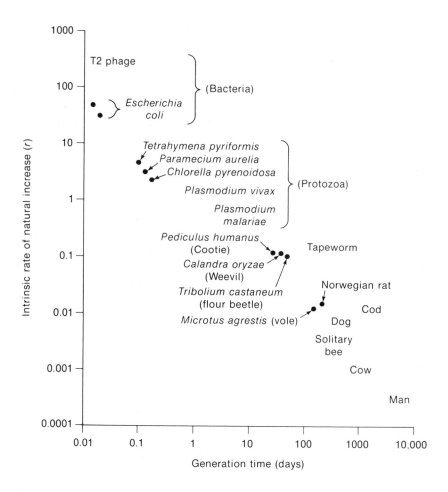

FIGURE 7-8
Inverse correlation between intrinsic rate of natural increase (*r*) and generation time in animals. (Data from Smith, 1954.)

that number of successional speciation events is less important than degree of divergence per event. In other words, evolution is most rapid early in radiation, when total number of speciation events per unit time is low relative to the overall rate later, when numerical diversity is greater (pp. 132–141). The idea that intense species selection coincides with episodes of quantum speciation amounts to a statement that rate of evolution increases with variability exposed to selection, according to something akin to a selection coefficient. This represents a general analogy with Fisher's Fundamental Theorem.

SUMMARY

Major trends in evolution are the result, not of phyletic transition, but of divergent speciation. Most are phylogenetic trends: net changes produced by multiple speciation events. The mechanisms by which phylogenetic trends form are analogous to processes operating within populations, but with species taking the place of individuals as fundamental units, and speciation and extinction replacing birth and death.

Phylogenetic drift, a chance process analogous to genetic drift within populations, is likely to produce significant trends only within small clades. The same is true of directed speciation, a bias in the direction of speciation that is somewhat analogous to mutation pressure within populations. Species selection is the dominant source of long-term phylogenetic trends within sizable clades. This process, which represents differential rates of speciation and extinction among species, is analogous to natural selection at the level of the individual. Divergent speciation generates the variability upon which species selection operates and is therefore somewhat analogous to mutation and chromosomal recombination at the level of the individual.

Within a sizable clade, it is impossible to predict when or in what environment or subpopulation speciation will occur. This random aspect of speciation largely decouples macroevolution from microevolution, even if one assumes that natural selection is the primary process by which new species arise.

Species selection is not equivalent to group selection of the sort advocated by Wynne-Edwards. The latter requires that new adaptations become fixed because they are of value to groups, while species selection carries no such implication, accounting only for the fate of adaptations after they are fixed within species.

Selection must occur not only among full-fledged species but also among small spatial or reproductive isolates. Isolate selection must, in part, amount to directed speciation because it favors the survival and expansion of isolates that differ from their parent species in a particular way. This kind of bias in the *direction* of speciation is not to be confused with variation in rate of speciation according to *position* in phylogeny (an aspect of species selection).

The agents of species selection are the factors that limit population sizes and cause extinction: competition, predation (including disease), habitat alteration, and random fluctuations of population size. Often, two or more agents conspire to terminate a lineage. Random fluctuations are likely to come into play only within small populations. Historically, the importance of competition in large-scale evolution has been exaggerated. As a result, predation has been granted much less significance than it deserves, especially within behaviorally simple, heavily preyed-upon taxa. Behaviorally advanced, weakly preyed-upon taxa, like many groups of top carnivores, are more likely to undergo species selection driven by competitive interactions. The directional nature of evolution makes biotic change a more likely agent of long-term species selection than is change in the physical environment (climatic change being the chief exception). Ironically, through the fragmentation of habitats, temporary deterioration of a particular kind of environment can select for species that occupy the very kind of environment that is deteriorating.

The punctuational model opposes the traditional notion that a major trend typically entails the driving of a lineage out of one mode of life into another (for example, fishes being forced from the water onto the land, as the first amphibians, or human ancestors being forced out of trees and into savannas). Rather, transitions are opportunistic in nature, reflecting the "experimental" nature of speciation. At least briefly, the ancestral lineage persists along with the new one.

Mass extinction, to the degree that it strikes randomly, is the antithesis of species selection. Sometimes, however, it operates with specificity, as in the differential mass extinction of large land vertebrates. In so doing, it represents a kind of catastrophic species selection.

In species selection, there is no precise equivalent to generation time, which is a determinant of rate of evolution within populations, but turnover time for lineages, which reflects rates of speciation and extinction, is the approximate analog. Differences in turnover time are largely responsible for major differences in rate of large-scale evolution between higher taxa like mammals and bivalves. Within a single taxon, however, as predicted by analogy with Fisher's Fundamental Theorem of Natural Selection, rate of large-scale evolution increases with phenotypic variability among species, which is generated by quantum speciation. Thus, rates of evolution are highest early in adaptive radiation, when degree of divergence per speciation event is high, even though the total number of events is low.

8

Why Sex Prevails

Traditional views of the function of sexuality hold that sex benefits the individual, population, or species by recombining and spreading genetic material and thus providing for effective evolutionary response to environmental change. Difficulties here are (1) there is no consensus among theoreticians that sex can generally accelerate evolution, and (2) a number of asexual clones of higher organisms can be seen to be flourishing in nature. Furthermore, punctuational arguments imply that phyletic evolution does not represent an effective enough response to environmental change to account for the near ubiquity of sex among higher organisms. Rather, sexual recombination has its primary effect in speciation. Except among a few simple, hardy, and widespread taxa, extinction rates are so high among higher organisms that rapid diversification is needed to offset them. The only adequate mode of diversification is divergent speciation, a process that is impossible for asexual taxa. Sex, then, is of greater value to clades (higher taxa) than to species, populations, or individuals. Asexual clones of higher organisms arise sporadically in the history of life but, unless they possess extraordinarily broad niches and enormous populations, disappear quickly, because they diversify much too slowly to overcome normal rates of extinction.

INTRODUCTION

On the face of things, sexual reproduction seems an unusual phenomenon. James Thurber and E. B. White wrote a book entitled *Is Sex Necessary?* Biologists also have difficulty explaining its value. By definition, natural selection favors those kinds of organisms that contribute a high percentage of genetic material to future generations. In many sexual species, however, in mating, a female sacrifices one half of her potential genetic contribution, with no apparent advantage to herself. As first considered in detail by Weismann (1889), a conventional argument used to dispel this paradox is that sexuality is of benefit not to individuals but to entire species (see, for example, Fisher, 1958, Chapter 6). The reshuffling of genetic material by sexual reproduction continually generates new combinations of genes. Some of these are favored and then, allegedly, spread throughout the entire species. Thus, sexuality is claimed to be of value in accelerating evolution: It is almost universally present in species of higher organisms because it enables them to succeed in a never-ending race of phyletic evolution against competitors, predators, and changes in the physical environment. The gradualistic nature of this standard explanation was noted in Chapter 2.

A variation on the preceding idea holds that sex is important not to the group, but, in a similar way, to the individual: Sex enriches the genetic legacy that the individual passes on by ensuring that the individual's genetic products will be combined with a variety of other genetic components (Ghiselin, 1974, p. 57; Lewontin, 1974, p. 196). Of course, this idea requires the presence of a changing environment and implies phyletic response. Sex is nearly ubiquitous among higher organisms. Therefore, like the idea of the preceding paragraph, this one is gradualistic in nature: It implies that successful phyletic response to environmental change is the rule.

Though widely adhered to, the idea that sexual recombination accelerates phyletic evolution is currently a matter of theoretical debate among geneticists. Moreover, even if sex does accelerate phyletic evolution to a degree, it does not necessarily follow that this effect is so important as to make sexual modes of reproduction a near necessity for the survival of lineages of higher organisms.

It should be appreciated that even without the female sacrifice noted above, the prevalence of sexuality would require an explanation. White (1978, p. 287) estimates that in the order of 1,000 animal species are asexual, which amounts to at most one in 1,000; yet many taxa clearly have the potential to evolve asexual species.

In the present chapter, the traditional view of sex will be challenged, largely on the basis of evidence already presented in favor of the punctuational model. The basic argument (Stanley, 1975b) can be expressed quite simply: Because species last much longer than has generally been appreciated, and because phyletic evolution advances far more slowly, there exists no gradualistic race of the sort that has commonly been envisioned. Most species, once established, survive with little change for long periods of time. Rapid evolution cannot be at such a premium as to account for the near ubiquity of sex. There is another side to this situation that, it seems to me, must largely account for the prevalence of sexuality. The punctuational model holds that rapid evolution is concentrated within speciation events, yet asexual taxa cannot speciate in the normal sense. As Mayr (1970, p. 374) has observed, "Without speciation there would be no diversification of the organic world, no adaptive radiation, and very little evolutionary progress."

The plight of asexual taxa in large-scale evolution can be viewed from a more quantitative perspective. In Chapters 5 and 9 of this book it is shown that higher taxa of particular kinds display characteristic rates of extinction. If a taxon is to survive, these rates must be offset by equal or greater rates of production of new species. Herein lies the problem of asexual taxa. Though unable to speciate, they are plagued by the same general rates of extinction as sexual taxa. It is no wonder that asexual modes of reproduction are relatively rare, even though many asexual "species" are individually quite successful. Thus, I suggest that the evaluation of sex be elevated to the level of the higher taxon. It is not primarily the species that benefits, but the clade. In effect, sexuality represents a virtual *sine qua non* for success in species selection. Before we embark on a more detailed presentation of the punctuational view of sex, it will be appropriate to consider more fully the conventional view.

GRADUALISTIC VIEWS OF SEX

Fisher (1930, 1958) and Muller (1932) claimed that the chief disadvantage of asexual species lies in their slow accumulation of valuable mutations. Each line of descent within a clone evolves independently. For this reason, many generations are required for several valuable mutations to appear sequentially within any one line of descent. Within sexual species, on the other hand, recombination permits the mixing of useful mutations so that a gradual, sequential origin within any line of descent is unnecessary and adjustment to environmental change is facilitated. Maynard Smith (1969, 1971) suggested that the early ideas of Fisher and Muller be modified; he concluded that sex accelerates evolution only when recombining genetic features that have evolved in different environments. On this point, Crow and Kimura (1969) took issue with Maynard Smith.

While most biologists believe that sexual reproduction predominates among higher organisms because it promotes variability and accelerates phyletic evolution, a few writers have taken adversary positions. An early source of some dissatisfaction was the observation already noted that sexuality represents a compromise for females of certain kinds of species: namely ones in which the male makes little paternal investment. The validity of this alleged sacrifice has been challenged (Barash, 1976) and defended (Maynard Smith and Williams,

1976). The error of the challenge, it seems to me, lies in failure to distinguish between realistic and unrealistic evolutionary options. Let us consider the possibilities for a sexual couple in which the male makes a paternal investment (perhaps providing food) that is equal to that of the female, allowing her to bear and raise n offspring when alone she could bear and raise only $\frac{1}{2}n$. The genetic contribution of each parent to the offspring will amount to the genetic complement of $\frac{1}{2}n$ diploid individuals. If such a couple could become asexual through conversion of both the male and female partners to asexual females, which would then bear $\frac{1}{2}n$ offspring each, neither gender would gain anything in the conversion. Given the evolutionary options in this particular case, sexuality entails no genetic sacrifice on the part of the female, but, in fact, this is not the way sex is lost. I have created an unrealistic scenario to expose the fallacy of the argument that sex entails no genetic sacrifice for the female. In reality, the female alone becomes asexual, and we must view the potential transition purely from her standpoint (as her evolutionary option). Furthermore, maternal investment in many sexual species heavily outweighs paternal investment. Let us assume, for simplicity, that there is no paternal investment; then the sexual female will be making the entire energetic investment but contributing only half the genetic contribution to n offspring. Through evolutionary transition to asexuality, she would still be able to bear and raise n offspring, but she would also be making the full genetic contribution to their progeny. Thus, in viewing evolution from the position of the female, who controls the eggs, we can see that she is indeed sacrificing her potential genetic legacy by breeding. The amount of sacrifice will vary with paternal investment.

The frequently enormous "cost of meiosis" to females has reasonably been viewed as paradoxical in light of the near ubiquity of sexuality among higher organisms. The common inference has been that there must exist some advantage of sexuality that is of greater magnitude (Maynard Smith, 1969, 1971; Williams and Mitton 1973; Wilson, 1975). It is therefore perplexing that geneticists have been unable to identify any advantage that now seems entirely satisfactory. Conflicting variations on the traditional idea were noted above.

Increasingly, in recent years, the traditional idea on the role of sexuality has itself been called into question on theoretical grounds. Using computer simulations to investigate the potential effects of genetic linkage, Thompson (1976) argued that sexual recombination has little effect on rate of evolution, except within extremely small populations, where few genotypes exist. In small populations, because of pronounced effects of linkage, recombination is seen as significantly increasing genotypic variety; but in large simulated populations little effect is observed. While introducing much greater sophistication, Thompson has returned to a view held by Darwin—that sexuality serves not to accelerate evolution but to slow and smooth out response to selection pressure. Thompson postulated that recombination prevents phyletic evolution from tracking short-term environmental change so closely as to endanger the survival of the species over the long term.

The conventional explanation for the value of sex, as articulated by Weismann (1889) and Fisher (1958), has also been opposed because it represents group selection, in the narrow sense of the phrase. In other words, sexual reproduction has traditionally been seen as prevailing because it is of value to the interbreeding unit (usually a species) rather than to the individual. The individual sexual female may, in fact, significantly sacrifice genetic contribution to future generations, as already noted. Fisher (1958, p. 50) singled out sexual reproduction as being the only kind of adaptation likely to have resulted from selection favoring the group rather than the individual. Williams has apparently looked beyond the traditional view because of his general rejection of selection of any sort above the level of the individual (Williams 1966, Chapter 4; 1975, Chapter 13). His alternative hypothesis is not that sex is of value for rapid phyletic evolution of the species, but that sex is favored because it provides for persistence of particular kinds of individuals under conditions of stringent selection—conditions which are taken to be the norm in nature. Williams' idea is that fitness is lognormally distributed within species and that only a few extremely fit individuals tend to survive. Individuals reproducing asexually can seldom be the winners here because they fail to yield diverse arrays of progeny. Sexual individuals will allegedly tend to produce a great variety of offspring, some of which will be the superior forms that form the next generation. This explanation can apply to the adaptive value of sexuality only in taxa with high birthrates, which has forced Williams (1975, pp. 102–103) to take the following position:

All organisms with really low ZZI [a measure of potential rate of population increase, combining fecundity and generation time] such as mammals, birds, and many insects have populations in which sexual reproduction must be consistently selected against. Their present exclusive reliance on sexual reproduction must be ascribed to inheritance from a high-fecundity ancestor in which the complete replacement of asexual with sexual reproduction was the evolutionary equilibrium. If and when any form of asexual reproduction becomes feasible in higher vertebrates, it completely replaces sexual. So in these forms sexuality is a maladaptive feature, dating from a piscine or even protochordate ancestor, for which they lack preadaptations for ridding themselves.

I find it difficult to accept such an argument. In the first place, many taxa with low rates of population increase have become asexual, yet asexuality has never taken hold to become widespread within them. There have, for example, been recognized several asexual species of lizards (Uzzell, 1970), the ancestors of which, like most lizards, were probably not particularly fecund, in comparison to many invertebrate taxa. Transition to asexuality has also occurred polyphyletically in the Insecta. I also find it difficult to believe that all taxa of high fertility are subjected to stringent selection during almost every generation. Species limited by wholesale juvenile predation, including many bivalve mollusks that I have studied, reveal little intraspecific variation that could be envisioned as providing a basis for such selection. If the human observer cannot detect such variation, it seems unlikely that the predator can either, at least with the great precision required by the hypothesis.

As noted earlier, Ghiselen (1974, p. 57) developed a hypothesis resembling that of Williams, suggesting that a sexually reproducing individual will leave more offspring than its asexual counterpart because the descendants of the sexual form will display greater variability in a changing environment.

ASEXUALITY AND ADAPTATION

Weismann is generally credited with introducing the idea that sex prevails because it accelerates evolution. He claimed that asexual species are certain to die out "because they are incapable of transforming themselves into new species or, in fact, of adapting themselves to any new conditions" (Weismann 1891, p. 298). This idea, in somewhat less extreme form, gained widespread acceptance and persists to the present day. I would suggest, however, that the idea remains primarily as a hypothesis to account for the rarity of asexual reproduction—a postulate with little empirical support. Dobzhansky (1970, p. 179) wrote:

> Sexual reproduction results in the formation of ever new combinations of heredi-
> tary determinants. ... Just how greatly evolution is speeded up by sex is controver-
> sial.

White (1978, p. 317) retains the view that asexual taxa are adaptively inferior because they are unable to evolve significantly: "But there is very little evidence that diversification and divergence of thelytokous [asexual] clones by mutation is likely to proceed to the point where they would be regarded as different species." He suggests that broadly adapted asexual "species" have inherited their genetic diversity polyphyletically from sexual ancestors. This condition may hold for some, but for many it stands as an untested hypothesis.

Cuellar (1977) also considers asexual taxa to be maladapted in the sense of being weak ecologic competitors. He cites evidence that many asexual forms are ecologic opportunists—the so-called "weeds" of ecosystems. Cuellar's suggestion is that asexual animals are unable to compete successfully with sexual forms and are consequently restricted to disturbed and newly formed environments. His primary example is of asexual lizards that occupy floodplains of rivers. The problem here is that competitive exclusion is far from universal in nature. A large percentage of the world's species have populations limited by predation and other forms of disturbance—far more species than are asexual. What, then, is holding back the general diversification of asexual forms? Furthermore, because competition is generally weak in disturbed habitats, it does not seem reasonable to envision a tendency to occupy these habitats as representing a severe restraint on the worldwide diversity of asexual "species." Quite apart from considerations of competitive ability, asexual taxa are inherently effective colonizers, by virtue of their rapid population growth. This trait derives automatically from the fact that all individuals are female, which doubles the basic (unrestrained) net birthrate for their populations. Given their intrinsic advantage for an opportunistic mode of life, it is hardly surprising that many turn out to be opportunists. There is no reason to assume that competition confines them to this ecologic role. In this light, it is important to note the frequently opportunistic nature of species that alternate between sexual and asexual modes of reproduction. These forms possess whatever phyletic advantage is conferred by sexuality and yet have nonetheless frequently become ecologic opportunists. The freshwater zooplankter *Daphnia* is a conspicuous example, populating newly formed bodies of water very rapidly by asexual reproduction and at times switching to sexual reproduction, which yields resting stages so resistant that they can endure desiccation and transport by wind.

In fact, there is evidence that in many taxa asexual reproduction is not maladaptive. Numerous asexual species seem extremely hardy and well adapted (Gustafsson, 1946–1947; Grant, 1971, pp. 345–348; Suomalainen *et al.*, 1976; Williams, 1975, pp. 161–162) The dandelion is perhaps the most famous of these. Solbrig (1971) has shown that the dandelion displays considerable genetic variability, and its adaptive success is legendary. The situation for plants in general has been summarized by Grant (1971, pp. 345–348). He suggests that some asexual forms seem to display limited variability, but that many do not seem to be short-lived in geologic time. The strongest evidence of adaptive success for asexual animals relates to weevils of the genus *Otiorrhynchus* (review by Suomalainen *et al.*, 1976). These forms display greater polymorphism than sexual taxa and have been more successful in adapting to new environments. Their genetic variability is apparently not inherited from sexual ancestors:

> Contrary to most earlier views, parthenogenetic insects are capable of evolution. There is considerable genetic differentiation within and between populations inhabiting different regions. At least a certain proportion of this variability is adaptive. Polyploid parthenogenetic genotypes seem to represent balanced and apparently successful genetic complexes. As such they are good examples of permanent heterozygosity. Their adaptedness and adaptability, as expressed by their ability to spread efficiently over large land areas, are, at least in part, attributable to heterosis. The differences between and within monophyletic populations are attributable to mutations. These mutations have occurred and established themselves in the populations since the origin of parthenogenetic reproduction. Populations of polyphletic parthenogenetic species are different by virtue of their different origin. They continue to diverge from each other through mutation. (Suomalainen *et al.*, 1976, pp. 248–249.)

CLADES VERSUS CLONES

It is evident that much ingenuity has gone into the devising of hypotheses to account for the prevalence of sex. The fact that such ingenuity has been required engenders the suspicion that something is amiss. Theoreticians have been reaching far and failing to agree or even take firm positions (see Maynard Smith, 1978, p. ix). The simple fact is that the gradualistic model has not been able, comfortably, to accommodate sexuality. All of the explanations outlined here offer potential reasons for sexuality being of some value. However, none reasonably accounts for its overwhelming dominance among eukaryotes, and none has passed empirical tests (few hypotheses have, in fact, been subjected to appreciable testing). Moreover, there is evidence that many asexual higher organisms are by no means maladaptive.

The point of departure for the following analysis is an observation by Stebbins (1950, p. 417) about the fate of apomictic (secondarily asexual) taxa of higher plants. Stebbins suggested that apomictic groups have tended to become extinct soon after forming, not only because of an inability to evolve rapidly in the face of environmental change, but also because of a poor capacity for diversification. He observed that they seem to form only incipient taxa, cropping up here and there in phylogeny but seldom diversifying enough to form new genera or subgenera. Van Valen (1975) further suggested that an equilibrium frequency of apomicts might result from a characteristic rate of production of these forms, in combination with characteristic disadvantages relative to similar sexual taxa, both in evolving phyletically and in speciating. In fact, as noted earlier, asexual taxa do not really speciate in the normal sense. According to the punctuational model, this, rather than slow phyletic evolution, must be their primary deficiency.

In comparing the diversification of sexual and asexual taxa, we are in effect comparing divergence within clades to divergence within clones. As noted earlier, an asexual clone expands by gradual divergence of individual lines of descent, as mutations accumulate. "Species" are recognized within sizable clones today because, as Mayr (1957) and Hutchinson (1968) have pointed out, differential extermination of certain forms has left discrete adaptive peaks surviving from what once more closely approximated an adaptive continuum. What is lacking from the expansion of a clone is quantum speciation. Whatever the precise genetic nature of this phenomenon turns out to be—and it may well vary considerably among taxa—there is no question that sexual recombination is essential for quantum speciation. For example, the regulatory gene model of Britten and Davidson (1969; 1971) for the origin of evolutionary novelties rests heavily upon genomic rearrangement. If the effects of recombination are concentrated within events of divergent speciation by way of small populations, then here too is where sex must play its major role.

It was noted in Chapter 3 that the evidence of weak gene flow among many conspecific populations in nature (Ehrlich and Raven, 1969) offers a potentially strong case against gradualism. If a beneficial new mutation or gene combination cannot spread throughout a species' entire range, the species cannot be expected to undergo rapid phyletic evolution, except by parallel evolution of separate populations, which is an unlikely development. Viewed in another way, sluggish or nonexistent gene flow deprives species of much of the alleged value of sex. We are left with the inference that sex can have a profound effect only within small interbreeding populations of the sort that are involved in speciation. A dramatically new adaptation is likely to become fixed within an entire species only by arising when the species forms, and only if the origin is by way of a small population.

It is perhaps relevant that Thompson (1976), in considering the importance of sex in terms of genetic linkage, concluded that "the effect [of sexual recombination] should be significant only in very small populations, smaller, perhaps, than are likely to be the rule in nature." Rather than being perplexed by this conclusion, we can welcome it as being thoroughly compatible with the punctuational model.

The plight of asexual taxa can be illustrated by diagrams resembling those of Figure 4-6, where a hypothetical clade characterized by quantum speciation is contrasted with one in which all evolution is phyletic. In the latter, speciation does nothing more than change the direction of phyletic evolution. A clone will evolve in a similar way. Each asexual line of descent can be compared to a lineage within a sexual clade. It was shown in Chapter 4 that established lineages like those depicted in Figures 3-8, 3-9, and 3-10 must have evolved very slowly. Certainly, if anything, asexual lines of descent within clones evolve even more sluggishly. The result will be very gradual diversification for newly formed asexual taxa.

Figure 8-1 portrays graphically the alleged fate in large-scale evolution of a typical asexual taxon of higher organisms. The figure contrasts asexual and sexual diversification, that is diversification within a clone and a clade. The kind of organism founding the hypothetical clone is meant to be identical to the species founding the clade in all respects except mode of reproduction. The clone expands slowly by the accumulation of mutations within individual lines of descent. The clade, in contrast, diversifies rapidly by divergent speciation. Limitations of the graphical mode of presentation rule out the depiction of speciation in which evolution is not accelerated, but this should not be taken to indicate that such events do not occur.

The key point illustrated in Figure 8-1 is that a normal incidence of quantum speciation yields much more rapid divergence than can occur within the clone. Inevitably, the clone will develop inhomogeneities through differential extinction among its members. As already noted, this process may produce large enough discontinuities for members occupying separate adaptive peaks to be recognized as discrete "species." For simplicity, this fragmentation of the clone is not depicted in Figure 8-1. A clone, unconstrained by interbreeding, in time is likely to develop a broader niche than that of a single, closely related sexual

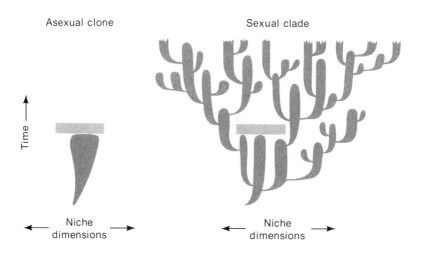

Asexual clone   Sexual clade

Time ⟶

← Niche →
  dimensions

← Niche →
  dimensions

FIGURE 8-1
Evolutionary success of a sexual clade relative to an asexual clone. The clade
and clone grow from species that are identical except for the presence or
absence of sexual reproduction. See text for explanation.

species. Pairs of asexual and sexual weevil species illustrating this condition
(review by Suomalainen, *et al.*, 1976) were mentioned earlier.

In Figure 8-1, an extinction event is shown as extirpating individuals within a
particular environmental hyperspace. The clone, which has diversified slowly,
goes extinct, while the clade, having expanded rapidly, persists by virtue of its
broad adaptive zone. In other words, the initial asexual "species," if it constitutes
most or all of the clone, has been quite as successful as an average species of the
clade. This condition is also consistent with previously cited empirical evidence
of the success of many asexual "species."

The crucial point is that almost every species is ephemeral in geologic time.
Even clades of three or four species, which might be regarded as comparable in
niche breadth to rather well-expanded clones, are short-lived. The impact of
extinction upon higher organisms is simply too great to be offset by clonal rates
of diversification. There is nothing to prevent asexual taxa from arising sporadically, but they will seldom persist as, in fact, Stebbins (1950) has observed of
their occurrence. It is not difficult to understand why nearly all asexual taxa have
close sexual relatives (Gustafsson, 1946–1947; Stebbins, 1950; Mayr 1963;
Grant, 1971). In an evolutionary sense, they simply never move very far from
their ancestors. The critical value of sex is in fostering the rapid formation of
discontinuities within phylogeny.

What the punctuational interpretation does, in essence, is to elevate the
paramount value of sexuality to a higher taxonomic level than traditionally

considered: It is not primarily the species that benefits, but the genus and all higher taxa within which genera are nested. The classic studies of Lewis on speciation in *Clarkia* were described earlier (pages 168 and 174). It is appropriate to quote again from this body of work to further illustrate the importance of speciational diversification:

> The occurrence of populations which are intersterile but morphologically similar, as well as adjacent populations that are morphologically distinct but apparently interfertile, suggests that differentiation which may lead to speciation in this genus may frequently occur rapidly. Most of the genetically isolated populations thus formed are probably short lived, but a few may become successful species. Gradual segregation of geographic races also occurs but the success of the genus seems not to lie so much in the gradual segregation of adaptive races as in the production of a multiplicity of locally adapted, genetically isolated and often ephemeral segregates. (Lewis, 1953, p. 19.)

In other words, it is not phyletic evolution that counts in *Clarkia*, but the casting off of isolates, some of which emerge as discrete new species, to contribute to the survival of the genus.

THE VALUE OF OUTBREEDING

Like many flowering plants, *Clarkia* possesses insect-pollinated flowers that seem to represent adaptations for outbreeding (Lewis, 1953). The apparent premium on outbreeding in nature has commonly been taken to indicate the importance of sexuality:

> Although the significance of sexual reproduction for securing genetic variability is still debated ... , naturalists ever since Darwin have been impressed with the lengths to which organisms will go to achieve cross-fertilization. Orthodoxy maintains that genetic variability for rapid evolution is the primary function of sexuality. (Uzell, 1970, p. 440.)

What seems often to have been overlooked is that the value of outbreeding is, at least in part, a result rather than a cause of the presence of sexuality. Given the presence of the diploid sexual genetic system, severe inbreeding quickly reduces heterozygosity. What is commonly called "inbreeding depression" has no counterpart in the transition to asexuality, however. In effect, inbreeding turns sexuality against the well-being of the organism, whereas loss of sexuality has no such immediate inimical effect. Thus, it is reasonable to suggest that outbreeding is to a considerable degree a secondary adaptive phenomenon, the value of which is greatly enhanced after sexuality is in place; it is selected for within species primarily because it is essential vis à vis inbreeding, not vis à vis asexual reproduction, which perhaps has the potential to foster considerable genetic variability.

It is important to recognize that we are confronted with a world in which eukaryotic asexuality is a secondary feature, and this historical condition (often entailing recent diploid ancestry) constrains the genetic prospects of the asexual forms that arise from time to time. It is a matter of conjecture just what asexual genetic system might exist if sexuality did not predominate, and if asexual organisms were farther removed from or unrelated to sexual forms. A system providing substantial genetic redundancy, perhaps akin to polyploidy but lacking chromosomes, might permit the maintenance of considerable genetic variability.

THE INITIAL IMPACT OF SEX

The arguments offered here relate to the prevalence of sexual reproduction, not to its origin. As with other adaptations that have been favored in species selection, the inception of sexual reproduction can be viewed as an experiment that fortuitously worked to the great advantage of its possessors. In fact, sex can be viewed as the first such experiment in the history of modern life, for until it evolved, there were no true species and there was therefore no species selection. The origin of genuine sexual reproduction has been suggested as having been a key factor in the sudden appearance of multicellular life in the late Precambrian (Schopf *et al.*, 1973). The primary effect of sexual reproduction would have been, not to accelerate phyletic evolution, but to provide for what must have been the world's first genuine adaptive radiation—adaptive radiation by speciation (Stanley, 1976).

Sex, then, is the creator and preserver of higher life. Its inception led to the origins of most eukaryotic phyla, and its presence has been necessary for their survival. Selection for sexual taxa must represent one of the most pervasive and stringent forms of selection among species or their equivalents. If many asexual species fare more or less as well as sexual species, as suggested by evidence cited earlier, then the critical component of selection here is not differential extinction, but differential rate of speciation. Even if the gradual expansion and fragmentation of a clone is viewed as a form of speciation, the result is equivalent to an extremely slow rate of sexual speciation.

PROMINENT ASEXUAL TAXA

There are important exceptions to the sexual imperative that seem clearly to support, if not to prove, the rule. Apparently since the Precambrian, the prokaryotes (blue-green algae and bacteria) have formed the largest group of asexual cellular organisms. In the traditional explanation for sexuality, these exceptional forms are seen as thriving with limited recombination by virtue of their short generation times. Generation time may indeed be a factor here, but in the punctuational scheme the success of prokaryotes must also be viewed in terms of resistance to extinction. Bacteria and blue-green algae are, in general, noted for their extremely broad niches. This condition, in combination with astronomical population sizes and widespread geographic distributions, would seem to dictate such low rates of extinction that rapid diversification is unnecessary.

Multicellular asexual groups that have had unusual evolutionary success offer more compelling evidence. The rotifer class Bdelloida is perhaps the most diverse living asexual metazoan taxon, containing more than 200 recognized "species" (Mayr, 1963, p. 433). The fact that all of these are parthenogenetic suggests that the class, and its asexual condition, are monophyletic, with the various "species" representing clonal fragmentation, as discussed above. The remarkable thing about particular kinds of bdelloids is that they are typified not only by great abundance, but also by extraordinary niche breadth. These tiny creatures are the denizens of minute bodies of water, like those at the bases of leaflets of damp moss. When conditions deteriorate, bdelloids can survive in frozen or desiccated states for years. They are also found in hot springs. Furthermore, like blue-green algae and bacteria, most varieties are cosmopolitan (Hyman, 1951; Hutchinson, 1968). Extremely low rates of extinction must characterize the Bdelloida. Extremely simple creatures, the bdelloids are likely to have arisen as early as the Paleozoic, so that their degree of divergence does not represent problematically rapid clonal expansion. If the group appeared 400 My ago, then doubling time for number of "species" would be about 50 My. In other words, an average clonal fragment, once recognized as a species, would expand and fragment into two comparable entities in 50 My. This is hardly an unreasonable rate for clonal evolution. Freshwater representatives of the Chaetonotoidea (Gastrotricha), another exclusively parthenogenetic group, are also abundant and cosmopolitan, thus conforming to the punctuational argument.

The species of most multicellular taxa are far more fragile, ecologically, than are members of the Bdelloida and Chaetonotoidea. In general, their numbers are far fewer, and their niches and geographic ranges, far narrower. These are traits that increase a species' vulnerability to extinction (Jackson, 1974). In general, it seems only through the rapid proliferation of such ephemeral entities by sexual speciation that eukaryotic higher taxa can survive for more than brief intervals of geologic time.

As an epilogue, I would like to point out that the punctuational analysis presented in the preceding section stands on its own. It does not, for example, rest upon the evidence that many asexual species are broadly adapted and genetically diverse, in violation of the traditional view that sexual recombination is required for lineages to survive. Similarly, the analysis does not depend upon arguments like those that Williams (1975) and Thompson (1976) have directed against the traditional view. Certainly, to the degree that independent lines of evidence oppose the gradualistic explanation for the prevalence of sexuality, the ideas advocated here gain strength. Nonetheless, the conclusion that clonal diversification is inadequate in the face of normal rates of extinction has not been conjured up to circumvent evident deficiencies of the traditional view. Rather, this conclusion flows directly from the punctuational model. Estimates of rates of speciation and extinction strengthen the argument. To the rates calculated in Chapter 5, others will be added in Chapter 9. These calculations provide a graphic picture of species turnover and, hence, of the inherent fragility of higher taxa existing at low diversity—a fragility that, it would seem, can only be offset by speciation.

SUMMARY

The ubiquity of sexual modes of reproduction among higher organisms has seemed especially noteworthy because the sexual female sacrifices as much as one-half of her potential genetic contribution to progeny, yet even without this genetic cost, the prevalence of sex requires an explanation. The traditional interpretation is strongly gradualistic. It is that sexual reproduction operates to the good of the group or individual by shuffling genetic material, spreading useful new genetic features, and placing alleles in varied genetic contexts. There has been much theoretical debate as to how, and to what extent, sexuality may actually accelerate evolution. There is, in addition, evidence that some asexual "species" are as abundant, genetically variable, and geographically widespread as similar sexual species. The apparently widespread premium on outbreeding in nature may, in large part, reflect the importance of avoiding inbreeding once sex is in place, rather than an inherent need for sexual generation of genetic variability.

Quite apart from these difficulties, the punctuational model opposes the traditional explanation for the prevalence of sexuality, implying that phyletic evolution is so slow that it can hardly be at a premium. Gene flow throughout many species is so sluggish that any potential value of sex for accelerating phyletic evolution (which by definition involves the entire species) is largely lost. Sexuality may be of some benefit to individuals or species, but is of primary benefit to higher taxa. What counts for survival is not rapid phyletic evolution, but rapid speciation. An asexual clone may actually diversify more rapidly than a single sexual species that is initially identical except in mode of reproduction, because variability in the sexual species is constrained by interbreeding. Species, however, are ephemeral in geologic time, and, by speciating, the sexual species can yield a clade that expands more rapidly than the asexual clone. For most higher organisms, speciation is necessary to offset normal rates of extinction. Most asexual forms are transitory because they cannot speciate in the normal sense of the term. The nature of exceptional groups—purely asexual taxa of high diversity, like the bdelloid rotifers—offers support for this conclusion. They possess such enormous populations and broad ecologic adaptations that their rates of extinction must be extraordinarily low.

9

Dynamics of Species Turnover: Variations Among Taxa

Calculations for a large number of taxa show that rate of speciation in adaptive radiation and rate of extinction are strongly correlated in the animal world. Because this relationship pertains to speciation during episodes of rapid diversification, it cannot be explained in terms of equilibrium linkage between speciation and extinction, whereby species are added only as rapidly as extinction makes room for them. Rather, the correlation seems to result fortuitously from the fact that the dominant factors controlling the frequency of speciation and extinction happen to affect the two processes similarly: (1) Rates of both speciation and extinction tend to vary inversely with the capacity for geographic dispersal, because although dispersal opposes the divergence of small populations, it also promotes widespread distribution, which reduces the likelihood of extinction. (2) Rates of both speciation and extinction tend to vary directly with level of behavior, because behavioral change is an effective isolating mechanism, and advanced, stereotyped behavior also represents a form of ecologic specialization that renders a species especially vulnerable to extinction. This dual role of behavior seems to explain why, when animal taxa are ranked according to their rates of evolutionary turnover, they form a hierarchy that crudely approximates a ladder of complexity, or scala naturae. *Various lines of evidence show that this hierarchy is not simply an artificial product of a bias in our perception of rate of evolution based on degree of morphologic complexity.*

INTRODUCTION

From the comparison of the Bivalvia and Mammalia presented in Chapter 5, there emerged three conclusions of general importance. One of them was that taxa tend to radiate at characteristic rates. Their net fractional rates of increase (values of R) reflect both rate of speciation (S) and rate of extinction, (E), suggesting that both components have rather consistent values for particular kinds of taxa. The second conclusion is that average rates of speciation and extinction vary greatly from taxon to taxon. The third is that rate of speciation and rate of extinction seem to be correlated: Mammals exhibit high rates, while turnover within bivalve families is relatively slow. This final conclusion, being based on figures for only two classes of animals, is in particular need of further evaluation. In this chapter, all three conclusions will be further explored and interpreted through the consideration of additional taxa. In the final chapter, several general implications of the exponential nature of diversification will be examined.

A valuable attribute of the exponential approach to the study of evolutionary rates is that it can embrace not only commonly studied fossil taxa, but also, to a degree, groups like insects and birds, which are of great ecologic and evolutionary interest but which in the past have seemed relatively inaccessible to macroevolutionary analysis.

Darwin referred to the most recent geologic period as the Age of Barnacles. I have been surprised to discover that this interval would be more appropriately labeled the Age of Snakes! The Colubridae, which include most living species of snakes, have been radiating at an extraordinary pace—probably more rapidly than any family of modern mammals, except possibly the murid rodents (Figure 5-2), which, perhaps not coincidentally, some of the snakes eat. Possibly, however, we live in the Age of Frogs! The Ranidae, the dominant frog family in North America, have been radiating rapidly, but their fossil record is too poorly known to reveal whether the hundreds of living ranid species of the world are the product of radiation since sometime in the Miocene or since a much earlier time. Snakes also eat frogs.

The sections that follow will document the estimation of values for average species duration and rates of adaptive radiation that are summarized in Figures 9-1 and 9-2. At the outset, a general explanation of procedure is called for.

The numbers plotted in Figures 9-1 and 9-2 for mammals and bivalves have been drawn from the estimates discussed in detail in Chapter 5. Figures 5-2, 5-6, and 5-8 display the extensive data employed here. With the exception of the gastropods, all other taxa depicted in Figures 9-1 and 9-2 are represented by less abundant data. This chapter amounts to a progress report in a nascent field of endeavor. Nonetheless, there is reason to believe that the estimates employed here are meaningful, and that the plots presented are unlikely to change greatly with the addition of data.

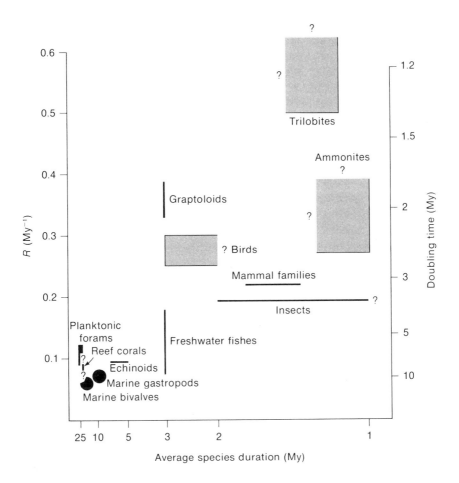

Figure 9-1
Plot of values of R estimated for major taxa versus average species durations, which are plotted inversely so as to be proportional to rates of extinction. Doubling times corresponding to values of R are displayed on the right. See text for sources of data.

THE CORRELATION OF S AND E

In Figure 9-1, values of R for taxa in the midst of rampant adaptive radiation are plotted against average species durations. Doubling time is inversely related to R and, being a more tangible variable, this too is labeled on the vertical axis. Figure 9-1 is plotted to test the previously mentioned conclusion that rate of speciation and rate of extinction are correlated. For this reason, because average species

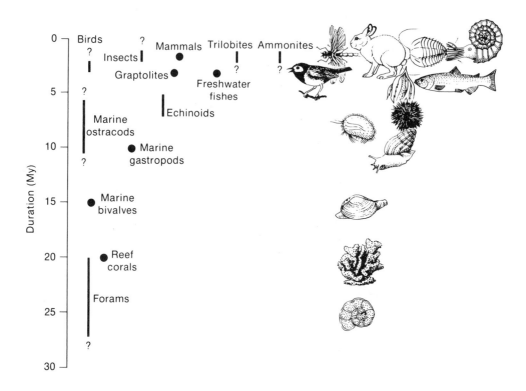

FIGURE 9-2
Hierarchy of average species durations estimated for major taxa. See text for sources of data.

durations are inversely related to rates of extinction, they are plotted against R on an inverse scale. Thus, in Figure 9-1 something roughly equivalent to a rate is plotted against an actual rate. A strong correlation between E and R is clearly indicated.

Why are rates of speciation and extinction not plotted directly in Figure 9-1? The principal reason is that a value of S can only be calculated by adding estimates of R and E. One problem here would be that, while E varies inversely with true species duration, the species durations that are employed here necessarily reflect not only termination of lineages but also pseudoextinction, and there is no way of evaluating the latter (although boundary conditions can be imposed, as done in Chapter 5). What is assumed in plotting Figure 9-1 is that the incidence of pseudoextinction is roughly the same from taxon to taxon (only an enormous disparity here could account for the correlation). Another problem would be that the calculation of S using values for R and E would render a plot of S versus E partially redundant from the standpoint of measurement (there would be a partial autocorrelation). Thus, one benefit of plotting average R versus average species duration stems from the fact that values of these variables are measured independently. Also, while there is no redundancy of mensuration, there is a mathematical redundancy that operates against the observed correlation and therefore strengthens the conclusion that a correlation exists: There is a partial negative autocorrelation between R and E, however they may be measured, because $R = S - E$. Thus, even if there were no correlation between S and E (for example if S were invariant) R would be negatively correlated with E. The positive correlation that appears in Figure 9-1 is robust with respect to this automatic bias.

It is important to bear in mind that the variables plotted in Figure 9-1 operate exponentially. An enormous disparity in arithmetic rate results from a two-fold difference in one of the exponential variables. What this means is that although many values that appear in Figure 9-1 are approximate, the substantial range of the values establishes an enormous variation in rates of turnover in the animal world. The observed variation can hardly be an artifact of procedure. The Mollusca partially illustrate this point.

It will be shown that for the Gastropoda and Bivalvia numerous data provide a consistent picture of longevities of species and rates of radiation (Figures 5-2, 5-8, 9-3, 9-4).* The ammonoid cephalopods, though well studied, are extinct and therefore cannot be expected to provide such accurate information. The techniques employed to estimate values of R, E, and S were developed to circumvent problems normally associated with analysis of fossil data at the species level and were meant to be applied to living clades, but the ammonites are extinct. Even so, the ammonites apparently differed so greatly from bivalves and gastropods in rate of evolutionary turnover that their fossil record, whose final chapter was written more than 60 My ago, clearly displays the disparity. Specialists in ammonite biostratigraphy have concluded that an average species lasted for perhaps a million years, fully an order of magnitude less time than the figure for bivalves and gastropods. Even the estimates derived from incomplete fossil data show that rates of adaptive radiation are correspondingly high for ammonites. During an Early Triassic interval of only 7 to 10 My, radiation from perhaps a single genus produced more than 150 species. Normal rates recorded for the Bivalvia and Gastropoda would produce less than a doubling of the initial number of species during so short an interval!

It should be appreciated that the measurement of rates by means of exponential techniques is by no means meant to imply that adaptive radiation follows a perfectly exponential course. Clearly, only net values of R are measured, glossing over irregularities in the pattern of diversification. The use of exponential measurement here reflects the recognition that diversification and decline are fundamentally geometric rather than arithmetic processes. Arithmetic rates have traditionally been employed with the same kind of simplification (Simpson,

---

* See appropriate section for figures referred to in this introduction.

1944; 1953). When we calculate number of new taxa arising or going extinct per million years, we are calculating an average arithmetic rate. If such rates have value, and indeed they have a history of widespread employment, exponential rates should have even greater merit. Certainly, there must be a general tendency for calculated values of R to represent underestimates of exponential rates, because some radiations will have followed distinctly sigmoid paths (Figure 5-1) during the interval evaluated and because some will have been set back by undetected pulses of extinction. That even characteristically low values of R do reflect approximately exponential increase, however, is illustrated for the Gastropoda and Bivalvia by the fact that for the radiating clades studied, semilogarithmic plots of number of species versus antiquity of radiation are rectilinear, as in Figure 9-4. (Also see Figure 10-5.)

The method by which data have been assembled to calculate values of R deserves more general explanation. Basically, clades (usually formal taxa) have been chosen which seem presently to be in the midst of unbridled radiation. Recognized levels of diversity through time often provide an indication here. If we consider the echinoids, for example, generic data presented in the *Treatise on Invertebrate Paleontology* make it quite evident that most families represented in the Recent are on the decline. The clypeasteroids, which in part represent expansion of the echinoid adaptive zone into the exposed sandy shore habitat (page 291), have proliferated genera at a much greater rate than have other Cenozoic groups, and it is no coincidence that they include many of the newest echinoid families. It is for these families that I have calculated values of R, which turn out to be quite similar (Table 9-1).

Here and there, higher taxa have evolved without leaving evidence of having ever radiated significantly. Such taxa have been avoided in the following compilation, along with those whose fossil records give clear evidence of a decline in rate of increase. Within these guidelines, I have attempted to be as objective as possible. For instance, within the Bivalvia and Gastropoda, data have been compiled for all apparently radiating taxa for which an initial interval of searching uncovered good estimates of both present diversity and time of origin. Only for very small, youthful clades (some genera), for obvious statistical reasons, is there a moderate amount of scatter in estimates of R.

In terms of measurement, it was noted in Chapter 5 that in the calculation of R, the value of N (number of species produced by radiation during time t) need only be approximately known because it enters the calculation logarithmically. Errors in estimation of t have a greater effect, introducing inversely proportional errors in the calculation of R. Since antiquity is usually underestimated, the direction of bias will be toward overestimation of R, but errors from this source must seldom exceed 10 or 15 percent, which, especially considering the direction of their bias, must be insignificant on a scale like that of Figure 9-1.

The estimation of an average species duration from a Lyellian Curve, or simply from the age of faunas in which 50 percent of species are judged to be extant, is also subject to errors introduced by pulses of extinction. Aspects of this problem will be discussed later in the chapter. It seems evident that, in general, the Lyellian technique, judiciously applied, is far superior to a simple compilation of apparent stratigraphic ranges—especially ranges for Paleozoic or Mesozoic species. Obviously such compilations consistently underestimate longevity. I have employed them only for planktonic forams and Pleistocene mammals (Figures 5-6 and 5-7), which have exceptionally good records, and for trilobites, ammonites, and graptolites, for which there is no alternative and for which durations were clearly very short.

A *SCALA NATURAE* OF DURATIONS?

It is of great interest that when the taxa evaluated here are ranked according to average species longevity, they form a crude approximation to a *scala naturae*—a ladder of adaptive complexity. The data that I have plotted in this way (Figure 9-2) are for groups represented in Figure 9-1 and for others for which it proved possible to estimate average duration but not R. Certainly, there are some exceptions to the trend, notably the graptolites, which were simple colonial animals (their status as hemichordates positions them in late chapters of textbooks on invertebrates, but would not seem to qualify them as advanced animals). In the final section of this chapter, I will suggest reasons for the rather striking general hierarchy. At this point, before treating particular taxa, it seems prudent to consider a potential source of bias that might lead some readers to question the reality of the apparent variation in rates of evolution that is depicted in Figure 9-1.

MORPHOLOGIC COMPLEXITY

It has been suggested that apparent variations among higher taxa in rates of evolution may be deceptive. They may simply reflect biases in our perception of evolution based on degree of morphologic complexity (Schopf *et al.*, 1975). The resemblance of Figure 9-2 to a *scala naturae* would seem to open the door to such an interpretation, which might then be applied to the range of measured values of R, and, thus, to the correlation depicted in Figure 9-1. The suggestion that procedural biases may determine measured rates (Schopf *et al.*, 1975) was directed toward higher level taxonomy, the idea being that number of morphologic characters used to divide a higher taxon into families represents an index of our bias. The present analysis, being conducted at the level of the species, is grounded in quite different morphologic criteria.

Certainly, there is no denying that species designations are more problematic for some taxa than for others, especially in the earlier portion of the Phanerozoic record. In some cases, morphologic bias may influence our judgment. The designation of chronospecies transitions is, of course, particularly susceptible to this kind of bias, but these transitions represent only one component in the determination of species longevity. Furthermore, as will be shown below, there are ways of testing the impact of these biases on longevity, and they can be shown not to account for the range of durations displayed in Figure 9-2.

Morphologic biases must be even less influential in the estimation of R. For one thing, these biases have a relatively weak effect on our recognition of living species, with which most values of R are calculated. Furthermore, as already noted, N enters the calculation logarithmically, which means that even a two or three-fold error in the tallying of living species will not greatly alter the position of a point in Figure 9-1. For some groups, like birds, mammals, freshwater fishes, echinoids, and planktonic forams, specific taxonomy for living forms is particularly well delineated.

Even if we restrict our focus to fossil data, which is not required here, there is a straightforward way to demonstrate the minor overall importance of morphologic complexity as an artificial control of our perception of species longevity, at least at the level of resolution required to demonstrate the spread of values in Figures 9-1 and 9-2. We can note that while a typical benthic foram test or bivalve shell is morphologically simpler than a mammalian skeleton, entire skeletons are unnecessary for the specific assignment of mammalian fossil material. In fact, single molars are commonly assigned unequivocally to particular species, and, if anything, these molars are less complex in the optically aided eye of the taxonomist than is an average foram test, especially since the advent of scanning electron microscopy. Hardly any molar from a Miocene deposit could be mistaken for that of a living species, yet most Late Miocene species of benthic forams are considered to be alive today. Thus, ten-million-year-old skeletons of mammals differ strikingly from living forms, not only in gross morphology but also in minute detail.

Another revealing comparison is of forams and nannofossils (coccoliths and similar algal forms). Nannofossils are smaller and, if anything, simpler in shape than most forams, yet, as will be discussed later in the chapter, display taxonomic durations that are an order of magnitude shorter than those of forams and rates of adaptive radiation that are strikingly higher.

A different kind of test of morphologic bias can be undertaken by comparing rates of turnover within taxa having entire skeletons of comparable complexity. Freshwater fish skeletons, for example, are generally as intricate as the skeletons of mammals, yet fossil data show freshwater fishes quite clearly to be of greater species longevity. About 40 percent of latest Pliocene freshwater fish species of North America are assigned to living species (Figure 9-12), while the percentage for mammals is an order of magnitude lower. As will be noted below, the taxonomic evaluation of the fishes is anything but shoddy, having been undertaken by neontologists who pioneered in the taxonomic study of living species using osteological features.

There is no denying that variation in morphologic complexity introduces artifacts into our estimates of faunal turnover. For example, this kind of bias may in part explain the fact that the 50 percent point of a Lyellian Curve lies at about 5 My before present for the Gastropoda (Figure 9-3), in contrast to its 7.5 My position for the Bivalvia (Figure 5-8,B). Gastropods are, on the average, slightly more complicated in external form than bivalves. On the other hand, I doubt that the apparent disparity in turnover rate is entirely spurious. Families in the two classes are comparable units that are, for the most part, discrete natural clades, and while about 20 living bivalve families range back into the Paleozoic, very few living gastropod families display such great longevity. Furthermore, there is much evidence that provinciality is a primary factor governing longevity of lineages, and it will be documented below that gastropods are more provincial than bivalves. It should be borne in mind that, in comparing these two classes, we are in effect splitting hairs in the context of Figure 9-1, where the points representing the classes are in close proximity relative to the general spread of data.

In summary, I am comfortable in the belief that the great range of mean values displayed in Figures 9-1 and 9-2 reflects real disparities. The following sections offer discussions of the data employed. (Some readers may choose to pass over these sections to the interpretations that follow.)

MARINE GASTROPODS

Figure 9-3 shows a Lyellian Curve for marine gastropods, representing a set of faunas nearly identical to that employed to plot Figure 5-8,B for bivalves. Interestingly, the curves are not congruent. Of the 28 faunas common to the two plots, 24 display higher extant percentages of bivalves than of gastropods. For some reason, although the great majority of molluscan faunas from other regions exhibit the same discrepancy, it has gone unnoticed or unmentioned by biostratigraphers, who have generally pooled data for the two classes in using percentages of extant species to date fossil faunas. For the gastropods, the 50 percent point is represented by faunas near the base of the Pliocene, having an age of about 5 My. This point yields a rough estimate of 10 My for mean species duration, about two-thirds the value for marine bivalves. It was noted in the preceding section that this difference may in part be an artifact of taxonomy, but probably reflects a real difference as well.

Rates of radiation may be slightly higher in the Gastropoda than in the Bivalvia. Taxa yielding an average value of R of 0.067 My^{-1} are represented in

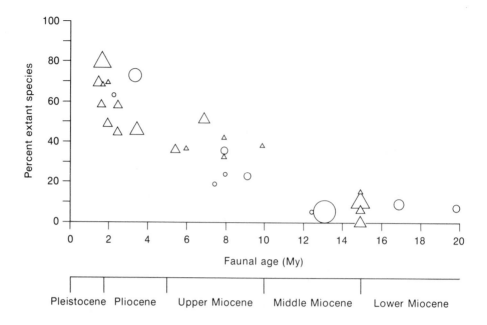

FIGURE 9-3
Lyellian Curve for temperate and subtropical gastropod faunas of California and Japan.
See Figure 5-8,B for explanation of symbols. (From Stanley *et al.*, 1979.)

Figure 9-4, which also displays the marked uniformity of rates of unbridled radiation in the Gastropoda. The data that are plotted represent all youthful, radiating taxa for which reliable figures for present diversity could be found in an interval of searching and for which Dr. Normal Sohl could provide me with an estimate of geologic age.*

Given rates of radiation in the Gastropoda at least as high as those in the Bivalvia, the higher rates of extinction in the Gastropoda suggest a slightly higher rate of speciation in adaptive radiation.

ECHINOIDS

For echinoids, our taxonomic knowledge of living species is excellent. Mortensen (1928–51) bequeathed to younger workers massive taxonomic monographs that are still regarded as nearly comprehensive and highly accurate at the species

---

* Norman Sohl, U.S. Geological Survey, 1978.

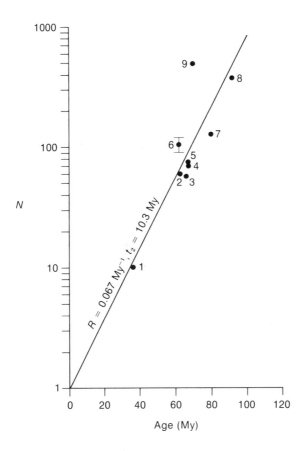

FIGURE 9-4

Plot of number of species (N) versus interval of radiation for extant radiating clades of the Gastropoda. Diagonal line depicts the rate of radiation (R) and doubling time (t_2) for an average clade. Taxa included: 1—*Harpa*, 2—Cassidae, 3—*Oliva*, 4—*Haliotis*, 5—Strombidae, 6—Turridae, 7—Turriculinae, 8—Mitridae, 9—Conidae.

level. Mortensen included information on the fossil occurrences of living forms, but even today the Cenozoic fossil record of echinoids remains poorly studied. Porter M. Kier has recently embarked on a study of these forms, which will no doubt yield excellent data for the kind of analysis undertaken here. Dr. Kier has informed me that very few Miocene species are considered to be extant.* In a

* Porter M. Kier, U.S. National Museum, 1978.

TABLE 9-1   Estimation of R for families of echinoids.

	t	N	R
Mellitidae	20	14	0.13
Clypeasteridae	41	35	0.087
Laganidae	47	26	0.069
		Mean	0.095

study of faunas of the Tamiami Formation of Florida, believed to be of Late Miocene age, Kier (1963) assigned only 2 of 9 species to extant species, and even these were placed in extinct subspecies. Cooke (1959) identified 18 species from Upper Miocene deposits of the southeastern United States, finding only one possible occurrence of a living species. The well-preserved fauna of the Yorktown Formation, now considered Pliocene in age, consists of 6 definitely extinct species (Kier, 1972). In the Lower Pleistocene Caloosahatchee Formation, 5 of 7 species are considered extant, but 3 of these are assigned to extinct subspecies (Kier, 1963). Later Pleistocene faunas consist largely of living species. It seems evident that the faunal datum at which 50 percent of species are extant lies somewhere in the Pliocene Series, but probably not near the base of the Pliocene because so few Upper Miocene species are extant. An estimate of 2.5 to 3.5 My, suggests an average species duration of 5 to 7 My. Thus, Durham's (1969) independent estimate of about 6 My, derived directly from stratigraphic ranges, seems reasonable. Clearly, species duration for echinoids is shorter than that for gastropods and bivalves.

Using the comprehensive monographs of Mortensen (1928–1951), I have calculated values of R for echinoid families radiating in the late Cenozoic. The Clypeasteroida (sand dollars and their relatives) is the only order that includes several families now in the midst of radiation. Estimates of R for three of these families are presented in Table 9-1. The Toxopneustidae, which are regular urchins belonging to the Echinacea, are also in the midst of their initial burst of evolution. Mortensen lists 34 living species for this family. If it arose in the Oligocene, which is suggested by fossil evidence but remains uncertain, this family has radiated with an R of 0.12 My^{-1}, which falls within the range of values for the other families (Table 9-1). All of these values are higher than the mean values for the Bivalvia and Gastropoda, averaging about 50 percent higher.

REEF CORALS

A few years ago, it would have been impossible to assess rates of turnover for species of reef corals. Now, fortunately, taxonomic problems for this group are becoming less clouded, and we can show unequivocally that its rates of speciation and extinction are rather low (similar to those of most other marine benthos). That great longevity typifies reef coral species can be seen simply by examining Figure 9-5, which displays stratigraphic ranges for members of the Caribbean Cenozoic fauna. Frost (1977), who published these ranges, noted that some lineages may have evolved in ways not evidenced by preserved skeletal morphology. From the standpoint of macroevolution, such cryptic change would, of course, have to be judged as minor. The Caribbean Cenozoic record is of high quality. Most of the 11 Recent Caribbean species unrecognized in a fossil state are of questionable taxonomic status. Despite the fact that species ranges in Figure 9-5 are terminated at the Recent and also at a point 15 My before the Recent, where the tabulation begins, most of the necessarily incomplete ranges displayed span between 8 My and 15 My. (Note that there was a pulse of speciation late in the Miocene, about 8 My ago, giving rise to many forms whose ranges abut against the Recent time plane.) Slightly more than 50 percent of the species recognized as existing 10 My ago are alive today, suggesting a mean duration in the vicinity of 20 My. Comparably great longevities are suggested by the stratigraphic ranges published by Chevalier (1968) for fossil reefs of New Caledonia, where, for example, 10 of 11 identified Upper Miocene species persist to the Recent.

Of all the substantial radiations of reef corals that have occurred during the Cenozoic, by far the most spectacular has been that of the genus *Acropora*. The value of R for the radiation of this genus should represent a good estimate of the maximum rate to be expected for an extensive radiation of reef corals. *Acropora* appeared in the Eocene. Its Recent representation has been greatly overestimated in the past. Dr. Carden Wallace, who is now studying the genus, has provided me with an estimate of about 70 species for its worldwide Recent diversity.* These data yield an estimate for R of about 0.09 My^{-1}, which resembles the rate for a typical radiation of echinoids. Presumably, rates for other sizable Cenozoic genera and families of corals have been slightly lower, resembling rates for the Gastropoda and Bivalvia. The coral genus *Fungia* may have undergone a dramatic radiation, but the fossil record is equivocal as to whether the rate has been truly exceptional. Wells (1966) recognized 16 living species and Judith C. Lang has provided me with an estimate of 30 for the maximum number that might exist.† The fossil record of *Fungia* is not good, but the genus may have appeared as late as earliest Miocene (Wells, 1966), which would give a value for R between 0.12 and 0.15 My^{-1}. An Early Oligocene origin would give a value between 0.08 and 0.10 My^{-1}, or about the same as for *Acropora*.

---

* Carden Wallace, Queensland Museum, Australia, 1978.
† Judith C. Lang, University of Texas, 1978.

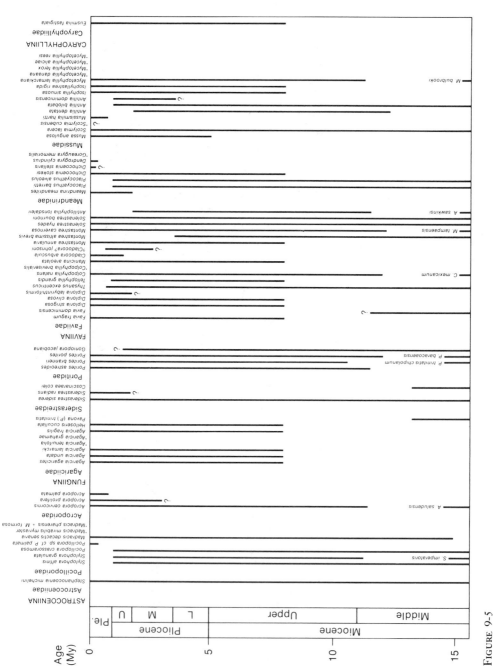

FIGURE 9-5
Stratigraphic ranges of Neogene reef corals of the Caribbean region. Asterisks denote species (mostly of uncertain taxonomic validity) that are unrecognized in the fossil record. (After Frost, 1977).

TABLE 9-2   Lyellian percentages estimated by Stach (1938) for marine Bryozoa.

Age	Extant percentage
Pleistocene (1.8 My–present)	70–100
Pliocene (5–1.8 My)	60–80
Miocene (23–5 My)	20–30
Oligocene (38–23 My)	7–15
Eocene (55–38 My)	2–5

BRYOZOA

Cenozoic Bryozoa, as now known, offer only limited possibilities for estimating species durations. One problem here is that mass extinction seems to have struck in some areas but not in others. Long ago, Stach (1938) published the percentages shown in Table 9-2, and these have been quoted by some later workers (Weisbord, 1967; Buge, 1972; Schopf, 1977). They must, however, be regarded as rough estimates that are outdated and lack substantiation. For Lower Pleistocene faunas of the Low Countries, Lagaaj (1952) found 70 percent of species extant. For slightly older, Upper Pliocene faunas embracing 124 species, the figure was only 40 percent. The great reduction in percentage surviving reflects the sudden, simultaneous extinction of many forms, apparently as a result of climatic cooling at the end of the Pliocene. Standing in sharp contrast to these data are those from samples taken off the west coast of Tasmania (Wass and Yoo, 1975). Here all 23 Upper Miocene species are also found in Upper Pleistocene sediments, which is virtually equivalent to being extant. Wass and Yoo concluded that "the Recent bryozoan fauna of the southern Australian continental shelf is a relict fauna lingering from Tertiary time." It would obviously be premature to attempt to assess rates of extinction for the Bryozoa.

The initial radiation of the cheilostomes seems to afford the best opportunity for estimating R for a major clade of bryozoans. Alan H. Cheetham informs me that there was a very low diversity of cheilostomes in the Upper Albian.[*] Possibly no more than 5 species existed in the entire world. For Upper Maastrichtian deposits, representing a time some 35 My later, Voigt (1930) reported on nearly 250 species in the area of the Baltic, northwest Germany, and Holland alone. Since this report, Voigt (1960), a reliable taxonomist, has stated that there are actually many hundred Maastrichtian species, a number of which remain to be described. A worldwide standing diversity in the late Maastrichtian of 300 species would give an R-value of 0.12 My^{-1} for the initial cheilostome

---

[*] Alan H. Cheetham, U.S. National Museum, 1978.

radiation. For $N = 1,000$, R would be 0.15 My^{-1}. These figures give us a lower limit for R, indicating that the cheilostomes underwent relatively rapid radiation for a marine invertebrate taxon (though not as rapid as radiations of other groups yet to be discussed).

PLANKTONIC FORAMS

The planktonic forams (Globigerinidae) are an unusual group, in that their fossil record at the level of the species is remarkably complete for at least the latter part of the Cenozoic. The tests of these animals rain down upon the seafloor in large numbers and have been widely employed in biostratigraphy. Figure 5-7,A displays species ranges from the thorough compendium that Blow produced for the latter 45 My of Cenozoic time. This histogram represents all species whose ranges intersect an interval of about 10 My, from the Upper Miocene to the Recent. As discussed earlier (page 109), rate of extinction must be calculated from a histogram representing an instant in time, which can be obtained by adjusting the frequency of species in each duration class in proportion to the duration represented. When this procedure is applied to the data of Figure 5-7,A, a mean duration of 15 My is obtained. Considering that more than half of the species are extant forms represented by partial ranges, the figure of 15 My must be regarded as an underestimate. The real figure must be substantially higher. The estimate is in reasonably good agreement with the estimate of 12 My that can be derived from the Late Miocene age of the zone in which 50 percent of the species recognized by Blow (1969) are extant. This 50 percent point falls in the upper part of what is known as zone 17, in sediments that according to Berggren and Van Couvering (1974) are about 6 My old. Blow is reputed to have been a taxonomic splitter, however, and Berggren (1969) plotted a Lyellian Curve that is perhaps more reliable; here the 50 percent point is represented by faunas 12 My old (Figure 9-6,A). Even here, the Lyellian approach probably yields a slight underestimate of species longevity, because in the latter portion of the Cenozoic the globigerines have been on the decline. Obviously, at times of stability or diversification, an average species has lasted at least 20 My.

The comprehensive data of Blow (1969) provide excellent opportunities to estimate R for planktonic forams. The entire planktonic family (Globigerinidae) has not been radiating significantly in the latter part of the Cenozoic. The genus *Globorotalia*, however, has diversified during this interval. Its members constitute more than one-third of all living species of planktonic forams. The radiation of *Globorotalia* occurred largely in the second half of the Cenozoic. Blow shows only three species of the genus as passing from zone 1 to zone 2 of the Late Oligocene. This boundary represents a time 30 My before the present. About 15.7 My later, in zone 9 (earliest Miocene), 15 species are present. (After this

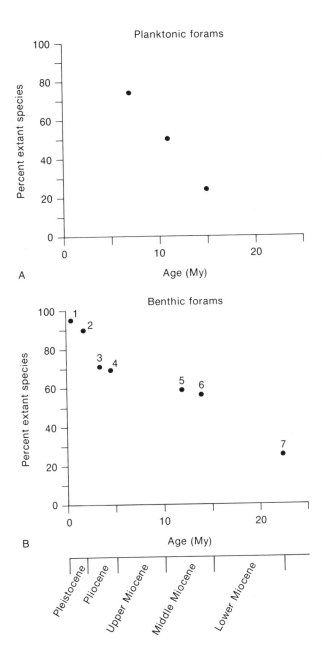

FIGURE 9-6
Lyellian Curves for foraminiferans. A: Plot for planktonic
forams of the world oceans. (Data from Berggren, 1969.) B:
Plot for benthic forams of the Atlantic Coastal Plain of the
United States (faunal size in brackets): 1—Late Pleistocene
[21]; 2—Early Pleistocene Croaton [52]; 3—Duplin [51];
4—Early Yorktown [72]; 5—St. Mary's [34]; 6—Calvert
[43]; 7—Silverdale [36].

time, the radiation of the genus decelerated.) The value of R for the radiation from the three species at the zonal boundary cited to the earliest Miocene is 0.10 My^{-1}.

It is also possible to evaluate R for the early portion of the Cenozoic radiation of the entire family. The globigerines suffered almost total annihilation in the mass extinction at the close of the Cretaceous, 65 My ago. Previously, it was thought that the Cenozoic radiation might have been monophyletic (Berggren, 1968), but Primoli Silva (1977) has presented evidence that it began with as many as five earliest Paleocene species. In the Middle Eocene, about 21 My later, 30 species were in existence. For a radiation from $N_0 = 5$, the value of R would have been 0.085 My^{-1}. To the degree that polyphyly is exaggerated here, the estimate of R would be reduced, but there seems to be a general agreement with the calculation of 0.10 My^{-1} for *Globorotalia*.

BENTHIC FORAMS

The record of benthic forams presents a similar picture. Thomas G. Gibson,* one of the few living workers who is expert in the taxonomy of both living and fossil faunas of a particular region, has kindly provided me with data that bracket the 50 percent point in the range of 11 to 15 My (Figure 9-6,B). This is quite similar to the determination for planktonic forams and leads to an estimate of mean species duration of 20 to 30 My.

It seems unwise, if not impossible, to calculate values of R for families of benthic forams. While species-level taxonomy is considered to be reasonably consistent and meaningful, families are not generally believed to be natural phylogenetic units, and few "families" have appeared since the early Cenozoic. Presently, radiating clades cannot be recognized with certainty.

TRILOBITES

In the trilobites, we encounter our first example of a taxon of invertebrates displaying rates of speciation and extinction comparable to those of mammals. Examples to follow will be the ammonites and the graptolites. It is no accident that all three of these invertebrate groups are widely used in temporal correlation. Their high rates of turnover offer excellent biostratigraphic resolution. Thus, despite the fact that all three groups are extinct, they are amenable to the kinds of analysis employed here. The Late Cambrian trilobites of North Amer-

---

* Thomas G. Gibson, U.S. National Museum, 1977.

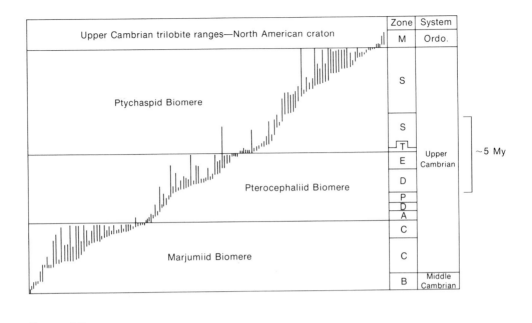

FIGURE 9-7
Composite stratigraphic range diagram for important species of three Upper Cambrian biomeres. (From Stitt, 1977.)

ica have been intensively studied during the past few decades by a number of workers. Many species have been collected from numerous stratigraphic sections, so that despite its great antiquity, the Late Cambrian record provides a realistic picture of species longevities. A distinctive pattern of evolution for Cambrian trilobites gave rise to the biomere concept of Palmer (1965a). A biomere is a biostratigraphic unit encompassing the fossil record of a radiation and bounded by what were probably slightly diachronous mass extinctions. Figure 9-7 depicts the three Late Cambrian biomeres. The extinctions were seemingly worldwide in scale. The species ranges plotted in Figure 9-7 do not include all of those documented for each biomere, but still serve to illustrate the evolutionary pattern. Appraisal of the dozens of ranges reported by Palmer (1965b), Longacre (1970), Stitt (1971, 1977), and others shows species to be of relatively short duration. Extremely few survived for as long as 3 My. If known ranges were highly fragmentary, they could not serve the function they do as guides for correlation. Figure 9-8,A displays the ranges recorded by Palmer (1965b) for the Pterocephaliid Biomere of the Great Basin. Only species recognized in four or more areas are included here. Errors introduced by the absence of a perfectly linear time scale are assumed to average out. The distribution for an instant in time, derived from Figure 9-8,A, is displayed in Figure 9-8,B; duration here is 1.1 My. The actual value cannot be much higher than this estimate, and certainly not above 1.5 My.

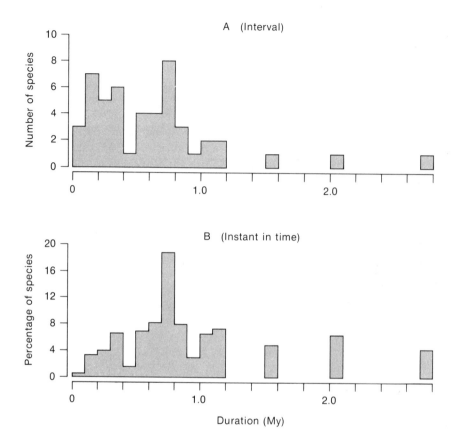

FIGURE 9-8
Species durations of trilobites of the Upper Cambrian Pterocephaliid Biomere of the Great Basin of the United States. A: Histogram for all species recognized in the interval. B: Histogram for an instant in time, derived from A by the method illustrated in Figure 5-5. (Data from Palmer, 1965b.)

The detailed biostratigraphic evaluation of Late Cambrian biomeres mentioned above also makes possible the evaluation of rates of adaptive radiation for trilobites. Stitt (1977) noted that in the early stage of diversification recorded by a biomere, only a few rather similar species existed (Figure 9-9). The exact number of species ancestral to each radiation can never be known, but was clearly quite low. The data of Palmer (1965b) show only one species at the very base of the Pterocephaliid biomere in the Great Basin, where he worked. Palmer shows a diversity of 27 species within the Pterocephaliid Biomere after about 2.7 My of radiation. If this radiation issued from a single ancestral species, the value of R would be 1.22 My^{-1}. If as many as five species were ancestral, R would still

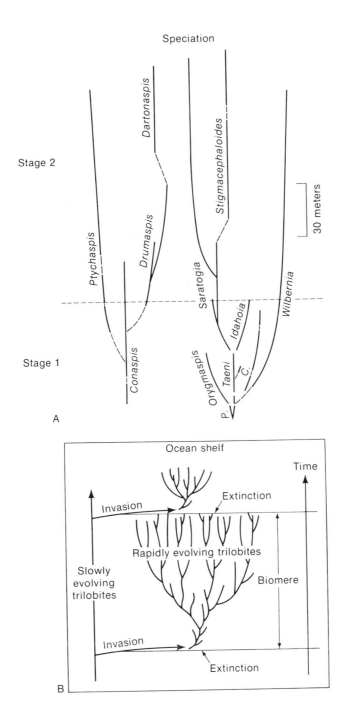

FIGURE 9-9
Phylogeny within a trilobite biomere. A: Relationship of genera within the *Taenicephalus* and *Saratogia* Zones of the Ptychaspid Biomere of Oklahoma. B: Diagrammatic representation of the origin, expansion, and termination of a typical trilobite phylogeny forming a biomere in an area of marine shelf deposition. (From Stitt, 1977.)

equal 0.62 My^{-1}, which is a very much higher value than can be found in any sizable radiation of bivalves, gastropods, planktonic forams, or echinoids.

Although values for average species duration and for R can be calculated only roughly, it is clear that the Ptychaspid Biomere (above—see Figure 9-7, which does not, in fact, display all known species) and the Marjumid Biomere (below) are characterized by values similar to those for the Pterocephaliid Biomere. Given the fact that the incompleteness of the fossil record must cause an under-estimation of R, remarkably high rates of speciation characterized the evolution of the trilobites.

BARNACLES

The barnacles are unusual arthropods in having a sessile habit, yet they also employ internal fertilization. Their fossil record is at present too poorly known for estimation of average species longevity. Rate of adaptive radiation is more amenable to study. As was discussed earlier (pp. 199–200), the Cthamaloidea, which appeared in the Cretaceous, seem to be on the decline, while the Balanoidea, which are known only from Eocene and younger deposits, are now radiating rapidly. Today the balanoids include 273 recognized species (Newman and Ross, 1976). If their radiation began 47 My ago, in the Eocene, the value of R would be 0.12 My^{-1}. Even if their record were pushed back to Paleocene time, 60 My ago, the value of R would be 0.093 My^{-1}, which is higher than for any family of bivalves that I have analyzed.

MARINE OSTRACODS

Given the existence of few living families of Cretaceous or younger age, there would seem to be little opportunity for evaluating values of R for marine os-tracods. Information bearing on average species duration is also somewhat in-consistent. Bold (1963) summarized data for Trinidad that show 16 of 36 Pliocene species (56 percent) extant and 13 of 29 Middle Miocene (Lengua Formation) species (45 percent) extant. In contrast, Hazel (1971) judged only about a third of all ostracod species of the Pliocene Yorktown Formation to be alive today. These figures place the ostracods somewhere in the range of marine echinoids, gastropods, or bivalves, but exactly where is uncertain. Imperfect knowledge of Recent species may depress some estimated percentages of extant species, but turnover rate in this group is obviously not nearly as high as in the trilobites, which are, of course, also arthropods.

AMMONITES

Kennedy and Cobban (1976) and Kennedy (1977) have assessed species durations of Mesozoic ammonites. Data are readily available, owing to the ammonites' great stratigraphic utility. Excellent resolution exists for Cretaceous forms of the Western Interior of North America, where beds of volcanic ash have been dated radiometrically. Endemic species of Mesozoic ammonites have commonly survived for intervals of only 0.5 to 0.7 My. More cosmopolitan species commonly span longer periods of time. The record duration seems to be that of *Phylloceras thetys*, which is considered to have existed for 25 My. Kennedy and Cobban, however, concluded that the lifetime of an average species was in the order of only 1 My. As evidence of this condition, regional correlation undertaken with ammonites is commonly accurate to between 0.2 My and 1 My.

For the ammonites, phyletic evolution is difficult to evaluate. The question is, to what extent do recorded species durations reflect phyletic transition, as opposed to termination of lineages? Kennedy and Cobban have noted that unusually long-ranging "species" are morphologically simple entities. Their lineages are difficult to subdivide. On the other hand, ammonite species with complex external features have been shown to display extraordinary intrapopulational variability of form (Figure 4-13). Many supposed phyletic trends may be spurious. The record of genera provides compelling evidence here. Among the many phyletic transitions that have been proposed, few valid transitions between genera have been recognized (Hallam, 1975). This means that most genera become extinct by termination of lineages. The fact that few genera pass from one stage to the next (see Figure 10-3) therefore indicates that lineages are indeed quite short-lived: An average Mesozoic stage spans only about 6 My. Clearly, short durations for species do not result simply from the fine subdivision of phyletic lineages.

Like the trilobites, the ammonites tended to radiate explosively following periodic mass extinctions. One of the most dramatic examples was the radiation of the Scythian (first stage of the Triassic), which followed the nearly total extinction of the ammonites at the close of the Permian. Possibly as few as three genera made the transition, and Kummel (1969) concluded that each of these but *Ophiceras* represented an evolutionary cul-de-sac. It was probably from this solitary genus that there issued a renewed and apparently unobstructed radiation of the ammonites. In surveying all known faunas of the final (*Prohungarites*) zone of the Scythian, Kummel provided ideal data for the estimation of R for this radiation. Owing to the imperfection of the record, the 154 species that Kummel recognized represent a minimum value for N. Radiation to this minimum diversity required between 7 and 10 My. The assumption of monophyly here yields a value of 0.50 to 0.72 My^{-1} for R, depending on the precise time of origin. If the radiation proceeded from as many as 10 species, R would have been 0.27 to 0.39 My^{-1}. As with estimates for other taxa, more exact data are unnecessary. Rates of radiation of ammonites were comparable to those of mammals and trilobites.

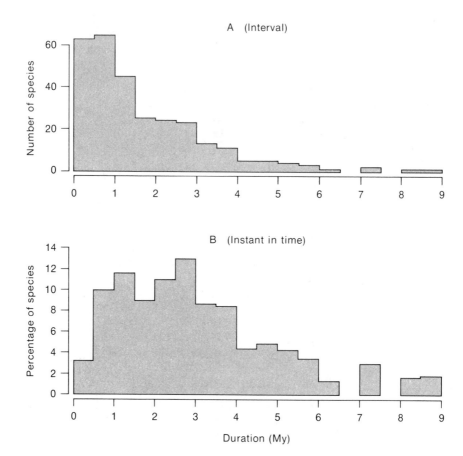

FIGURE 9-10
Species durations of Silurian graptoloids of the British Isles and Eire. A: Histogram for all recognized species of the interval. B: Histogram for an instant in time, derived from A by the method illustrated in Figure 5-5. (Data from Rickards, 1977.)

GRAPTOLITES

In part because of their planktonic mode of life, graptolites of the order Graptoloidea serve as useful index fossils for the Ordovician and Silurian Systems. Also, as with the trilobites and ammonites, rapid taxonomic turnover is a trait conferring biostratigraphic utility. Rickards (1977) estimated a mean duration of 1.9 My for Silurian species of graptolites. In Figure 9-10, Rickards' data are converted to a histogram for an instant in time, in the manner employed for other taxa. Mean duration here is about 3 My. The exact shape of the histogram is open to question because of crude temporal resolution, especially for short ranging species, but the data clearly show graptolites to be of short mean longevity.

Estimates of R can also be made from Rickards data. He recorded apparent trends in diversity during two episodes of worldwide explosive diversification. Like those recorded in trilobite biomeres, each of these followed a ubiquitous mass extinction. Because both apparent radiations are documented by widespread geographic evidence and span several zones, they must be real events rather than artifacts reflecting some kind of preservational bias. During 5 My, beginning in the earliest Llandovery, species diversity expanded from 12 known species to 59; during 3 My beginning in the *Nassa* zone of the Wenlock, the number increased from 7 species to 22 (Figure 9-11). These radiations reflect relatively high values of R: 0.32 My^{-1} and 0.40 My^{-1}, respectively. Thus, rates of species turnover within the graptoloids was unusually high for the marine realm, resembling those of trilobites and ammonites.

FRESHWATER FISHES

Thanks to the assistance of Gerald R. Smith,[*] it has been possible to obtain what seems to be a reliable estimate of mean species duration for late Cenozoic freshwater fishes of North America. The taxonomy of these forms has been established by Dr. Smith and others who have pioneered in the use of osteological features in the diagnosis of living fish species. For this reason, the data now available, though relatively sparse, are consistent and meaningful. Figure 9-12 depicts percentages of living species occurring within known fossil faunas. Faunas of the Late Pliocene and earliest Pleistocene, averaging about 2 My in age, contain about 38 percent living species. Only about 12 percent of mammals of this age are extant (Figure 5-8, A), indicating that species of freshwater fishes tend to be longer lived. The 50 percent point must occur within fish faunas having an age of about 1.5 My. Mean species duration must be in the neighborhood of 3 My.

The implication that species of freshwater fishes typically survive longer than species of mammals gains support from the great longevity of certain wellstudied species of fishes. The North American catfish *Ictalurus punctatus* has survived essentially without change since the Middle Miocene (about 15 My). Note also that species of *Amia* have existed for several million years (Figure 5-13).

Unfortunately, origins of presently radiating clades are poorly documented. Late Cenozoic radiations within large lakes have occurred at astronomical values

* Gerald R. Smith, University of Michigan, 1977.

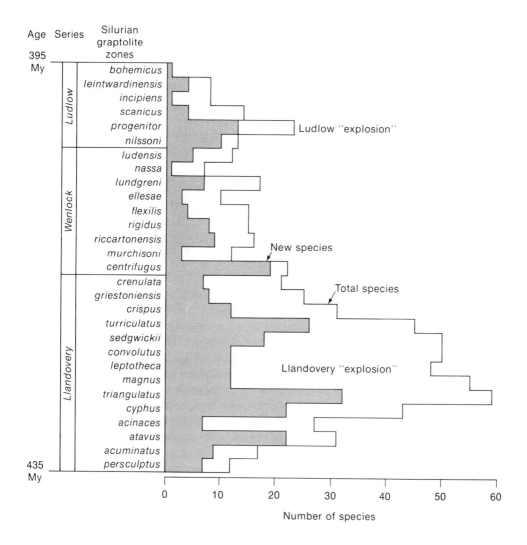

Figure 9-11
Diversification and decline of graptoloids during the Silurian in the area of the British Isles. (Data from Rickards, 1977.)

of *R*. The cichlid radiation of Lake Victoria yielding about 170 species from a small number of progenitors during a maximum of 750,000 years was discussed in Chapter 3. We seek here more typical values, representing radiation in larger geographic regions. The North American catfishes from an apparently monophyletic family (Ictaluridae), which, with 43 species, is the largest endemic North American family of freshwater fishes; the oldest known fossil representative is of Late Paleocene age (Lundberg, 1975). *Ictalurus*, the most species-rich

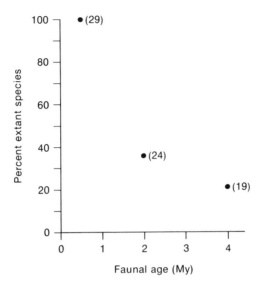

FIGURE 9-12
Lyellian plot for freshwater fishes of North
America. Numbers in parentheses represent
number of species forming each sample. (Data
from Uyeno and Miller, 1963; Eshelman, 1975;
Neff, 1975; and Smith and Lundberg, 1972.)

genus of the family, has its earliest known record at the base of the Oligocene, and in the subsequent 38 My has radiated to a present diversity of 18 species, giving a value for R of 0.076 My^{-1}. It seems likely that groups of smaller fishes, like the Cyprinidae (minnows) have expanded more rapidly. The North American cyprinids seem to have radiated more or less monophyletically to a diversity of about 245 species since being introduced from Eurasia perhaps no earlier than the Oligocene. The most ancient American cyprinids now known are about 31 My old (Cavender, 1968). Thus, the value of R may have been as high as 0.18 My^{-1}.

While much uncertainty remains, the preceding calculations suggest that average rate of radiation for freshwater fishes is higher than that for marine bivalves and gastropods, but lower than that for mammals. Note that the estimated species duration (3 My) is also intermediate.

INSECTS

Perhaps more than any other group, the insects illustrate the utility of the techniques employed here for evaluating rates of speciation and extinction. The insects are generally considered to have an unusually poor fossil record, yet their fossil occurrences show that their faunal turnover resembles that of birds and mammals. Only insects entombed in amber provide sufficient morphologic detail for comparison with living species. Frank M. Carpenter, a leading authority on fossil insects, informs me that he knows of only one definitely pre-Pleistocene form assignable to a living species. This is *Fannia scalaris* Fabricius, which has been carefully identified by Willi Hennig (1966), the foremost taxonomic expert on the muscid flies, of which *Fannia* is a member. As evidence of the quality of detail provided by preservation in amber, Hennig's identification is based on the morphologic pattern of tiny hairs. It seems apparent that additional living insect species existed before the Pleistocene, but the percentage cannot be great.

It seems evident that species durations for insects must resemble those for mammals. Rates of radiation are consistent and are only slightly lower than mammalian rates. Dr. Carpenter has provided me with geologic ages for currently radiating taxa of the Diptera (true flies) and Formicidae (ants).* He has also supplied estimates of modern diversity for the former, and Dr. W. L. Brown has provided comparable data for the latter.† Table 9-3 presents these data and calculations of R. The mean value (0.19 My^{-1}) is quite high, though slightly lower than for mammals.

BIRDS

There has been a widespread misconception that birds have evolved very slowly. There are indeed some Miocene lineages that seem to have survived to the present, but there is also evidence that many groups of birds have undergone rapid evolutionary turnover.

Brodkorb (1959) and Fisher (1967) have expressed the view that no Pliocene fossil bird belongs to a living species. This, however, seems impossible in light of the positively skewed shapes of histograms of species longevities for other taxa (Figures 5-6, 5-7, and 9-10). Furthermore, both authors assigned a Pleistocene age to the Blancan Stage, now considered to range well back into the Pliocene, spanning the interval from 4 My to 1.5 My before the present. Brodkorb (1959) concluded that the oldest known living species are Blancan, and compilations by Fisher (1967) suggest that about 40 percent of Blancan species are extant. It would be hazardous to attempt to extract an estimate of average species duration from the rather uncertain fossil species designations. Birds lack teeth, which are useful in mammalian taxonomy, and species are often identified on the basis of fragmented skeletal evidence. Nonetheless, it seems likely that the figure exceeds that for mammals, approaching that for freshwater fishes (about 3 My).

---

* Frank M. Carpenter, Harvard University, 1978.
† W. L. Brown, Cornell University, 1978.

TABLE 9-3   Estimates of R for presently radiating taxa of insects.*

	Age	N	$R(My^{-1})$
(Order: Diptera)			
Tipulidae	Eocene (47 My)	3,800	0.18
Bombyliidae	L. Oligocene (35 My)	4,000	0.24
Asilidae	Eocene (47 My)	4,700	0.18
Syrphidae	Eocene (47 My)	5,000	0.18
(Family: Formicidae)			
Ponerinae	Eocene (47 My)	2,000	0.16
Myrmicinae	Eocene (47 My)	10,000	0.20
Formicinae	Eocene (47 My)	4,000	0.18
		Mean	0.19

* Data from Brown (1973) and from personal communication—Frank M. Carpenter, Harvard University, and W. L. Brown, Cornell University, 1978.

Rates of radiation are more easily evaluated, despite the fact that times of origin of many modern families are unknown. In particular, the passerine birds (small perching forms, including song birds) are radiating rampantly today. The subgroup known as the nine-primaried oscine passerines began radiating in North America not appreciably earlier than the Miocene (Feduccia, 1977; Storrs L. Olson*). Today they include more than 800 species. This estimate gives a value of R of about 0.30 My^{-1}. The size and complexity of the radiation and the reliability of the data here demonstrate that rates of speciation comparable to those of mammalian families must be common for birds. Even if speciation was accelerated during Pleistocene climatic changes (page 205; see also Mengel, 1964), rapid radiation is indicated. If, for example, the Pleistocene yielded an eight-fold increase in diversity of nine-primaried oscines (and this would seem an extreme figure) the preceding radiation to a diversity of about 100 species at the start of this interval would represent a value of R of at least 0.25 My^{-1} for the earlier radiation. While these calculations represent but a single radiation, the clade considered includes about 10 percent of living species of birds. Clearly, a high rate of radiation is indicated for the class.

REPTILES

Because of climatic cooling, reptiles in many regions have not been flourishing during the Late Cenozoic. Nonetheless, Auffenberg and Milstead (1965) reported knowing of no North American genus that became extinct during the

* Storrs L. Olson, U.S. National Museum, 1978.

Pleistocene. This fact, together with the observation of the same authors that most living species existed in the Early Pleistocene, indicates that average species longevity is greater than in the Mammalia (see Figure 5-8,A). To minimize Pleistocene climatic effects, given the ectothermic nature of reptiles, I have sought more specific data from low latitudes. Of 28 species of snakes recognized by Auffenberg (1963) in the Pleistocene of Florida, 27 are considered to be extant. The problem here is that faunal ages are not precisely known. Many collected fossils may be Late Pleistocene in age. One the other hand, direct comparison with mammals and birds is possible, and this demonstrates a greater longevity for snakes (although some mammalian extinctions may have resulted from human interference):

> The most striking fact concerning the Late Pleistocene snake fauna of Florida is its similarity to the modern one. The great degree of extinction witnessed among the mammals and birds in Pleistocene and post-Pleistocene time apparently is not present in the reptiles. Evidently only one Pleistocene Florida species of snake has become extinct, *Crotalus giganteus*. (Auffenberg, 1963, p. 211.)

While it seems evident that, on the average, reptilian species survive longer than mammalian species—a condition that is consistent with the greater longevity of reptilian genera (Estes, 1970)—mean longevity for species of reptiles remains uncertain.

As noted near the beginning of this chapter, the Neogene Period could be appropriately labeled the Age of Snakes. Most living snakes—the exceptions include the boids—belong to the family Colubridae (caenophidians). There are about 1,400 species of colubrids (Porter, 1972), yet the known fossil record of this family in North America extends back only about 13 My to the Middle Miocene; all older species belong to more primitive families, and the few known colubrid species of the Middle Miocene appear to represent an ancestral grade of organization for the family (Holman, 1976). A Middle Miocene origin for the colubrid radiation would yield a value for R of 0.56 My^{-1}, which is apparently higher than the value for any living family of mammals (Figure 5-2). Even if the origin of the colubrid radiation were pushed back to a date 25 My before the present, the value of R would become 0.29 My^{-1}, which would represent very rapid radiation for a family of mammals. The modern snakes have entered a new adaptive zone, which they are expanding at a remarkable rate. Because of uncertainty as to mean species duration, no point for colubrid snakes is plotted in Figure 9-1. It may be, however, that for colubrids the ratio $[S/E]$ is unusually high.

REASONS FOR THE CORRELATION

The correlation displayed in Figure 9-1 seems to establish three rules of macroevolution. One of these is *that rate of speciation in adaptive radiation and rate of*

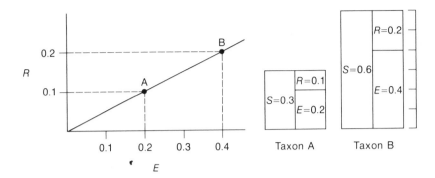

FIGURE 9-13
Illustration of the fact that an increase in S and E by a given proportion results in an increase in R by the same proportion. In this hypothetical example, all rates are twice as high in taxon B as in taxon A.

extinction are strongly correlated in the animal world. It will be recalled that the horizontal axis of the graph indirectly depicts rate of extinction and that the documented correlation with R then establishes a correlation with S (because $S = R - E$, S must increase as R and E increase).

The second rule is *that R tends to increase in proportion to S and E.* (This is implicit in what has just been stated because of the procedure followed, but is not a corollary of the first rule; the first rule would not preclude R being constant throughout the animal world, for example). In other words, as rate of extinction increases across a spectrum of taxa, rate of speciation in adaptive radiation tends to increase *proportionally,* so that R grows in proportion to these two rates (see Figure 9-13).

Why rates of speciation and extinction behave in such a way as to yield the first two rules forms the subject of most of the remainder of this chapter. This question relates to the third rule, which is *that rate of evolutionary turnover tends to increase among taxa in association with adaptive complexity.* This was discussed earlier with respect to rate of extinction, for which more data have been considered (Figure 9-2) than are represented in Figure 9-1.

The most obvious factor potentially contributing to the correlation of S and E is dictated by simple arithmetic. A taxon that for some reason is characterized by a high rate of extinction will not survive unless it happens also to be possessed of a high rate of speciation. This relationship cannot account for the apparent absence of taxa with low rates of extinction and high rates of speciation, however, and can therefore form only a part of any general explanation.

It may be that under certain circumstances rate of speciation is controlled by rate of extinction—that competition may reduce rate of speciation to a degree that new species are likely to be added only when established species die out. Under such circumstances, rate of speciation would track rate of extinction, which might be expected to vary systematically. This kind of relationship is an underlying premise of the theoretical work of Van Valen (1973). A less stringent connection, in which competition tends to equalize the two rates, although they still fluctuate significantly, has been assumed in the simulation of diversification and decline of taxa by Raup *et al.* (1973) and Gould *et al.* (1977). Rate of speciation may sometimes be linked to rate of extinction in such a manner, but it is very important to appreciate that this kind of relationship cannot explain the correlation displayed in Figure 9-1 because this plot represents adaptive radiation, in which rate of speciation greatly exceeds rate of extinction. Many of the clades considered, such as those containing sand dollars, snakes, carnivorous snails, and acroporid corals, represent the invasion of entirely new adaptive zones for the classes to which they belong. Clearly, we must seek other, less direct kinds of linkage between S and E. Two factors would seem to be of great significance here: dispersal ability and level of behavior. The effects of these variables will now be explored in some detail.

DISPERSAL ABILITY

Dispersal ability seems to play a major role in determining both species longevity and rate of speciation within higher taxa (Mayr, 1963 and 1970, Chapter 18; Jackson, 1974). First, let us consider longevity. Cosmopolitan species, being represented by numerous populations, would be expected to survive longer than endemic species, and there is empirical evidence that the predicted correlation does indeed exist for bivalve mollusks (Jackson, 1974) and ammonites (Kennedy and Cobban, 1976; Kennedy, 1977). There is a potential problem of circularity, however. Does a species tend to persist because it is broadly distributed, or, as J. C. Willis (1922) asserted in his book *Age and Area*, is it broadly distributed because it has survived for a long period of time? We might answer this question using species for which excellent fossil data show a pattern of early dispersal to maximum range. At present, we are faced with a compelling hypothesis: Endemic lineages in general, should, tend to be short-lived.

The case of the Cretaceous brachiopod *Terebratulina chrysalis* represents a supporting example. In a detailed study of the remarkably well preserved brachiopods of the Cretaceous Danish chalk, Surlyk (1972) found *T. chrysalis* in all of the 150 rock samples that he disaggregated. In life, if seems to have occupied all kinds of substrata available in the deep-water environment of deposition, and it is the only species of Surlyk's study that is also found in adjacent nearshore facies. Thus, it was both broadly adapted and widespread, and Surlyk noted that the lineage to which it belonged seems to have survived to the present time, with little evolutionary change.

The history of the bivalve species *Hiatella arctica* is another case in point. This species today occupies an unusually wide variety of substrata and ranges from the arctic region to the tropics. One of the oldest of living bivalve species, this form has a fossil record extending back at least 50 My to the Eocene. Presumably, its survival, like that of the *Terebratulina chrysalis* lineage, has partly been a function of its broad geographic distribution.

The observation that bivalve species have a slightly greater average longevity than gastropod species (compare Figures 5-8, B and 9-3) may in part be a function of greater taxonomic splitting of gastropods, as noted earlier, but may also reflect a real biologic contrast. Having assembled data for 1,137 species of benthic mollusks from Florida, Nicol (1977) showed that a higher percentage of gastropods than bivalves are endemic to Florida (12.7 percent versus 3.6 percent) and that a higher percentage of bivalve species are wide-ranging. These contrasts seem in part to reflect a higher percentage of planktonic larvae in the Bivalvia.

We must be careful not to overgeneralize in this kind of analysis. Taking the other side of the argument, we could cite the long persistence of some species in localized refugia. If protection from potential predators and competitors is most likely to be found in small, protected ecologic "islands," then a considerable fraction of extremely long-ranging species might be expected to have narrow rather than broad geographic ranges. Widely heralded examples include the survival of *Metasequoia* in a small area of China, of *Sphenodon* on a few islands off the coast of New Zealand, and of numerous lemurs and other primitive mammals on Madagascar. It seems likely, however, that for the vast majority of species, which are not confined to protective refugia, there is a correlation between geographic range and geologic longevity. Widespread species have not put all their eggs in one geographical basket and can tolerate the kind of local habitat destruction that extirpates endemic forms. Quite apart from the question of the number and size of populations, Mayr (1963 and 1970, Chapter 18) and Jackson (1974) have emphasized the notion that physiologically tolerant species tend to be widely distributed and that this condition should enhance their inherent tendency to resist agents of extinction.

Species that disperse readily have been postulated to speciate readily as well (Slobodkin and Sanders, 1969). I would agree with Mayr and Jackson, however, that for such forms geographic isolation, and even steep genetic gradients, will not be maintained easily; hence, speciation should be rare. This view is evident in the long held belief that speciation is more frequent on land than in the sea (Day, 1963; Mayr, 1963).

The alleged terrestrial-marine contrast is confirmed by Figure 9-1, which shows mammals, birds, and insects to have short durations and high values of R. In the marine realm, trilobites, ammonites, and graptolites exhibit exceptionally high rates of turnover. These groups will be evaluated below.

A kind of test within a test is possible for some terrestrial groups. The effects of unusually weak powers of dispersal can be seen, for example, in *Melanoplus*, a genus of grasshopper. Sections of the genus that contain flightless individuals are deployed as numerous, narrowly ranging species. Those including fully winged forms consist of small numbers of widespread species, which clearly are the product of lower rates of speciation (White, 1973, p. 370). Mayr (1963, pp. 581–582) reviewed several similar examples. The genus *Busycon* provides a comparable illustration within the marine Gastropoda. Species of this genus have no free larval form but produce largely immobile egg cases. Populations are localized in areas of shallow water, and gene flow among them tends to be weak. As discussed by Pulley (1959), species of this genus show enormous morphologic variability (Figure 9-14), reflecting the partial isolation of many populations.

The general effect of the nonmarine habitat upon rates of speciation and extinction can also be seen in natural experiments of the geologic past. During the Pliocene, in the Caspian region, a large lake known as the Pontian Sea became isolated from the world ocean system. This lacustrine body had a complex history that is poorly understood, but its waters seem to have been brackish. A population of the marine cockle *Cerastoderma* was apparently isolated in the Pontian Sea, and from it there issued an incredible adaptive radiation. It is uncertain exactly how many species were formed here, but a new family, the Limnocardiidae, evolved and is now divided into five subfamilies and nearly forty genera. The morphologic diversity of the Limnocardiidae was enormous (Figure 5-11), and a few species survive today (Gillet, 1946). The radiation, which proceeded for no more than about 3 My and was perhaps much shorter, resembled those that have taken place more recently in the Great Lakes of Africa. What has become of *Cerastoderma* elsewhere? In the marine realm, its representatives have continued to speciate at rates that are typical for marine Bivalvia. Its precise phylogeny is unstudied, but the genus arose in the Late Oligocene and today contains very few species (for example, two in Britain and one in the Eastern United States). Clearly, environmental conditions not characteristic of the marine realm can induce rapid speciation in those marine taxa that normally speciate at lower rates. The role of geographic isolation in the Pontian radiation is unclear, but geographic and microgeographic heterogeneity in space and time are suggested by the complex history of the inland sea. The isolated Paraná Basin of South America was the site of a similar but less dramatic molluscan radiation in Permian time (Runnegar and Newell, 1971). The Pontian and Paraná examples illustrate the importance of geographic factors in the control of radiation. The sluggish multiplication of marine bivalves species seems to be an inherent aspect of bivalve biology only insofar as the group is fundamentally marine.

It would be most interesting to examine geologic longevities of species of terrestrial snails, but unfortunately this does not appear to be possible at present. Data are available for freshwater snails, and these reveal longevities commonly as great as those of marine forms. Freshwater fishes also show rates of turnover much lower than those of terrestrial taxa (Figure 9-1). Such low rates are not surprising in light of the great powers of dispersal that characterize many freshwater taxa.

Even so, dramatic endemic radiations are quite common in lakes and river

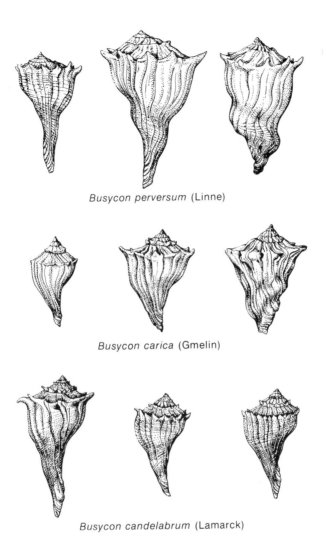

Busycon perversum (Linne)

Busycon carica (Gmelin)

Busycon candelabrum (Lamarck)

FIGURE 9-14
Morphologic variability in three species of the marine snail
Busycon. (From Pulley, 1959.)

systems, as has been discussed earlier (pages 45 and 103). These radiations usually follow the origin of new bodies of water, like the Mekong River System or the Great Lakes of Africa. The endemic, monophyletic radiation of triculine gastropods generated at least 88 species in a maximum of 12 My (Davis, 1978), giving a minimum value for R of 0.37 My^{-1}, which is equivalent to that of an unusually rapidly radiating family of mammals.

In general, the instability of freshwater habitats fosters great variation in rates of speciation and extinction. The late Cenozoic history of North American mollusks illustrates this condition. Taylor (1968) contrasted the fauna of the Rexroad Formation of Kansas with that of the Tulare Formation of California. Both are of Blancan age (mid-Pliocene to early Pleistocene). Of the 31 freshwater mollusks recognized from the Tulare, only 10 (32 percent) are assigned to living species, compared to 13 of the 16 freshwater snails (81 percent) of the Rexroad fauna. Note that in Figure 9-3, which depicts marine snails, no two points for contemporaneous faunas differ so greatly. Taylor observed that the living Rexroad species have remained widely distributed in a tectonically stable area. Members of the Tulare fauna, in contrast, occupied an unstable area, in which tectonic movements and volcanism have occurred, filling basins and changing drainage patterns. The extinction of most Tulare species, many of which were lacustrine forms confined to single basins, is hardly surprising.

Of the more than 60 total nonmarine molluscan species of the Blancan Rexroad and Saw Rock Canyon faunas of the continental interior, about 75 percent are extant. This percentage is similar to those for marine gastropods and bivalves (Figures 5-9 and 9-3) of the same age (about 2.5 My), suggesting that in the absence of tectonism and other modes of habitat destruction, freshwater snails and bivalves resemble marine snails and bivalves in rate of turnover.

It is hardly surprising that terrestrial faunas also exhibit occasional pulses of radiation entailing extraordinary rates of speciation. Examples are ecoinsular radiations, like the one that has produced about 238 species of *Drosophila* on the Hawaiian islands during about 5.6 My ($R = 0.98$ My^{-1}), with even higher rates characterizing single islands, like Hawaii. Local Pleistocene radiations of birds (page 205) may represent additional examples.

The marine realm is more uniform, but is of course not entirely without spatial and temporal heterogeneities that cause some variation in rates of speciation and extinction. It was observed earlier that late Cenozoic environmental deterioration seems to have caused mass extinction of European bryozoans, while Australian faunas have remained quite stable (page 243). In the next chapter, it will be suggested that increasing endemism and habitat destruction accelerated extinction within late Cenozoic faunas of marine mollusks in the tropical Americas. It was also noted earlier that volcanic islands, like Cocos Island, may represent sites of speciation for shallow-water benthos (page 170).

It is uncertain with what frequency ecoinsular or other radiations occasionally proceed at exceptionally high rates in the marine realm. Because of an absence of precise data, I have not attempted to assess quantitatively the notable radiation of the Oligopygoida, an order of echinoids (Kier, 1967). This distinctive group is known only from the Middle and Upper Eocene of the Caribbean and eastern Gulf of Mexico. Porter Kier informs me that not even fragments of oligopygoid tests can be found in Paleocene sediments.* Poor stratigraphic resolution

---

* Porter Kier, U.S. National Museum, 1978.

obscures standing diversity for any slice of Late Eocene time, but at least 25 valid species are recognized for the total Late Eocene, which together with the Middle Eocene spanned only about 10 My. Are we confronted here with a burst of marine speciation whose rate rivals the rates of typical terrestrial radiations? Possibly, more detailed assessment of this or some other evolutionary episode will reveal an extraordinary marine adaptive radiation that requires special evaluation. In other words, while the values of R summarized earlier in the chapter are relatively consistent for radiations within a particular taxon, we cannot rule out the possibility that exceptions will appear.

BEHAVIORAL COMPLEXITY

To explain the presence of rapid turnover rates within some marine taxa, like trilobites and ammonites, it would seem necessary to look beyond environmental controls. I would suggest that level of behavioral interaction plays an important role here. A rather remarkable, previously noted feature of Figure 9-2, which lends itself perhaps too readily to simplistic interpretation, is that high rate of speciation and extinction correlate strongly with position in the scale of life, or with level of biologic organization. Lyell (1830) was aware of this relationship, as was Darwin, who also recognized the terrestrial-marine contrast:

> The productions of the land seem to change at a quicker rate than those of the sea. ... There is some reason to believe that organisms, considered high in the scale of nature, change more quickly than those that are low; though there are exceptions to this rule. (Darwin, 1859, p. 313.)

In comparing terrestrial and marine life, Darwin was concerned not with differences in potential for dispersal, but with level of biologic complexity. As it happens, conspicuous terrestrial taxa are more advanced animals than are most marine taxa. Characteristically, Darwin offered a gradualistic hypothesis. His idea was that competition, which he saw as the primary driving force of selection, is especially intense within and among advanced taxa. Without rapid change, he believed, these forms would become extinct. This allegation is reminiscent of the gradualistic explanation for the dominance of sexual modes of reproduction described in the previous chapter.

Competitive interactions among advanced animals are to a considerable degree behavioral in nature. In a punctuational context, advanced behavior takes on different meaning. It seems evident that level of ethological complexity must play a role both in differential rate of extinction and in differential rate of speciation. Let us first consider rate of speciation. Here two assertions in the writings of Ernst Mayr have important implications:

A shift into a new niche or adaptive zone is, almost without exception, initiated by a change in behavior. The other adaptations to the new niche, particularly the structural ones, are acquired secondarily. (Mayr, 1963, p. 604, citing Mayr, 1958, 1960.)

Ethological barriers to random mating constitute the largest and most important class of isolating mechanisms in animals. (Mayr, 1963, p. 95.)

These are broad generalizations, which like all others are subject to refinement. In this chapter, we are examining differences among taxa, and clearly there exist major discrepancies in level of behavior. We can predict from Mayr's statements that advanced creatures with complex patterns of behavior should differ from simple forms in both mode and frequency of speciation.

In studying burrowing bivalves, I have been struck by the great simplicity and uniformity of the behavioral program that controls locomotion (Trueman, 1966). The clam, in fact, has no head! It does not defend a territory, nor does it sense its neighbors with any degree of sophistication, if at all. It does not engage in courtship. It does not even choose a mate, but impersonally broadcasts its gametes into the surrounding water mass, to an uncertain fate. The anthropomorphic flavor of this description is intentional. It does not fit the primitive subject, which is utterly lacking in ethological traits of a human quality. Birds, mammals, and to a lesser extent, other animals are much closer to humans in the nature and complexity of their behavior.

Certainly, timing of reproduction can represent a behavioral isolating mechanism, if reproduction is seasonal and if timing is classified as an aspect of behavior. Nonetheless, simple organisms must, on the average, possess fewer ethological isolating mechanisms than complex organisms. Lacking significant agents of character displacement (Brown and Wilson, 1956), they may exhibit a weaker degree of divergence per speciation event (Stanley, 1973c). More important in the present context is the obvious point that primitive organisms should speciate at rates lower than those of advanced organisms. Some invertebrates are virtually lacking what Mayr has considered, in general, to be the most important kind of isolating mechanism in the animal world. In contrast, for some higher animals, a minor change in some ritual of courtship can lead to immediate and total isolation of a very small subpopulation. It will be recalled that Bush (1975) and Wilson *et al.* (1975) have proposed that advanced behavior tends to increase frequency of speciation for the different reason that it favors the formation of small, inbreeding social groups.

Complex behavior, especially in courtship, would seem to account for the high rates of speciation of the trilobites and ammonites, which were advanced, somewhat vertebrate-like animals in level of sensory perception and behavior. Marine mammals offer possibilities for similar tests. Apparently, the Delphinidae (dolphins) today represent the most rapidly expanding family of the order of whales (Cetacea). Today, after radiation since Late Oligocene or Early

Miocene time (approximately 22 My), about 62 species are in existence, giving an R-value of 0.18 My^{-1}. This is a relatively high rate for the marine realm (see Figure 9-1), perhaps supporting the idea that advanced behavior promotes speciation.

An interesting parallel exists in the plant world, where pollination by animals can be viewed as a behavioral isolating mechanism (Grant, 1949; 1963, pp. 363–366). We would expect rates of speciation to be more rapid in such plants than in wind pollinated forms.

Just as advanced behavior accelerates speciation, it would seem to increase rate of extinction because, in effect, it represents a form of ecologic specialization. Although most evaluations of niche breadth have entailed the measurement of environmental resources, behavioral adaptation greatly constrains the volume of environmental hyperspace occupied by a species. In short, complex behavior is often a form of ecologic specialization. The historical fleeing of many species of mammals, particularly predators, from advancing human civilization represents a familiar example. Often, a species is disturbed behaviorally and vacates territory well before its proximate needs can no longer be met. Within a few taxa, like our own species, advanced behavior contributes ecologic flexibility, but I am referring in this discussion to the more common, sterotyped kinds of behavior that delicately balance an organism with its environment, increasing the number of things that can go awry.

Thus, it seems likely that rate of speciation and rate of extinction increase with behavioral complexity, just as they decrease with dispersal ability. What seems remarkable about these paired relationships is that they are fortuitious—that is, it just happens that each of the two controlling factors has the same effect on speciation that it has on extinction. Because of this coincidence, some taxa can flourish at one end of the ethological or dispersal spectrum, while other taxa flourish at the other end. High rates of turnover will characterize some species and low rates, others. This relationship will be more fully explored in the following chapter.

Certainly, dispersal ability and behavioral complexity may not account fully for particular rates of speciation and extinction. Why the graptoloids, which were colonial planktonic animals, underwent rapid turnover is uncertain. Although the fossil record reveals considerable provinciality for some faunas, others were cosmopolitan (Skevington, 1973; Berry, 1973), and weak dispersal would not seem to account for the generally high rates of speciation and extinction of graptolites. Cenozoic nannoplankton have also speciated and gone extinct at remarkably high rates (Hay, et al., 1967). Coccolithophores nearly disappeared at the end of the Cretaceous, but re-expanded to full diversity by the mid-Paleocene, only about 5 My later! Further testimony to the remarkable rates of speciation in these forms is the observation that living floras are dominated by species that have appeared only within the last 100,000 years. These rates contrast sharply to those for planktonic forams (Figure 9-1). It may be that certain modes of reproduction promote speciation, for example, by raising the probability of selfing, or inbreeding. Possibly, the fact that the balanoid barnacles have diversified more rapidly than any family of Cenozoic bivalves relates the the barnacles' reproductive behavior, which involves direct, internal fertilization.

RATES OF PHYLETIC EVOLUTION

Another interesting condition is what seems to be a correlation between rate of speciation and rate of phyletic evolution. This correlation can be seen, for example, in the comparsion of Figure 2-2, which depicts chronospecies of scallops, with Figure 4-11, which depicts much shorter-ranging chronospecies of elephants. More generally, mean duration for chronospecies of mammals is in the order of a million years (Figure 4-7). The simple fact that typical species of bivalves, including those disappearing as lineages are terminated, last on the average, an order of magnitude longer demonstrates the great disparity in rate of phyletic transformation. None of the various kinds of animals which have speciated at low rates and whose species have lasted for several million years (Figure 9-1) can have undergone phyletic evolution as rapid as that of the Mammalia. The strength of the apparent correlation remains uncertain for lack of information. The only other group of rapidly speciating taxa in which phyletic evolution has been rather extensively studied is the ammonites. In this rapidly speciating group, as in the mammals, phyletic evolution seems to have been relatively rapid. It is not uncommon for measurable phyletic change to have occurred within spans of 1 to 2 My (Brinkman, 1929, 1937).

Why do taxa whose lineages multiply and terminate at high rates also undergo rapid phyletic change? The answer would seem to lie primarily in the important role of population size, which, as Mayr (1967; 1970, p. 345) has emphasized, exerts a profound control over rates of evolution (see also Boucot, 1975, pp. 59–62, who seems, however, to blur the distinction between phyletic extinction and termination of lineages). The apparent confinement of quantum speciation to small populations is the extreme condition here. In Chapter 6, a case was made that extremely rapid change must occur almost entirely within small, diverging populations because total rates of extinction (numbers of species going extinct per unit time) are so low in the animal world that rapid divergence by bottlenecking and rebirth of lineages must be an exceedingly rare phenomenon. In other words, it is very seldom that an established lineage takes a rapid evolutionary step in some new direction. Still, although phyletic evolution seems almost universally to be slow, it does occur and its rate should vary with population size. Taxa with small, panmictic populations can be subjected to uniform selection pressures more readily than widespread species composed of disjunct demes occupying varied habitats. Thus, we should expect taxa whose species have weak powers of dispersal to exhibit not only relatively high rates of speciation and extinction, but also relatively high rates of phyletic evolution. It will be recalled that the most rapid phyletic evolution now recognized among late Cenozoic Mammalia is for elephants (page 85), and this perhaps constitutes evidence that population size has a stronger influence than generation time.

THE FUTURE

This chapter has been concerned with the gross comparison of major taxa (classes, or groups of similar rank). As more complete sets of data become available, it may become feasible to contrast subsets of species within these groups. Are S and E similar for infaunal and epifaunal bivalves? For passerine and nonpasserine birds? For monocots and dicots of the plant world? Are values of these parameters higher for gastropods with internal fertilization than for other members of the class? These are intriguing questions that must be addressed with caution but should be considered in future work. Many of the questions relate to what must be subtle disparities, so that a partitioning of the data now available would reduce sample sizes to a point where results would be inconclusive. As noted earlier, because of the imperfection of the fossil record, there is also great danger in the simple, direct enumeration of species longevities from stratigraphic data. Consider, for example, the idea that species of small population size tend to have short geologic durations. Even if apparent stratigraphic ranges seem to support this relationship, we must ask whether we are not simply confronted with an artifact of preservation: The smaller a species' population size, the less fully, on the average, will we document its actual vertical range.

Unfortunately, for plants it is difficult to evaluate longevities of species. The fossil record of pollen, which offers promise, in general cannot yet be studied at the level of the species. On the other hand, values of R should be calculable and may prove interesting.

For animals, there remains the question of how the incidence of quantum speciation relates to other biologic features. For example, apart from the fact that mammals speciate rapidly, is their evolution characterized by a high *incidence* of strongly divergent speciation? If this trait is present, is it because (1) mammals tend to be highly competitive, owing in part to advanced behavior, and to undergo character displacement frequently (Stanley, 1973c), (2) they form inbreeding social groups more readily than do behaviorally simple animals (Wilson *et al.*, 1975), or (3) they have complex endocrinological systems that present many opportunities for rapid epigenetic restructuring (Løvtrup, 1976)?

SUMMARY

Rate of speciation (*S*), rate of extinction (*E*), and net rate of diversification (*R*) in adaptive radiation vary greatly in the animal world, but have characteristic values for particular taxa. These variables tend to be strongly correlated, all having relatively high values within groups like the mammals, birds, insects, trilobites, ammonites, and graptolites and having relatively low values within groups like the bivalves, gastropods, echinoids, reef corals, and forams. Apparent differences in rates of turnover reflect a real evolutionary disparity. They are not artifacts that reflect biases in taxonomy related to variations in morphologic complexity.

The correlation of rates of speciation and extinction may, in part, reflect the obvious impossibility of a taxon's surviving with a high value of *E* and a low value of *S*, but this cannot explain the absence of the reverse condition. Also, because the correlation has been established by the evaluation of rapidly radiating taxa, it cannot result from competitive equilibrium, in which rate of extinction dictates rate of speciation. The correlation seems primarily to represent fortuitous linkages: (1) Rates of speciation and rates of extinction both vary inversely with dispersal ability. Thus, terrestrial animals, in which populations are often readily restricted by geographic barriers, exhibit high turnover rates. Freshwater animals, which typically have excellent powers of dispersal, commonly exhibit lower rates than terrestrial groups. Both terrestrial and freshwater faunas typically display greater variability than marine faunas in rates of turnover, owing to the temporal and spatial heterogeneity of nonmarine environments. (2) Rates of speciation and extinction increase with behavioral complexity because advanced behavior constitutes an important isolating mechanism but also represents a form of specialization that increases probability of extinction. Thus, a hierarchy of taxa arranged according to rates of speciation and extinction crudely resembles a *scala naturae*, with adaptively complex taxa near the top. Pollination specificity in plants may resemble advanced behavior in accelerating speciation and extinction.

Rates of phyletic evolution seem to vary among taxa in much the same manner as rates of speciation and extinction, probably because phyletic evolution accelerates with a decrease in population size, which, in general, accompanies a decrease in dispersal ability.

10

Speciation, Extinction, and Diversity in Time and Space

The analysis of exponential rates of change has many applications to the study of taxonomic diversity. Species selection tends to increase rate of speciation (S) and to decrease rate of extinction (E). As shown in the previous chapter, however, the two rates happen to be causally linked and are therefore positively correlated. Species selection strains this linkage, but has failed to develop relatively high [S/E] ratios except in a few "supertaxa." Given this limitation, potential "strategies" for maintenance of diversity relate largely to position in a spectrum ranging from high values of S and E to low values. Taxa characterized by high rates have the advantage of evolving rapidly but the disadvantage of being intrinsically unstable and vulnerable to mass extinction.

Taxa characterized by severe interspecific competitive interactions tend to be elastic, in that a sudden decline in their diversity, if not too severe, removes the primary restraint on their diversification and is automatically followed by a rebound in diversity when normal conditions are restored.

Spatial gradients in diversity may often develop because of the relative difficulty of speciation into rigorous habitats. The slow start that characterizes exponential increase automatically accentuates diversity gradients. Evidence is equivocal as to whether S, E, and R tend to be higher in tropical than in nontropical taxa, but S is probably higher because of the multiplicative effects of biotic diversity. In general, however, geographic configuration has more profound effects upon rates of diversification and decline than does climate. Very large regions divided into many semi-isolated habitats represent crucibles of diversification, in that they foster high rates of speciation and low rates of extinction.

INTRODUCTION

Rates of speciation and extinction carry many implications not brought to light in previous chapters, especially with regard to the numbers of species existing in particular places at particular times. These rates are not solely responsible for determining momentary diversity at any location (emigration and immigration are also important), but they play a part that has perhaps been too little appreciated.

EVOLUTIONARY STABILITY AND MASS EXTINCTION

The fact that rates of speciation and extinction vary markedly among higher taxa has great significance in the study of mass extinction. A simple mathematical consequence of the possession of high rates of speciation and extinction is a general instability in large-scale evolution (Stanley, 1977). In other words, taxa that have high rates of speciation and extinction are inherently vulnerable to mass extinction. This point is perhaps most effectively made by means of two examples, which are illustrated in Figure 10-1.

We have documented a general correlation in the animal world between mean rate of extinction and mean rate of speciation in adaptive radiation. Accordingly, let us postulate the existence of two radiating higher taxa in which S is 1.25 times as high as E. In one of these taxa, however, we will let the values of both S and E be four times as high as in the other. This means that R will also be four times as high. Diversification of the hypothetical taxa is shown in Figure 10-1,A. After a period of radiation, we impose a 3 My interval of environmental deterioration that has the same proportional effect upon each taxon. During this interval, let us assume that for each taxon rate of extinction is doubled and rate of speciation is cut to one-half its original value. (This choice is arbitrary; any of an infinite variety of effects could be postulated, as long as their impact on the two taxa is equivalent, or proportional to S and E.) As shown in Figure 10-1,A, the result for the taxon with high original values of S, E, and R is a disastrous decline to extinction. (Note that exponential decline must terminate at a diversity of one species, but we can assume that the final species goes extinct quickly.) For this taxon, a high positive value of R is converted to a high rate of decline. In the other taxon, a low positive value of R becomes only a low rate of decline, and this species survives the 3 My catastrophe with modest reduction in diversity, to radiate anew, at the original value of R, when the interval of environmental deterioration comes to an end.

A simpler way of appreciating how high values of S and E yield instability is to consider the extreme case in which speciation is altered by 100 percent—

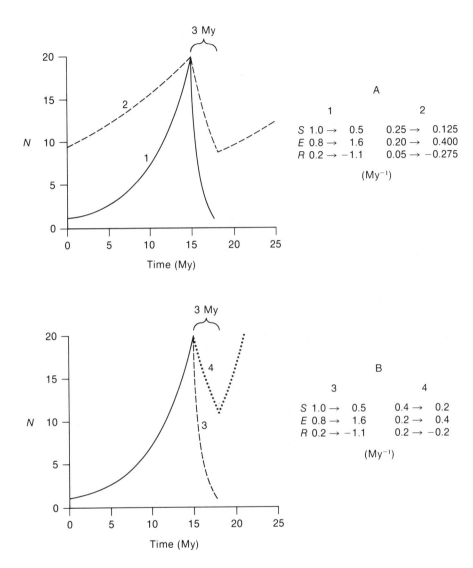

FIGURE 10-1
Evolutionary instability attending high rates of evolutionary turnover of species (high values of S and E). A: A hypothetical example in which initial values of S, E, and R are proportional in two taxa (four times as high in taxon 1 as in taxon 2). B: A hypothetical example in which initial values of R are identical, but result from high values of S and E in taxon 3 and low values in taxon 4. In both A and B after 15 My there occurs a 3 My interval of environmental deterioration, during which rates of speciation are lowered and rates of extinction are elevated in proportion to their original values. See text for fuller explanation.

speciation is shut off entirely. Then even if rate of extinction remains un-changed, a taxon with a high rate of extinction will die out quickly, whereas a taxon with a low rate will weather a brief catastrophe.

Another kind of comparison also illustrates how high turnover rate confers instability. This point relates to the fact that the correlation among values of R, S, and E cannot be perfect. To the degree that the correlation is imperfect, a characteristic value of R in adaptive radiation may result from high values of S and E or low values. A hypothetical example is represented in Figure 10-1,B, where we see that as in Figure 10-1,A, the taxon with high rates of speciation and extinction suffers a terminal decline when subjected to the effects of envi-ronmental deterioration. In this example, environmental deterioration results in a doubling of rate of extinction and a halving of rate of speciation. The other taxon radiates with the same value of R, but with this value resulting from lower rates of speciation and extinction. Though subjected to proportional changes in rate of extinction and rate of speciation, this taxon survives the crisis.

The analogous question in demography of how stability of population size should relate to birth and death rates seems to have received little attention in population ecology, although MacArthur and Wilson (1967, Chapter 4) have shown that probable time of extinction of a propagule colonizing an island increases with the ratio of birthrate to death rate. This conclusion is analogous to that of the preceding paragraph, where because R is held constant, the ratio $[S/E]$ increases with a decrease in the values of S and E.

It is certainly true that the ultimate explanation of vulnerability to mass extinction is to be found in an understanding of the controls of high rates of speciation and extinction, as discussed in the previous chapter. On the other hand, *a priori* knowledge that a group has a certain rate of speciation *or* a certain rate of extinction, even if this rate is to some degree understood in ecologic terms, tells us nothing. In order to predict the relative stability of a taxon, we must also know how its characteristic values of S and E relate arithmetically to each other and to characteristic values for other taxa.

We can extend this approach to nonradiating taxa. If, for example, we con-sider a taxon that has stagnated after radiation, so that rate of extinction equals rate of speciation, the higher these rates happen to be (the higher the turnover rate is), the more unstable will be the diversity of the taxon. If complete ecologic saturation (a condition never truly realized in nature) has resulted from competi-tive interactions, then new species will appear only when others die out, unless new habitats form or niches are constricted. In other words, extinction rate will govern speciation rate. Then, if characteristic extinction rate for a taxon is high, turnover rate will be high and the taxon will be inherently unstable, just as it will have been during its radiating phase.

Evidence of the fossil record is in excellent accord with such analysis. Of the groups portrayed in Figure 9-1 that would be expected to exhibit instability by virtue of possessing high rates of speciation and extinction, the birds and insects have too poor a fossil record for evaluation. The other such groups, however—mammals, trilobites, ammonites, and graptolites—are all famous for their high

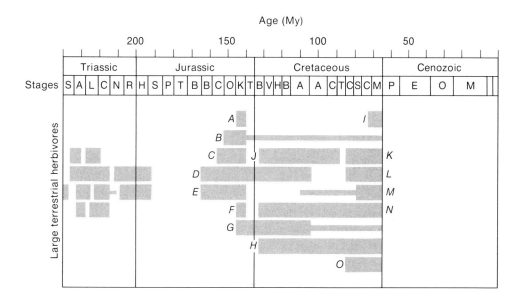

FIGURE 10-2
Ranges of Mesozoic families of large, terrestrial herbivores. Stages of each Mesozoic period are indicated by their first letters. Narrow segments of bars represent intervals in which families are present but very rare. *A*, camarasaurids. *B*, diplodocids. *C*, stegosaurids. *D*, brachiosaurids. *E*, cetiosaurids. *F*, camptosaurids. *G*, hypsilophodontids. *H*, panoplosaurids. *I*, ceratopsids. *J*, iguanodontids. *K*, hadrosaurids. *L*, protoceratopsids. *M*, titanosaurids. *N*, pachycephalosaurids. *O*, euoplocephalids. (From Bakker, 1977.)

frequency of mass extinction. We can take the mammals here as being representative of certain kinds of terrestrial tetrapods, and Bakker (1977) has presented evidence that tetrapods of large body size (heavier than 10 kilograms) suffered major extinctions at the ends of the Triassic and Jurassic periods, as well as during the well-known latest Permian and Cretaceous crises (Figure 10-2). He informs me that a mid-Cenozoic mass extinction of mammals can also be recognized.* The biomere pattern of trilobite evolution (Palmer, 1965a; Stitt, 1977), which was discussed earlier (Figures 9-7 and 9-9), is one of repeated worldwide mass extinctions at intervals of 5 My or so. Also as predicted, the history of the ammonites was punctuated by mass extinction. Figure 10-3 depicts instability for this group at the generic level. A quotation from Kennedy and Cobban (1976, p. 80) encapsulates the pattern:

> The evolutionary history of ammonites is depicted as one of alternate episodes of "explosive" radiation, continued evolution, and widespread extinction. These latter

---

* R. T. Bakker, Johns Hopkins University, 1978.

episodes generally correspond approximately with major stratigraphic boundaries, as at the Permain-Triassic, Triassic-Jurassic, and Jurassic-Cretaceous boundaries. The Cretaceous-Paleocene boundary, it may be argued, is a fourth such event, which the group did not survive.

The Graptoloidea (planktonic graptolites), which range from lowest Ordovician to mid-Devonian, exhibit a similar periodic pattern of explosive radiation and mass extinction. The most notable nonterminal mass extinctions seem to have come at the end of the Ordovician and roughly 12 My later, in the early part of the Wenlock (Rickards, 1977). In this group, the terminal mass extinction came quickly, after only about 130 My of evolution.

Standing in sharp contrast to these unstable groups, are the bivalve and gastropod mollusks, in which the famed Permian and Cretaceous marine extinctions took only moderate tolls, and in which no severe overall decline has occurred during an interval of about half a billion years. There can be no question that the biologic traits fostering low rates of speciation and extinction in these taxa have endowed them with considerable stability.

It should not be overlooked that what we call "mass extinction" can result from a decrease in rate of speciation, just as it can result from an increase in rate of extinction. In purely mathematical terms, if the two rates are equal, then change in each by a particular percentage will have the same effect on rate of diversification. Just as Thorson (1950) concluded that the larval stage of a marine invertebrate is the most vulnerable in ontogeny, we might argue, following a loose analogy, that speciation should be the most sensitive stage in the "life history" of a species (Stanley, 1977). A small population of the sort that might evolve into a new species is far a more fragile entity than a fully established, widely deployed species (page 198). This view is supported by data for the modest mass extinction of bivalves at the close of the Triassic (Nakazawa and Runnegar, 1973) and for the mass extinction of terrestrial tetrapods in general (Bakker, 1977). Compilation of stratigraphic ranges for genera in these groups reveals a marked failure in the production of new lineages at the critical times, but little increase in rate of extinction. Lowered rates of production of new genera (and, we may infer, species) seem also to have contributed heavily to mass extinctions of ammonites (Figure 10-3). Perhaps the particular sensitivity of speciation is exemplified by the history of North American reptiles during the late Cenozoic, when climates were cooling. Auffenberg and Milstead (1965) concluded that virtually no new reptile species evolved during the Pleistocene: rate of speciation declined nearly to zero. On the other hand, termination of long-ranging species was clearly a major component of mass extinctions of trilobites (Figure 9-6).

Certainly, additional factors must be considered in the evaluation of catastrophic extinction. The planktonic forams, which exhibit low rates of speciation and extinction (pp. 243–245), suffered almost total extirpation at the ends of the Cretaceous Period and Oligocene Epoch (Cifelli, 1969). In each case, this decline

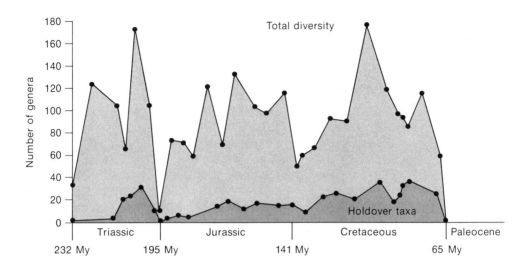

FIGURE 10-3
High rate of generic turnover in the evolution of the Ammonitina. (From Kennedy, 1977.)

was but one aspect of a major biotic breakdown in the pelagic realm, evaluated in a highly significant paper by Fischer and Arthur (1977). The fact that planktonic forams are vulnerable to mass extinction seems to relate to their occupancy of a habitat that is subject to periodic deterioration. Fischer and Arthur offer evidence that the pelagic ecosystem has collapsed many times during the Phanerozoic, with a periodicity averaging about 30 My.

That evolutionary instability is an unavoidable consequence of high rates of speciation and extinction is, I believe, one of the most important conclusions of the kind of exponential analysis advocated here. It is no accident that, in basic pattern of macroevolution, the ammonites, trilobites, and graptolites resemble terrestrial tetrapods of large body size. Here, too, we may find a hint as to why many primitive taxa persist over long intervals of geologic time. Their low rates of speciation and extinction endow them with an inherent resistance to mass extinction.

Confirming Cope's venerable observation, Bakker (1977) has shown with much more thorough analysis that in the terrestrial realm small tetrapods have tended preferentially to pass through the filter of mass extinction (Figure 10-2). The preceding analysis suggests at least a partial explanation. First, we can recall that all species of European mammals considered to have records extending back more than about 3 My, into the Astian, are very small forms (page 79), despite the fact that large mammals tend to be better preserved. This finding represents strong evidence that geologic longevity is inversely related to body size: Taxa of small animals normally exhibit low rates of extinction. The high value of R for the murid rodents (Figure 5-2) suggests that small mammals may also, on the average, speciate more rapidly than large mammals. (In fact, a difference in S would exist even if R were the same for small and large mammals, so long as E is higher for the former.)

Thus the ratio $[S/E]$ is elevated for clades of small mammals and must render them relatively stable. This condition can be interpreted in light of the geographic controls of S and E evaluated in the previous chapter. Among species of mammals in general, as body size increases, average population size decreases quite markedly. This relationship is partly a simple consequence of the existence of the trophic pyramid of numbers (although not all small mammals occupy low trophic levels) and partly a result of the increased requirements for food and space that accompany increase in body size. Given the obvious importance of population size to lineage survival, it is no surprise that small mammals exhibit lower average rates of extinction than large animals. On the other hand, we would expect rate of speciation to be higher for small animals because of weak dispersal. McNab (1963) showed that home range size of mammals increases, on the average, with body size raised to the power 0.63. He related this condition to metabolic requirements. In fact, large mammals experience the environment in a fine-grained fashion, facing relatively few barriers to dispersal. Thus, a typical species of large body size will contain relatively few individuals, but they will disperse rather readily over large areas. While its small numbers will render such a species vulnerable to extinction, its ease of dispersal should suppress rate of speciation. In contrast, while an average species of small body size should be relatively resistant to extinction by virtue of large total population size, its demes will occupy more restricted regions. In a complementary way, these requirements would seem to explain why subpopulations of small mammals tend rather easily to become segregated sufficiently for speciation.

In short, we would predict that, with an increase in body size, rate of speciation and rate of extinction will tend to converge (or at least the ratio $[S/E]$ will decrease). This prediction may account for the tendency of higher taxa of large body size to decline in diversity suddenly in the event of environmental deterioration.

It should be borne in mind that mass extinction actually represents one end of a continuum. After diversifying, a clade may decline slowly or rapidly. Very rapid decline constitutes mass extinction, but the phrase remains ill-defined. If, as must often happen, conditions deteriorate slowly, as with the gradual proliferation of new predators that inhibit speciation and accelerate extinction, the temporal pattern of diversity will resemble that represented in Figure 5-15,A, rather than in Figure 10-1, in which an abrupt reversal of conditions is depicted. We can envision this more subdued pattern of decline as resulting from a progressive decay in the rate of speciation and/or growth in the rate of extinction, as opposed to the sudden switch in values depicted in Figure 10-1. The assumption of the foregoing analysis is that, whether their onset is sudden or gradual, changes in rates will be proportional to original values.

RANDOM FACTORS

Recently a group of workers (Raup *et al.*, 1973; Gould *et al.*, 1977) has conducted computer simulations of the diversification and decline of clades. Basically, the computer program employed permits artificial clades to expand and contract by the random fluctuation of rates of branching and extinction about mean values (probabilities). This research group finds some discrepancies between real and randomly generated artificial clades. Their general conclusion, however, is described as follows (Gould *et al.*, 1977, p. 32): "How different, then, is the real world from the stochastic system? ... The answer would seem to be 'not very'—the outstanding feature of real and random clades is their basic similarity." This similarity is not taken as a definite indication of stochastic origins for real patterns, but even the alleged resemblance is open to question.

A problem with the simulations published to date is that they have not been scaled to match the real clades to which they have been compared. For example, certain artificial clades may resemble real clades for genera within orders of brachiopods (Gould *et al.*, figure 1), but the shapes of the artificial clades have resulted from extremely low standing diversities (averaging only about 4 taxonomic units per ordinal clade), which have tended to undergo rapid fluctuations. In fact, standing diversity for genera within a real order of brachiopods has averaged about 25. In stochastic processes, "population" size is of prime importance, and much less severe random fluctuations are to be expected for a mean standing diversity of 25 genera than for the mean simulated diversity of only 4. Similarly, artificial clades may resemble real clades for families within orders of reptiles (Raup *et al.*, 1973), but here the shapes of the artificial clades have resulted from unnaturally high probabilities of branching and extinction (0.037 My^{-1} compared to a real value that can be estimated to be approximately 0.020 My^{-1}). A third problem is that the use of higher taxa (genera and families) as units of diversity has also introduced artificial instability to simulations by making large steps of change more probable than if species had been used as units. Simulations that two colleagues and I have conducted at the species level using empirically scaled parameters show much less dramatic fluctuations than are found in the random simulations published to date. These considerations leave unshaken my belief that rapid radiation and decline of large clades normally relate to nonrandom causal factors: adaptive innovations and the agents of species selection.

OPTIONS FOR MAINTAINING DIVERSITY

The fact that rates of speciation and extinction vary markedly among taxa raises a question as to whether something akin to the so-called life-history "strategies" of interbreeding populations may develop at the higher level of species selection.

I will conclude that no close analogy exists, but consideration of the matter raises interesting points about the nature of selection among species.

A traditional view of population biology has been that some species are "*r*-strategists," which survive opportunistically by virtue of high fecundity, and others are "*K*-strategists," which sustain their populations near the saturation point by virtue of superior competitive abilities. The idea has been that energetic limitations tend to prevent species from combining high fecundity with competitive superiority. One theoretical problem with seeking an analogy here is that no energetic trade-off exists at the level of the clade. There is no reason to assume *a priori* that a taxon in the early stages of radiation cannot be characterized by both rapid speciation and long mean lineage duration (Gould and Eldredge, 1977). Another problem with seeking an analogy here is that in recent years the concepts of *r*- and *K*-selection themselves have been widely ridiculed (see review by Stearns, 1977). The more appropriate way to view selection for particular values of life history parameters is simply to recognize that fitness will be maximized. Depending upon such things as the value of parental investment and the pattern of mortality during ontogeny, high or low fecundity may be favored, as may concentration of reproductive effort either early or late in ontogeny.

At the level of species selection, "strategies" are less complicated. Once species are well established, it is unlikely that *on the average*, their speciational "fecundity" changes greatly until extinction approaches. The same can be said for probability of extinction; we do not know how this changes during the existence of an average species (Raup, 1976b; Sepkoski, 1976), but once a species is fully established, the probability is not likely to vary enormously (Van Valen, 1973), at least in comparison to the markedly nonlinear patterns of individual survivorship that typify many species.

A general limitation of the analogy is that a typical clade is characterized by many fewer speciation and extinction events during its lifetime than the number of births and deaths that occur within a species. This condition, when combined with the weakly varying probability of extinction and rate of species production during the lifetime of a fully established species, must preclude the evolution at the cladal level of anything equivalent to the complex and subtle life history "strategies" that are observed within populations. What remains is quite simple: We should expect species selection to increase the value of R. This, in fact, is the fundamental tendency of species selection, which, by definition, tends to increase the equivalent of fitness within a clade (Figure 7-3): *It maximizes rate of speciation and minimizes rate of extinction*, to the degree that these rates are heritable (to the extent that they are determined by genetically controlled phenotypic features).

As was indicated in the previous chapter, rate controls do indeed seem to relate to inherent biologic traits. On the other hand, it was concluded that the very traits that tend to increase S also tend to increase E. In other words, S is seldom relatively large for a taxon if E is relatively small (Figure 9-1), even though there is no energetic trade-off precluding the evolution of divergent values. To the extent that steady-state saturation by species exists, there is direct linkage between S and E, but I have argued that such linkage is minimal for some taxa. Certainly, it is especially weak for radiating clades, upon which Figure 9-1 is based. In the preceding chapter, it was suggested that much of the linkage of S and E is indirect and fortuitous, reflecting the fact that speciation and extinction happen to be controlled by similar factors, including dispersal ability and behavioral complexity.

The coupling of rates of speciation and extinction obviously has a considerable damping effect on the tendency of species selection to maximize S and minimize E. There must be some play in the system, however, and taxa in which the linkage has somehow been weakened—taxa in which a relatively high $[S/E]$ ratio has evolved—we might term **supertaxa.** As discussed above, groups of small mammals may qualify for such recognition. Possibly, a high ratio also obtains for the modern snakes (Colubridae), which are radiating within a new adaptive zone at an extraordinary rate. Although mean species duration is not known precisely, it must be greater than in the Mammalian, and yet R is higher in the Colubridae than in an average family of mammals (page 258). Snakes resemble small, fossorial mammals, and perhaps are amenable to the kind of explanation offered above for small mammals. The cichlid fishes may also represent a supertaxon. Karel Liem informs me that not only do cichlids tend to speciate at extraordinary rates after gaining access to vacant habitats (see page 45),* but there is some evidence that they also tend to displace established species of other families when introduced to occupied territory. It seems likely that the remarkable adaptive flexibility of the pharyngeal jaw apparatus of the cichlids (page 103) confers upon this group both a general competitive superiority and the capacity for rapid evolutionary expansion. Whether competitive dominance actually extends the longevity of cichlid lineages is uncertain, but it seems probable that the group's average rate of extinction is at least no higher than the average for other fishes, indicating that its $[S/E]$ ratio should be unusually large.

Given the general correlation between S and E, the primary variable for cladal "strategy" will be the magnitude of these coupled variables. In other words, the primary question is whether turnover rate will be high, low, or intermediate. A number of factors must be considered. An advantage of possessing high values of S and E, as we have seen (Chapter 5), will be greater frequency of adaptive breakthroughs and, more generally, rapid evolution. I would suggest that a purely numerical advantage of the same kind may have been a factor in the Late Cretaceous and Cenozoic ascendancy of the animal-pollinated angiosperms. The formation of ethological barriers represents a major isolating mechanism for angiosperms (Grant, 1971, pp. 74–75). It must be true that plants which are pollinated and dispersed by animals have been able to occupy habitats unavailable to wind-pollinated taxa (Regal, 1977), but the simple numerical advantage conferred by accelerated rates of speciation must not be overlooked.

As already discussed, a disadvantage of a "strategy" involving rapid turnover will be inherent vulnerability to mass extinction. The consequences may be seen

---

* Karel Liem, Harvard University, 1976.

in the history of the trilobites, graptolites, and ammonites, all of which suffered enormous cladal instability and, finally, went extinct.

The kind of trade-off that typifies the evolution of high or low turnover rates may be illustrated by the relative merits of planktonic and nonplanktonic larval development in the macroevolution of the Gastropoda. Hansen (1978) has uncovered evidence that within the Eocene Volutidae, species with nonplanktonic larvae had narrower geographic ranges than species with planktonic larvae. As would be predicted from the discussion of dispersal ability in the previous chapter (Jackson, 1974), they seem also to have undergone more rapid speciation and extinction (Figure 10-4). Yet since the Eocene, nonplanktonic larval development has become fixed in the Volutidae: All species now have it! One potential explanation for this trend would be that it developed by species selection. Another would be that it resulted from phyletic evolution (transition *within* lineages). Still another would be that directed speciation yielded a general shift to the nonplanktonic mode. (It was easier to shift from a planktonic state to a nonplanktonic state than vice versa.) The prevalence of planktonic species in many other families argues against the third possibility: There seems to be no inherent directional tendency at work in this aspect of gastropod speciation. A way of testing ideas of this sort would be to determine the detailed pattern of phylogenetic trends. For directional speciation or phyletic evolution to have produced the general pattern, switching from one mode of development to the other would have had to be frequent. We would therefore expect to find many clades (genera) displaying the transition. On the other hand, if few genera altered to the nonplanktonic condition, and the trend was accomplished by the differential expansion of clades with nonplanktonic larvae, then species selection would be indicated.

What would, of course, be generally implied by the fixation of the nonplanktonic mode through species selection is that the difference in net rate of speciation, which favored nonplanktonic species, more than offset the difference in rate of extinction, which would have worked in the opposite direction. The eventual elimination of the planktonic mode would indicate that net rate of extinction here exceeded net rate of speciation. Figure 10-4 hints that this form of species selection may indeed have been operating. Numbers of species and species ranges are, of course, incompletely known, but for at least the latter part of the interval depicted, taxa with nonplanktonic larvae increased in diversity (note that seemingly noncontemporaneous clusters of species of the Upper Eocene must, in fact, have overlapped in time). Among lineages with nonplanktonic larvae, speciation seems to have done little more than offset extinction throughout the interval depicted.

It is reasonable to inquire why there has clearly been a net increase in behavioral complexity in the history of life if, indeed, either high values of both S and E or low values will suffice for survival of a taxon. Perhaps the key factor is that species interact during evolution, and in doing so alter each others' rates of speciation and extinction. Often, a relatively primitive taxon that had earlier radiated successfully has declined because of an unfavorable competitive or

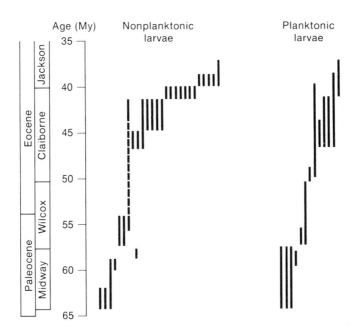

FIGURE 10-4
Recognized stratigraphic ranges of species of volutid gastropods
in the lower Cenozoic of the Gulf Coast of the United States.
(After Hansen, 1978.)

predatory interaction with other, behaviorally more advanced taxa. In the long
run, the cumulative effect of events of this type has probably been to increase the
average values of S and E in the animal world, but this has happened because
advanced behavior has won out, not because high (coupled) values of S and E
have been advantageous in and of themselves. If pollination by animal vectors
could be shown to accelerate both speciation and extinction in angiosperms, a
parallel argument might emerge for plants.

While there has been a general tendency for coevolution to advance the level of
behavioral complexity, many ethologically simple taxa survive today. The presence of some forms of this type is to be expected, however, because not all
evolutionary solutions to problems imposed by the behavioral advancement of
interacting taxa will be behavioral, and even if they are, they may not represent
the complication of behavior. Species selection operates on a wide range of traits.

ELASTICITY AND HIERARCHIES OF CLADES

What happens to a taxon after being decimated by mass extinction? In Chapter
7, it was pointed out that the nature of species selection must differ among

higher taxa according to the relative impact of competition and predation upon component species. These interactions should also exert an influence upon evolutionary response to mass extinction. The basic point here is that for a taxon in which interspecific competition is severe, mass extinction removes the major constraint on diversification. Following a sudden extinction, we would expect diversity to rebound rapidly. I have referred to such a taxon as having an **elastic** quality (Stanley, 1977). This metaphor would seem to fit the scheme of MacArthur (1969) who observed that there should be no true saturation of habitats in nature. Rather, he suggested, diversification should proceed at progressively lower rates, as if filling a balloon.

In Chapter 7, birds and mammals (especially top carnivores) were described as taxa that biologists have viewed as undergoing intense competitive interactions. These groups may be hypothesized to have diversified elastically. We can also predict the presence of an elastic quality within invertebrate groups that are characterized by (1) advanced behavior, which tends to intensify competitive interactions or (2) a position high in the food web, which decreases the likelihood of being heavily preyed upon.

In contrast, for a taxon in which predation and other forms of disturbance represent primary controls of population density, mass extinction will not necessarily be followed by an increase in rate of speciation. If the primary predatory groups and agents of disturbance remain, it will be no more likely that new species will take hold after a mass extinction than before. Taxa of this sort can be viewed as inelastic, in that before and after major extinctions, they tend to expand at some typical value of R. Though they cannot be perfectly inelastic, they may never have reached a condition in which elastic resistance to continued radiation is particularly noticeable. Their expansion will have been less sigmoid (more purely exponential) than that of elastic taxa.

It is difficult to test the idea of elasticity in the fossil record. One problem is that mass extinctions have seldom been confined to particular taxa. Unfortunately, the marine fossil record in the temporal vicinity of mass extinctions is relatively poor, owing to the coincidence of the extinctions with regression of the world oceans from continental shelves. Thus, for some taxa it is difficult to assess trends in diversity through critical intervals of time. For the Bivalvia, which I would predict to have evolved inelastically (page 200), Nakazawa and Runnegar (1973) report that at least 46 genera existed in the early part of the Early Triassic, 19 being known directly from Lower Triassic fossil occurrences and the rest being found in both older and younger deposits, which establishes their presence early in the Triassic. The meaning of the increase to at least 146 genera in the Late Triassic, representing radiation during 25 My, is difficult to assess. First, considerably more than 46 genera may have existed at the outset. Second, reliable data at the species level are inaccessible. Finally, important predatory groups may also have gone extinct in the crisis, and the possible effects of such eliminations on speciation, being unknown, cannot be distinguished from any effect of release from competition.

The ammonites, having been predators with advanced behavior, would be expected to have radiated elastically. They nearly disappeared at the end of the Permian. The extraordinary rebound of this group from perhaps three or four species to 154 during the 7 to 10 My Scythian Stage (Early Triassic) might be seen as bearing out the prediction of elasticity. On the other hand, the high value of R here (page 266) may simply reflect normally rapid speciation associated with complex behavior because, as noted in the previous chapter, behavior constitutes an important isolating mechanism. In other words, intense competition may have a compound effect, operating both on the mechanism of speciation and on the opportunity for speciation. There is no simple way to separate the two components.

There is, however, an entirely different approach that may shed light on elasticity, while revealing other interesting features of large-scale evolution as well. This is the comparison of net rates of diversification at various taxonomic levels within a higher taxon. The Mammalia and the bivalve mollusks serve as contrasting examples. Recall that the mean value of R is much higher for families of mammals than for families of bivalves, and there is no overlap in values of R for the two groups (Figure 5-2). Let us consider clades within either class. From equation 5.1, one can see that if all clades have radiated with the same value of R, a semilog plot against geologic age of the present diversities of clades will yield a straight line. Figure 10-5 represents such a plot for genera and families of bivalves. Note that the familial data conform rather well to a straight line and that generic data fall along the same line.

Figure 10-6 is the corresponding graph for mammals. The points for all families but the murid rodents (Old World rats and mice) form a rather linear plot. The murids have radiated with exceptional rapidity. What is striking is that many single genera of mammals have speciated much more rapidly even than the murids. Points for several newly radiating genera of the Bovidae (cattle, antelopes, and the like) fall well above the line for families. The ages of the genera are not precisely known, but the plotted figures cannot be grossly in error. Of course, other extant genera have radiated less rapidly. Some are monospecific. Generic clades are not well enough documented that it is generally possible to discriminate between living genera of low diversity that have radiated slowly and genera that are on the decline after an initial radiation. Even so, it is clear from Figure 10-6 that many genera of mammals have radiated very rapidly in comparison to rates for entire families. That this contrast is not simply a matter of stochastic fluctuation reflecting the small size of generic clades is indicated by the absence of a comparable disparity in the Bivalvia (here only modestly greater scatter is observed). I can find no evidence that any genus of marine bivalves has radiated to a diversity of 10 species or so at a rate several times higher than the normal rate for families. Many nonbovid genera of the late Cenozoic obviously fit the pattern shown in Figure 10-6, although some have suffered recent extinctions. Examples are *Felis*, *Canis*, and *Hyaena*.

A bit of reflection will reveal that the difference in values of R at the genus

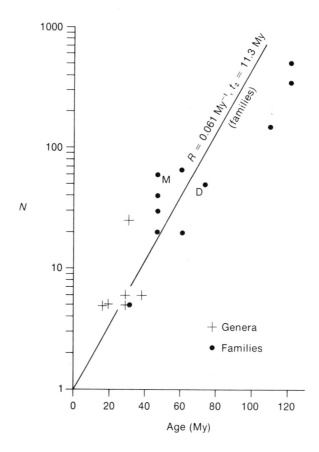

FIGURE 10-5
Plot of number of species (*N*) versus interval of radiation
for extant radiating clades of the Bivalvia. Diagonal line
depicts the rate of radiation (*R*) and doubling time (t_2) for
an average familial clade. M—Mesodesmatidae;
D—Donacidae. (Data from Stanley and Newman, 1979.)

and family level for mammals must reflect a basic inhomogeneity in phylogeny.
As depicted schematically in Figure 10-7,A, many genera must represent rather
discrete clades—pulses of rapid radiation within the phylogeny of families.
Values of *R* are much lower for families because barren areas exist between
generic clades. The implication is that mammalian genera are often natural
entities. This condition, in fact, seems manifest in the general taxonomic agreement that exists among experts who classify mammals into genera.

The marine Bivalvia present a different picture altogether (Figure 10-5). The
comparison with mammals in particularly interesting because the two groups

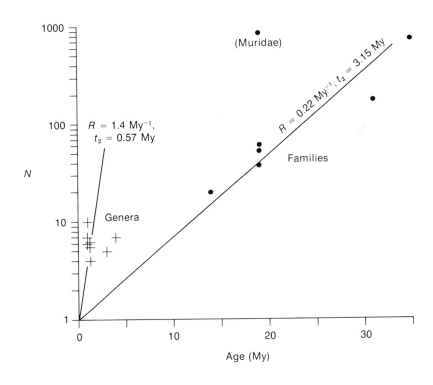

FIGURE 10-6
Plot, like Figure 10-5, for mammals. Generic data are for rapidly radiating
bovid genera. (See Stanley, 1977.) Data for families from Figure 5-2.

contain roughly the same number of living species (about 3,900 species of placental mammals compared to about 5,300 species of marine bivalves) and the
species in each class are lumped into roughly 100 families. In each class it is
quite clear that families are for the most part discrete, natural clades. An average
family of bivalves, however, lacks a hierarchical structure comparable to that of
mammals. Here, the fact that values of R for radiating genera resemble those for
families implies that the fabric of phylogeny within a familial clade is relatively
homogeneous (Figure 10-7,B). This condition is reflected in the confused state of
generic taxonomy, even within the living Bivalvia. Presumably, then, many
formally recognized genera are polyphyletic. This condition would bias values
of R in an upward direction, which strengthens the empirical observation. (Note
that the assumption in all calculations has been that $N_0 = 1$.) The gastropods
resemble the bivalves, both in similarity of R-values for genera and families
(Figure 9-4) and in muddled generic taxonomy.
    It is not immediately obvious why within the evolution of mammals there
occur distinctive pulses of radiation embodied in genera. A reasonable hypothe-

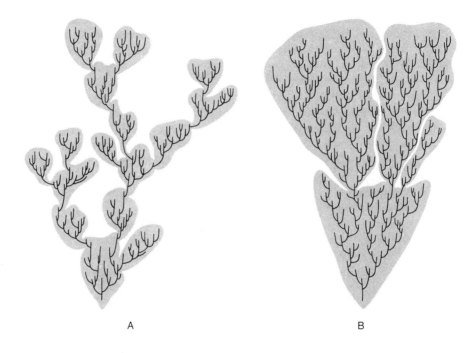

A   B

FIGURE 10-7
Inferred patterns of phylogeny. A: Pattern for a typical family of mammals. Most genera (outlined) represent discrete pulses of radiation with values of R that are higher than the net value for the family. B: Pattern for a typical family of the Bivalvia. Most genera do not represent very discrete clades and do not represent pulses of radiation.

sis would be that, owing to behavioral interactions, particular geographic regions can accommodate only small numbers of congeneric species. Today an average genus contains only three or four. Kohn and Orians (1962) noted that there are 34 genera of North American mammals in which all continental species are allopatric, and that extensive abutments of species' geographic ranges exist in 20 additional genera, suggesting competitive exclusion. Perhaps what typically happens within a family is that when a distinctive new genus evolves, it speciates initially at an extremely high rate, until interactions (especially interference competition) impede further diversification. Such a pulse should occur only occasionally, when a distinctive adaptation opens a new adaptive zone, producing what we recognize as a new genus. This explanation is based in part on the weak dispersal abilities of mammals in the heterogeneous terrestrial realm that confine many generic radiations to small geographic areas. The pattern displayed in Figure 10-6 is consistent with the conclusion that most new genera of mammals arise rapidly, by divergent speciation (Chapter 4).

Presumably, the relative weakness of cladal clustering within marine bivalve families reflects, in part, the typically great powers of dispersal of bivalves. This condition contrasts to that suggested for families of mammals, which seem to spread haltingly, as component genera radiate within relatively small regions bounded by geographic barriers. In addition, I would suggest that competition within the Bivalvia is too weak for saturation often to be approached even where dispersal is somewhat restricted (page 200). In other words, discrete small bursts of radiation are not to be expected within large clades of marine Bivalvia. Bivalve families are generally discrete natural clades. What typically has led to the decline of certain other families is uncertain, but my view would be that evolutionary advances in predation have been significant (page 200). For some groups, mass extinction by other agencies may also have had an impact.

Further evaluation of the bivalve record at more than one taxonomic level yields an important conclusion concerning the meaning of changes in diversity. This is that within a class, cessation in the diversification of subclasses and orders is by no means an indication that diversity has stagnated at the level of the genus or species. A prime example would be the leveling off of numbers of orders of marine invertebrates at the end of the Ordovician (Sepkoski, 1978). It is relevant here that all of the extant subclasses of the Bivalvia, as delineated by Pojeta (1971), were established early in the Paleozoic. Even so, families (Figure 5-10,B), and probably genera, have continued to proliferate to the present day. I would suggest that the early cessation of subclass production was largely a matter of morphologic specialization: Once in place, essential, primitive features were not easily altered. Since the early Paleozoic, the bivalve adaptive zone has expanded in some remarkable ways, as with the origin of the coral-like rudist bivalves and the huge tridacnids, but enough fundamental features of ligament, shell, and hinge construction have persisted that no new subclass is recognized. It would seem that complex, integrated morphologic systems have not easily broken down even during enormous and persistent radiations.

DIVERSITY GRADIENTS

Much has been written about the origin and maintenance of taxonomic diversity gradients. My aim here is not to review this entire subject but to inject into it some of the techniques and inferences of the preceding pages. Ultimately, diversity is the product of rate of speciation and rate of extinction, which makes it amenable to the kinds of analysis presented in this book.

Diversity gradients, whether latitudinal or local in scale, have traditionally been evaluated in two different ways. The two approaches are not mutually exclusive, and, in fact, are not clearly separable. Both are evident, for example, in the analysis of Fischer (1960). The first might be termed the saturation approach. Here it is asserted that diversity gradients reflect the fact that more species can exist in one area (for example, the tropics) than in another (for example, the arctic). The other approach emphasizes realized rather than potential diversity. Here the balance between rate of immigration or speciation and rate of emigration or extinction is considered.

It is the second approach that will be pursued here. It is important to recognize, however, that the saturation approach can be viewed as a variant of this one if it is accepted that complete saturation never occurs. We can recall here MacArthur's analogy between diversification within a habitat and the filling of a balloon. This view emphasizes constraints upon diversification but does not exclude evaluation in exponential terms. In population biology, the analogy would be with the substitution of a sigmoid logistic curve, describing the damped exponential growth of a population under conditions of crowding, for a purely exponential curve reflecting unbridled increase. In the past, the fundamentally exponential nature of diversification has been overlooked by many workers. The ideas to be discussed here relate to increase that can be approximated by purely exponential equations. In other words, the concept of saturation is not considered.

Slobodkin and Sanders (1969) have presented an important model for the origin of diversity gradients that has perhaps been too little appreciated. In part, their scheme depicts low rates of speciation into hostile environments, such as estuaries with fluctuating salinities or high-latitude regions with cold temperatures. Speciation from hostile habitats to equable habitats occurs at a higher rate. The imbalance contributes to the existence of lower diversity within the hostile regions. This is not the only feature of the Slobodkin-Sanders model, but it is the one that will be examined here.

In particular, I will consider an environment that is familiar to nearly everyone—the marine sandy beach habitat, by which I mean the wave-ridden shoreface that lies exposed to the open sea, and also the contiguous zone that extends to a depth of a meter or so below the low tide mark. C. M. Yonge (1950) observed the great difficulty of living here. Invertebrates must find protection within sand, but this is not enough. They must be particularly rapid burrowers for efficient reburial, because frequent dislodgment by waves is inevitable and the results can be disastrous. Waves carry small objects up onto the beach and, at least in the modern world, predators are quick to victimize exposed individuals.

The marine sandy beach today supports only a sparse variety of macroinvertebrate species. Clearly, the habitat has the potential to support many species, yet much evidence suggests that existing communities represent early stages of diversification. It is significant that populations on the shoreface come and go. At any time, long stretches of sand are largely barren of certain species, while others support dense populations. Local degree of colonization often changes from year to year. This condition suggests that nothing approaching evolutionary saturation has occurred. Furthermore, throughout the world only a few families of invertebrates are represented. Most of these, it can be shown, are in the early stages of generally unbridled radiation.

Consider, for example, the Gastropoda. Most snails of the sandy shore are elongate forms, like terebrids and olivellids. They are chiefly mobile neogastropod predators, which have radiated only during the last 150 My or so. Following the loss of the right gill (Linsley, 1977), the evolution of an elongate shell aperture seems to have triggered the radiation of elongate neogastropods (Vermeij, 1977).

The sandy beach is also occupied by only a few varieties of bivalves, all of which are modern (post-Jurassic) taxa. The most prominent are wedge clams (the Donacidae and the Mesodesmatidae, which are closely convergent with them) and, lower on the shoreface, the surf clams (certain mactrids and the convergent venerid *Tivela*). There is no sign that the radiation of such forms has abated. The Donacidae and Mesodesmatidae have multiplied at rates closely resembling those of other presently radiating bivalve taxa (Figure 10-5). The evolution of efficient burrowing required a long earlier interval of bivalve evolution. A critical step was the evolution of mantle fusion and siphons, which was hardly underway before the late Paleozoic (Figure 10-8). It was apparently not until the Cretaceous that bivalves conspicuously occupied the shoreface. *Tancredia*, the initial invader, may have made the transition as early as the Jurassic.

Among the echinoids, the only sandy shore dwellers are the familiar sand dollars. These did not evolve until the beginning of the Cenozoic. Their rates of radiation, which are slightly higher than those of bivalve and gastropod taxa, are depicted in Table 9-1.

Conspicuous macrofaunal arthropods of the modern sandy shore are also of relatively modern origins. The dominant form is *Emerita*, the mole crab, which cannot have originated much before the Cenozoic. *Callianassa* and its close relatives have formed conspicuous fossil burrows in sandy shore deposits only since the Jurassic or Cretaceous.

It is important to recognize that morphologic features of the sort that characterize modern denizens of the sandy shore are unknown in groups of the Bivalvia, Gastropoda, and Echinoidea that lived before mid-Mesozoic times. All of these skeletonized classes required almost half a billion years to evolve the requisite adaptations. Furthermore, preserved burrows are quite rare in sandy shore deposits older than mid-Mesozoic. Possible exceptions are the *Skolithos* burrows that were formed as long ago as the earliest Cambrian or latest Precambrian. Some of these occur in deposits that may represent exposed sandy shores.

Considering the incipient stages of radiation represented by clades of the modern sandy shore, there is no question of habitat saturation, or equilibrium, at the present time. Given the rates at which species are being added and the generally exponential nature of increase, it seems evident that upon returning to earth 10 or 15 My hence, we would find species diversity of these clades to have approximately doubled.

Incipient stages of radiation must be represented today not only within newly

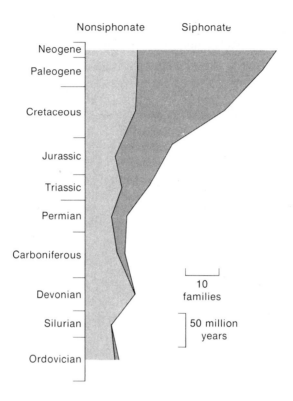

FIGURE 10-8
Increase in number of siphonate families of the Bivalvia since late Paleozoic time. (From Stanley, 1977.)

invaded habitats, but also within habitats that in the recent past have been partly
vacated by mass extinction. Both on the land and in the sea, major mass extinctions have commonly taken place at intervals in the order of 30 My (Fischer and
Arthur, 1977; Bakker, 1977). This number is not presented here as a necessarily
consistent value, but as an indication of order of magnitude. Taxa like bivalves,
gastropods, echinoids, and planktonic forams, with doubling times in the order
of 10 My, if affected by such extinctions might be expected never to diversify to
the point where exponential increase is slowed by competitive interactions, at
least by such interactions within these groups themselves. Taxa like mammals
are more likely to approach some sort of equilibrium. Webb (1969, 1976) has
argued in favor of widespread equilibrium for late Cenozoic mammals. Whether
or not this condition has obtained, here again we seem to observe the elastic
pattern of the large-scale evolution of mammals.

What at least partly offsets the contrast between explosively radiating and sluggishly radiating taxa is the differences in stability evaluated in the previous section. Because explosively radiating taxa are inherently unstable, or vulnerable to mass extinction, even though they have the potential to approach equilibrium rapidly, their diversification is frequently interrupted by mass extinction. The point is well illustrated for trilobites in Figure 9-7. In the vicinity of the Late Cambrian, they suffered worldwide mass extinctions at intervals of about 5 My! Can we reasonably entertain the idea that the trilobite faunas terminated at the upper limits of biomeres were approaching equilibrium? The absence of any such condition is suggested by the occurrence of progressive diversification throughout most of the interval spanned by a typical biomere.

It is also important to bear in mind that even where evolutionary stagnation of a taxon can be documented, there is not necessarily any reason *a priori* to invoke competitive interactions as a cause. Predation can also serve as a braking mechanism. Many higher taxa that have radiated initially in the absence of severe predation (and, in fact, may have radiated in part because of this condition) may later, in effect, be overtaken by newly evolving predators that suppress speciation and/or accelerate extinction. Earlier in the chapter, this kind of interaction was invoked as a partial explanation for the decline of certain bivalve taxa. Still, it should not be forgotten that less severe predation can have the opposite effect, relieving competition and fostering diversification (page 204).

By no means do I mean to argue that diversification is never slowed by internal crowding, but it seems evident that equilibrium has received more theoretical emphasis than it deserves, given our knowledge of the frequency of extinction and of the rates of radiation that typify major taxa. It has been widely debated whether worldwide diversity of marine benthos has changed greatly since the mid-Paleozoic (Valentine, 1970, 1973; Raup, 1972, 1976a; Bambach, 1977). Given the occurrence of Mesozoic and Cenozoic mass extinctions at intervals in the order of 30 My, we must ask whether even a demonstrable appearance of constant diversity would have significance with respect to questions of species packing or ecologic equilibrium.

Under similar scrutiny, the origins of latitudinal diversity gradients may be somewhat less puzzling. Regions at high latitude are equivalent to sandy beaches and estuaries in their tendency to act as speciational filters for taxa that evolve in warmer regions, yet as Slobodkin and Sanders (1969) observed, invasion by way of speciation in the opposite direction is less rapid. It should be noted that the interesting scheme proposed by Stehli, Douglas, and Kafescioglu (1972) for the origin of latitudinal diversity gradients in modern planktonic forams amounts to this kind of partial application of the Slobodkin-Sanders model. The idea here is that new kinds of planktonic forams have evolved in the tropics and only with difficulty have then taken speciational steps in the direction of the poles.

There is, however, another important factor. Fischer (1960) plotted diversification within a latitudinal belt as a convex-upward curve, reflecting a decrease in rate with time. If we alter such a curve so that it has a sigmoid shape reflecting exponential diversification, then a taxon in a cold region may never move beyond

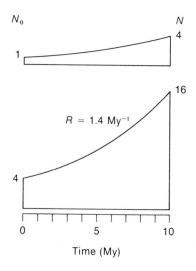

FIGURE 10-9
Diversification of two faunas at the
same exponential rate (R = 1.4
My^{-1}), producing a much higher
arithmetic rate of increase (12
species in 10 My for a fauna initially containing 4 species) than the
arithmetic rate (3 species in 10 My)
for a fauna initially containing only
1 species.

the "tail" of the potential curve, in which increase is roughly exponential but,
arithmetically, very slow (Figure 10-9). What keeps the taxon at this incipient
stage of radiation is in part reflected in the relative youth of modern temperate
and arctic biogeographic regions. We now know that during much of the
geologic past, tropical and subtropical conditions have covered a very high
percentage of the earth's surface. The vast cold-temperate and arctic regions of
the present world are the product of late Cenozoic cooling. Thus, the time-
honored view that regions of warm climate have served as centers of radiation
(Darlington, 1959) tends to lose meaning: During much of geologic time, it has
been warm nearly everywhere.

Later in the chapter, it will be shown that geographic factors unrelated to
temperature have tended to govern rates of extinction and, probably, speciation
for Cenozoic marine Mollusca. If this conclusion proves to hold for other taxa as
well, then an unusually high rate of speciation may not be the primary cause of
high diversity in the tropics. Here, of course, we are considering high *fractional*
rate (S). What the preceding discussion asserts is that total arithmetic rate (*num-*

ber of species added per unit time) is high within the tropics, partly because the tropics, unlike colder climatic zones, never shrink drastically. Our tendency to think in arithmetic terms has perhaps caused us, at times, to view radiations actually characterized by equal values of R (or doubling times) as having occurred at highly disparate rates (Figure 10-9).

Certainly, also, one of the difficulties of immediate, wholesale occupancy of newly formed cold regions is the delay caused by the difficulty of evolving fundamental adaptive specializations required to live there. Before or soon after certain breakthroughs have evolved under cold conditions at high latitudes, these regions must often have turned largely warm-temperate, shifting biotas poleward and stifling incipient cool temperate and arctic radiations at early stages of exponential increase.

The preceding analysis concerns the impact on diversification of the timing and frequency of entry into a habitat. In effect, for equation 5.2 (page 104) only N_0 and t have been considered. The question remains whether the value of R, reflecting fractional rates of speciation and extinction, may differ significantly from habitat to habitat for a given kind of taxon. For instance, does fractional rate of increase in number of species contribute to diversity gradients between tropical and temperate or boreal regions? This question can be approached, without regard to the relative influence of S and E, by calculating values of R for similar taxa restricted to different latitudes. Study of the ways in which mean values of these parameters may vary along latitudinal gradients represents a potentially fruitful application of exponential analysis. One obvious problem is that many taxa are not sufficiently restricted by latitude to be of value here. Nonetheless, some provide suitable data. One prediction seems secure (MacArthur, 1969; Mayr, 1969a). At low latitudes the high diversity of certain groups of organisms must accelerate the diversification of others. The provision of substratal space by tropical plants is an example, and so is the propagation of tropical diversification upward through food webs. Here we are not considering N_0 for the taxon in question; we are considering how niche partitioning within the taxon can be affected by the diversity of other taxa.

Using meager data, I have as yet uncovered no dramatic latitudinal disparities in values of R. Of the bivalve taxa for which data are plotted in Figure 10-5, *Mya*, a cold-water genus, has radiated unusually slowly, producing approximately the same number of species as have evolved within the tropical genus *Tridacna* in half the time. On the other hand, the cold-water genus *Neotrigonia* has diversified more rapidly in Australia, having produced about six species in a time interval that is uncertain but cannot exceed about 20 My. *Conus* is a predatory genus of tropical gastropods, species of which specialize on particular kinds of prey (Kohn, 1959). The great variety of potential prey species in the tropics may have accelerated speciation in *Conus* by increasing the probability that small populations would expand into new species. On the other hand, among the Insecta, the Myrmicinae, a subfamily of ants restricted to the largely nontropical Palearctic and Nearctic regions (Brown, 1973), appear to have radiated more rapidly than any other large group of ants or than any radiating family of dipterans (Table 9-3).

Care must be taken in comparing values of R for high and low latitudes because adaptive radiation is not perfectly exponential. Calculations therefore yield minimum values. If rate of diversifications has actually slowed significantly, the calculated value will then reflect the carrying capacity of a region (whatever the restraint) and will be spuriously high for a region for which carrying capacity is great (perhaps the tropics, in situations where niche partitioning depends upon complexity of the biotic environment).

The effect of latitude on rate of extinction can also be studied, although also not definitively at present because of the sparseness of data. The Lyellian Curves used to estimate rates of extinction for bivalves and gastropods (Figures 5-8 and 9-3) were based on temperate and subtropical data. In Figure 10-10, data for huge faunas of the tropical Americas are added to these plots. Note that the added points fall below the band of points representing nontropical faunas. This distribution would seem to suggest that tropical molluscan faunas may turn over much more rapidly than faunas of cooler regions. This hypothesis can be tested by calculating Lyellian percentages for molluscan faunas of the tropical Pacific (Stanley, 1979). Ladd (1966, 1972, 1977) has provided data for the gastropods of the sizable South Pacific area extending from the Mariana Islands and Palau to Tonga. As shown in Figure 10-10, this South Pacific fauna has been remarkably stable. Evidently, during the late Cenozoic there has been a much higher incidence of extinction in the tropical American faunas than in either temperate or tropical faunas of the Pacific. Thus, latitude (or temperature) does not seem to have been the critical factor. For some other reason, during the Pleistocene Epoch the tropical Americas suffered a mass extinction that eliminated at least a thousand (and perhaps several thousand) more species than would have been expected to have died out during such an interval under normal conditions. As a result, the present tropical American fauna is markedly depauperate. Only a thousand or so shelled species inhabit the Caribbean.

The fact that many tropical American faunas for their age exhibit unusually small percentages of living species (Figure 10-10) has escaped the recognition of some taxonomists, who continue to use Lyellian percentages to date these faunas. The situation is perhaps epitomized by the fauna of the Cantaure Formation of the Paraguaná Peninsula of Venezuela (Jung, 1965), now believed to be of late Early Miocene age (Díaz de Gamero, 1974). This fauna contains a number of endemic species, all of which seem to have become extinct quickly. It is of great interest that Petuch (1976) has discovered a highly endemic living assemblage of mollusks that also occurs along the Paraguaná Peninsula, in a small bay. Several of the species seems to be direct descendants of forms known from the Upper Miocene and Lower Pliocene of Ecuador and Peru. Whether the Paraguaná location is coincidental is uncertain, but this kind of relict fauna may offer clues as to the nature of late Cenozoic extinctions in the tropical Americas.

A parallel can be found in the history of the reef corals of the Caribbean

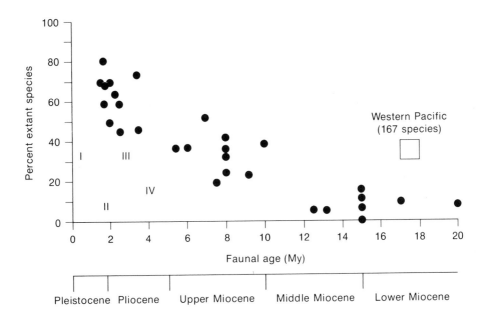

FIGURE 10-10
Effects of endemism on faunal turnover (rate of extinction). Dots define a Lyellian Curve for gastropod faunas of western North America and Japan (see Figure 9-3 for details). Roman numerals represent large tropical American faunas: I—Caloosahatchee Fauna, Florida (344 species); II—Bowden Fauna, Jamaica (406 species); III—Mare Fauna, Venezuela (270 species); IV—Esmaraldas Fauna, Ecuador (117 species). Large square represents the stable western Pacific fauna studied by Ladd (1966, 1972, 1977). (From Stanley, 1979.)

biogeographic province. It has long been recognized that these forms suffered dramatic decline following tectonic isolation of the region, first from the Mediterranean region early in the Miocene and later from the Indo-Pacific region with the uplift of the Isthmus of Panama (Wells, 1956). During the same interval, the tropics have shrunk toward the equator. Today there are not many more than 60 species of Caribbean reef corals (Goreau and Wells, 1967), and the number of Indo-Pacific species is nearly an order of magnitude larger. In the mid-Cenozoic, the disparity was not nearly so pronounced. So brief has been the interval of decline in the Caribbean that a reduction in rate of speciation is not likely to be fully responsible. Apparently, rate of extinction has also increased.

A significant extinction of Caribbean reef corals seems to have resulted from the onset of climatic cooling at the end of the Pliocene (Frost, 1977). At this time, the narrowly distributed eastern Pacific coral reef fauna may have been entirely eliminated (Dana, 1975). In each of these extinctions, not only temperature change but also sea-level fluctuation may have played a destructive role. For mollusks, while temperatures should have had similar effects throughout the Pacific and Atlantic, sea-level fluctuation may have been more disastrous on both coasts of the tropical Americas, where populations were much less widespread. The degree to which Pleistocene events were responsible for late Cenozoic mass extinction of tropical American mollusks can perhaps be determined in the future by detailed study of Pliocene and Pleistocene faunas.

The remarkably stable western Pacific island faunas for which data are plotted in Figure 10-10 represent a sample of a vast geographic region in which, at any time during the interval considered, myriads of volcanic islands and numerous larger landmasses have supported shallow-water species. Nothing like the periodic constriction in the tropical Americas has occurred. It seems reasonable to draw the general inference that degree of provinciality exerts a stronger influence over rate of extinction than does position with respect to the equator.

I would venture to suggest that the tropical Pacific marine realm approaches the ideal geographic configuration for adaptive radiation of shallow-water faunas. It is an enormous area divided into numerous semi-isolated subregions. Its size and heterogeneity must depress rates of extinction by favoring the establishment of large populations, segments of which occupy subregions whose environmental histories are largely independent of one another (Figure 10-11). When some subregions are being destroyed or rendered inhospitable, others should support thriving populations. At the same time, environmental heterogeneity on a vast scale should foster the frequent divergence of local populations: Rate of speciation should be high. The endemic shallow-water species of Cocos Island (page 170) perhaps forms an example. If these inferences are correct, the combination of high rates of speciation and low rates of extinction should yield relatively high values of R.

Regions having biogeographic configurations like that of the Indo-Pacific shallow-marine realm may represent great crucibles of diversification, where a disproportionately high percentage of taxa make their appearance. In essence, the advantages that accrue to their inhabitants resemble those of groups that I have called supertaxa. Interestingly, the characteristics of these regions are essentially like those envisioned by Wright (1931, 1977) as providing optimal conditions for rapid evolution within a single species. The arguments of the preceding paragraph retain the idea that such regions are favorable to species, but only because they promote survival, not because they accelerate phyletic evolution. I have, in fact, argued that such regions are especially stifling to phyletic change (pp. 47–51). They enhance rates of evolution not at the level of the species but at the level of the higher taxon.

Parenthetically, I would note that while draining of the continental shelves has commonly been viewed as an agent of mass extinction for shallow marine benthos (Chamberlain, 1898; Schuchert, 1914; Stille, 1924; Newell, 1952; Schopf, 1974), we must not ignore the role of island archipelagos, which generate and sustain enormous biotas. Plate tectonic movements, including subduction of the seafloor, have operated continuously during the Phanerozoic. The

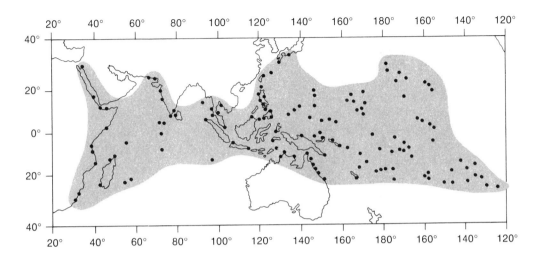

FIGURE 10-11
Pan-Pacific distribution of the marine gastropod species *Rhinoclavis sinensis* (Gmelin). (From Houbrick, 1978.)

island arcs that result represent a dynamic but collectively persistent physiographic feature of great biogeographic significance—they harbor representatives of most families of marine benthos. Consequently, the role of constriction of continental shelves in mass extinction at the family or order level is in doubt. Certainly, the shifting of a subduction zone, leaving island arcs to erode, may cause regional extinction, as may the draining of a continental shelf.

SUMMARY

The pronounced variation among taxa in rate of speciation and rate of extinction has important consequences. Assuming that fluctuations in these rates are proportional to original values, we can predict that taxa characterized by high rates will exhibit phylogenetic instability (frequent mass extinction). This prediction is born out by the high incidence of mass extinction for terrestrial tetrapods, especially those of large body size, and for trilobites, ammonites, and graptolites. Bivalves and gastropods, which have much slower turnover rates, have suffered far less from mass extinction.

Variation in rates of speciation and extinction might seem to open the way for phylogenetic options analogous to the life history "strategies" of populations. There are several problems here. For instance, probability of extinction and speciation do not vary during the lifetime of an average fully established species nearly as greatly as do fecundity and probability of death during the lifetime of an average individual. This condition and the relatively small number of speciation and extinction events in adaptively discrete clades make the evolution of complex macroevolutionary "strategies" improbable. Also, in macroevolution there is no energetic trade-off equivalent to that recognized in the traditional concepts of r- and K-selection within populations. Species selection automatically tends to increase speciation rate and to decrease extinction rate (it tends to maximize R). This tendency, however, is opposed by other factors. There is a general correlation in the animal world between rate of speciation and rate of extinction, which results, in part, fortuitously from the control of the two variables by the same biologic factors. The linkage between rate of speciation and rate of extinction is constantly being strained, but is not easily broken. Still, some "supertaxa" may possess relatively high rates of speciation and low rates of extinction. Fossil data suggest that groups of small mammals, like the murid rodents, are examples. The explanation may be that large population sizes reduce rate of extinction for small mammals, while limited powers of dispersal accelerate rate of speciation. The cichlid fishes, which radiate unusually rapidly and also seem to enjoy success in competition, may be another kind of "supertaxon."

In macroevolution the primary equivalent of a life-history "strategy" relates to the more flexible variable: rate of turnover of species. The question is, will the semilinked rates of speciation and extinction within a taxon be high or low? Turnover rate will, to a degree, be fixed within discrete taxa by the limited flexibility of relevant biologic traits, but in the long run, a high rate will offer the advantage of fostering rapid large-scale evolution and the disadvantage of increasing the impact and frequency of mass extinction. Low rate of turnover will have the opposite effects. With the general increase in behavioral complexity of animals (and in equivalent traits of vascular plants), there may have been a general increase in average turnover rate during the Phanerozoic, but this, in large part, represents selection for the traits themselves rather than for the higher turnover rates that they happen to produce.

Animal taxa displaying advanced behavior and occupying high positions in food webs would seem to exhibit elastic evolution, which means that (1) competitive interactions tend to brake their radiations rather abruptly and (2) because their diversity is self-limiting, their fractional rate of speciation accelerates markedly following mass extinction.

Within a particular class of animals, a hierarchy of clades may exist. This condition is revealed by the presence of different values of R at different taxonomic levels. For the Mammalia, values of R for many genera are much higher than values for families, indicating that genera represent pulses of rapid evolution and that they approximate natural clades. No such condition exists within the Bivalvia or Gastropoda, in which, not surprisingly, genera are taxonomically problematic. The mammalian condition perhaps results from weak dispersal and competitive braking, with a rapid pulse of radiation occurring within a family when a distinctive new genus appears, but slowing quickly as competitive interactions set in within the limited area to which the genus has access.

As a class diversifies, a leveling-off in the numbers of component subclasses or orders does not necessarily indicate an approach to competitive saturation. In the Bivalvia, where number of families has continued to increase to the present time, production of new subclasses ceased early in the Paleozoic, apparently reflecting the inflexibility of fundamental morphologic specializations.

Species diversity in many habitats today does not simply reflect saturation or rates of immigration and emigration (or local extinction). The antiquity of endemic radiation also plays an important role. The wave-ridden, sandy shore is poorly inhabited by marine macrobenthos, not because food or space is limiting but because invasion has been sporadic and, in many cases, geologically recent. (The few taxa living here are radiating at rates resembling those of non-beach-dwelling taxa of the same classes.) Many biotas of cool, high latitudes are also of relatively recent origin. In the punctuational scheme, the invasion of hostile regions is seen as being sporadic because of a catch-22. Taxa can only invade by speciating and yet they can only do this if their populations have already invaded. Whether values or R within a class tend to be relatively high in the tropics is not yet known, but such a tendency seems likely for some groups because of compounding effects: High tropical diversity of taxa providing nutrition or substrates promotes rapid diversification of dependent taxa.

Local geographic history seems to be of greater importance to rate of extinction than is latitudinal position. During the late Cenozoic, for example, tropical molluscan species of the Western Pacific seem to have been long-lived relative to those of cooler regions, whereas faunas of the tropical Americas have suffered unusually high rates of extinction because of extensive destruction of shallow marine habitats. In fact, for shallow-water benthos, the Indo-Pacific realm seems to represent the optimal kind of region for radiation—a crucible of diversification. Its insular geographic texture makes speciation likely, while its enormous size favors the development of large, scattered populations that should resist extinction. High net rates of radiation should result.

Bibliography

Auffenberg, W. (1963) The fossil snakes of Florida. *Tulane Stud. Zool.*, *10:* 131–216.

Auffenberg, W., and Milstead, W. W. (1965) Reptiles in the Quaternary of North America. *In* Wright, H. E. and Frey, D. G., eds. *The Quaternary of the United States.* Princeton, N.J., Princeton Univ. Press, pp. 557–568.

Avise, J. C. (1977) Is evolution gradual or rectangular? Evidence from living fishes. *Proc. Nat. Acad. Sci. U.S.A.*, *74:* 5083–5087.

Ayala, F. J. (1975) Genetic differentiation during the speciation process. *Evol. Biol.*, *8:* 1–78.

Baker, H. G. (1963) Evolutionary mechanisms in pollination biology. *Science, 139:* 877–883.

Baker, H. G., and Hurd, P. D. (1968) Intrafloral ecology. *Ann. Rev. Entomol.*, *13:* 385–414.

Bakker, R. T. (1977) Cycles of diversity and extinction: A plate tectonic/topographic model. *In* Hallam, A., ed. *Patterns of Evolution.* Amsterdam, Elsevier, pp. 431–478.

Bambach, R. K. (1977) Species richness in marine benthic habitats through the Phanerozoic. *Paleobiology, 3:* 152–167.

Barash, D. P. (1976) What does sex really cost. *Amer. Nat., 110:* 894–897.

Bateson, W. (1894) *Materials for the Study of Variation.* New York, Macmillan, 598 pp.

Berggren, W. A. (1968) Problems of some Tertiary planktonic foraminiferal lineages. *Tulane Studies in Geol., 6:* 1–22.

Berggren, W. A. (1969) Rates of evolution in some Cenozoic planktonic foraminifera. *Micropaleont., 15:* 351–365.

Berggren, W. A., and Van Couvering, J. A. (1974) *The Late Neogene.* Amsterdam, Elsevier, 216 pp.

Berry, D. L., and Baker, R. J. (1971) Apparent convergence of karyotype in two species of pocket gophers of the genus *Thomomys* (Mammalia, Rodentia). *Cytogenetics, 10:* 1–9.

Berry, W. B. N. (1973) Silurian-Early Devonian graptolites. *In* Hallam, A., ed. *Atlas of Palaeobiogeography.* Amsterdam, Elsevier, pp. 81–87.

Birky, C. W., and Gilbert, J. J. (1971) Parthenogenesis in rotifers: The control of sexual and asexual reproduction. *Amer. Zool., 11:* 245–266.

Blow, W. H. (1969) Late Middle Eocene to Recent planktonic foraminiferal biostratigraphy. *In* Brönnimann, P., and Renz, H. H., eds. *Proc. First Conf. Planktonic Microfossils.* Geneva, 1967, Leiden, E. J. Brill, pp. 199–421.

Bock, W. J. (1970) Microevolutionary sequences as a fundamental concept in macroevolutionary models. *Evolution, 24:* 704–722.

Bock, W. J. (1972) Species interactions and macroevolution. *Evol. Biol., 5:* 1–24.

Bold, W. A. van den. (1963) Upper Miocene and Pliocene Ostracoda of Trinidad. *Micropaleont., 9:* 361–424.

Boreske, J. R. (1974) A review of the North American fossil amiid fishes. *Mus. Comp. Zool. Bull., 146:* 1–87.

Boss, K. J. (1971) Critical estimate of the number of Recent Mollusca. *Occas. Papers on Mollusks, Mus. Comp. Zool., Harvard Univ., 3:* 81–135.

Boucot, A. J. (1975) *Evolution and Extinction Rate Controls.* Amsterdam, Elsevier, 425 pp.

Bowen, Z. P., McAlester, A. L., and Rhoads, D. C. (1974) Marine benthic communities of the Sonyea Group (Upper Devonian) of New York. *Lethaia, 7:* 93–120.

Bretsky, P. B., Flessa, K. W., and Bretsky, S. S. (1969) Brachiopod ecology in the Ordovician of eastern Pennsylvania. *Jour. Paleont., 43:* 312–321.

Briggs, J. C. (1974) *Marine Zoogeography.* New York, McGraw-Hill, 475 pp.

Brinkmann, R. (1929) Statistisch-biostratigraphische Untersuchungen an mitteljurassischen Ammoniten über Artbegriff und Stammesentwicklung. *Göttingen, Abhandlung Gesell. Wiss.,* vol. 13, no. 3.

Brinkmann, R. (1937) Biostratigraphie des Leymeriellestammes nebst Bemerkungen zur Paläogeographie des nord-westdeutschen Alb. *Mitt. Geol. Staats Inst. Hamburg, 16:* 1–18.

Britten, R. J., and Davidson, E. H. (1969) Gene regulation for higher cells: A theory. *Science, 165:* 349–357.

Britten, R. J., and Davidson, E. H. (1971) Repetitive and nonrepetitive DNA sequences and a speculation on the origins of evolutionary novelty. *Quart. Rev. Biol., 46:* 111–133.

Brodkorb, P. (1959) How many species of birds have existed? *Bull. Fla. State Mus., 5:* 41–53.

Brown, J. H. (1971) The desert pupfish. *Scient. Amer., 225* (November): 104–110.

Brown, W. L. (1973) A comparison of the Hylean and Congo-West African Rain Forest Ant Faunas. *In* Meggars, B. J., Ayensu, E. S., and Duckworth, W. D., eds. *Tropical Forest Ecosystems in Africa and South America.* Washington, D.C., Smithsonian Inst., pp. 161–185.

Brown, W. L., and Wilson, E. O. (1956) Character displacement. *Syst. Zool., 5:* 49–64.

Buettner-Janusch, J. (1973) *Physical Anthropology: A Perspective.* New York, John Wiley, 572 pp.

Buge, E. (1972) Remarques sur les méthodes d'utilisation stratigraphique des Bryozoaires postpaléozoiques. *Mem. Bur. Rech. Géol. Min. France, 77:* 55–58.

Bulman, O. M. B. (1970) Graptolithina with sections on Enteropneusta and Pterobranchia. *In* Moore, R. C., ed. *Treatise on Invertebrate Paleontology,* 2nd ed. Part V. Lawrence, Kansas, Univ. Kansas and Geological Society of America. 163 pp.

Burt, W. H. (1957) A subspecies category in mammals. *System. Zool., 3:* 99–104.

Bush, G. L. (1975) Modes of animal speciation. *Ann. Rev. Ecology and Systematics, 6:* 339–364.

Bush, G. L., Case, S. M., Wilson, A. C., and Patton, J. C. (1977) Rapid speciation and chromosomal evolution in mammals. *Proc. Nat. Acad. Sci. 74:* 3942–3946.

Cameron, A. W. (1958) Mammals of the islands in the Gulf of St. Lawrence. *Nat. Mus. Canada Bull., 154:* 1–165.

Campbell, B. G. (1974) *Human Evolution.* Chicago, Aldine, 469 pp.

Carlquist, S. (1965) *Island Life*. Garden City, N.Y., Natural History Press, 451 pp.

Carson, H. L. (1968) The population flush and its genetic consequences. *In* Lewontin, R. C., ed. *Population Biology and Evolution*. Syracuse, N.Y., Syracuse Univ. Press, pp. 123–137.

Carson, H. L. (1975) The genetics of speciation at the diploid level. *Amer. Nat.*, *109:* 83–92.

Carson, H. L. (1970) Chromosome tracers of the origin of species. *Science*, *168:* 1414–1418.

Cavender, T. M. (1968) Freshwater fish remains from the Clarno Formation, Ochoco Mountains of North-Central Oregon. *Ore. Bin.*, *30:* 125–141.

Chamberlain, T. C. (1898) The ulterior basis of time divisions and the classification of geologic history. *Jour. Geol.*, *6:* 449–462.

Chevalier, J. P. (1968) *Expédition Française sur les récifs corallieus de la Nouvelle Calédonie*. Volume Troisième. Paris, Éditions de la Fondation Singer-Polignac, 155 pp.

Cifelli, R. (1969) Radiation of Cenozoic planktonic Foraminifera. *System. Zool.*, *18:* 154–168.

Clark, J. D. (1976) African origins of man the toolmaker. *In* Isaac, G. L., and McCown, E. R., eds. *Human Origins*. Menlo Park, Calif., W. A. Benjamin, pp. 1–53.

Clarke, B. C., and Murray, J. J. (1969) Ecology, genetics, and speciation in land snails of the genus *Partula. Biol. Jour. Linn. Soc. London*, *1:* 31–42.

Cloud, P. E. (1948) Some problems and patterns of evolution exemplified by fossil invertebrates. *Evolution*, *2:* 322–350.

Connell, J. H. (1961a) The effects of competition, predation by *Thais lapillus*, and other factors on natural populations of the barnacle, *Balanus balanoides. Ecol. Monogr.*, *31:* 61–104.

Connell, J. H. (1961b) The influence of interspecific competition and other factors on the distribution of the barnacle *Chthamalus stellatus. Ecology*, *42:* 710–723.

Cooke, C. W. (1959) Cenozoic echinoids of the Eastern United States. *U.S. Geol. Surv. Prof. Paper 321*, 106 pp.

Corbet, G. B. (1964) Regional variation in the bank-vole *Clethrionomys glareolus* in the British Isles. *Proc. Zool. Soc. London*, *143:* 191–217.

Corbet, G. B. (1975) Examples of short- and long-term changes of dental pattern in Scottish voles (Rodentia; Microtinae). *Mammal. Rev.*, *5:* 17–21.

Crow, J. F., and Kimura, M. (1969) Evolution in sexual and asexual populations: A reply. *Amer. Natur.*, *103:* 89–90.

Cuellar, O. (1977) Animal parthenogenesis. *Science*, *197:* 837–843.

Dana, T. F. (1975) Development of contemporary eastern Pacific coral reefs. *Mar. Biol.*, *33:* 355–374.

Darlington, P. J. (1959) Area, climate, and evolution. *Evolution*, *13:* 488–510.

Darwin, C. R. (1859) *On the Origin of Species*. London, John Murray, 490 pp.

Darwin, C. R. (1871) *The Descent of Man, and Selection in Relation to Sex*. London, John Murray, 2 vols.

Davis, D. D. (1964) The giant panda: A morphological study of evolutionary mechanisms. *Fieldiana Mem. (Zool.)*, *3:* 1–339.

Davis, G. M. (1978) The origin and evolution of the gastropod family Pomatiopsidae, with emphasis on the Mekong River Triculinae. *Acad. Nat. Sci. Philadelphia*, Monogr. 20, 120 pp.

Day, J. H. (1963) The complexity of the biotic environment. *System. Assoc. Publ.*, *5:* 31–49.

De Beer, G. R. (1940) *Embryos and Ancestors*. Oxford, Clarendon, 108 pp.

Delamare-Deboutteville, C., and Botosaneanu, L. (1970) *Formes Primitives Vivantes*. Paris, Harmann, 232 pp.

De Vries, H. (1905) *Species and Varieties, Their Origin by Mutation*. Chicago, The Open Court, 847 pp.

Díaz de Gamero, M. L. (1974) Microfauna y edad de la Formación Cantaure, Península de Paraguaná, Venezuela. *Asoc. Venez. Geol. Min. y Petrol., Bol. Inf., 17:* 41–47.

Dice, L. R., and Blossom, P. M. (1937) Studies of mammalian ecology in Southwestern North America, with special attention to the colors of desert mammals. *Carnegie Inst. Washington, Publ. 485:* 1–129.

Dobzhansky, T. (1956) What is an adaptive trait? *Amer. Nat., 90:* 337–347.

Dobzhansky, T. (1970) *Genetics of the Evolutionary Process*. New York, Columbia Univ. Press, 505 pp.

Dobzhansky, T. (1972) Species of Drosophila. *Science, 177:* 664–669.

Doyle, J. A., and Hickey, L. J. (1976) Pollen and leaves from the mid-Cretaceous Potomac Group and their bearing on early angiosperm evolution. *In* Beck, C. G., ed. *Origin and Early Evolution of Angiosperms*. New York, Columbia Univ. Press, pp. 139–206.

Durham, J. W. (1969) The fossil record and the origin of the Deuterostomata. *North Amer. Paleont. Convention, Chicago, 1969, Proc.,* pp. 1104–1132.

Durham, J. W. (1978) The probable metazoan biota of the Precambrian as indicated by the subsequent record. *Ann. Rev. Earth Planet. Sci., 6:* 21–42.

Ebersin, A. G. (1965) Sistema: Filogenia Solonovatovodnikh Kardiid. Molloski Voprosy Teorii Prikladnoi Malokologii. Tezisy Dokl., Sb. 2, *Izd. Zool. in-ta AN SSSR*.

Ehrlich, P. R., and Raven, P. H. (1964) Butterflies and plants: A study in coevolution. *Evolution, 18:* 586–608.

Ehrlich, P. R., and Raven, P. H. (1969) Differentiation of populations. *Science, 165:* 1228–1232.

Eldredge, N. (1971) The allopatric model and phylogeny in Paleozoic invertebrates. *Evolution, 25:* 156–167.

Eldredge, N., and Gould, S. J. (1972) Punctuated equilibria: an alternative to phyletic gradualism. *In* Schopf, T. J. M., ed. *Models in Paleobiology*. San Francisco, Freeman, Cooper, pp. 82–115.

Emry, R. J. (1970) A North American Oligocene pangolin and other additions to the Pholidota. *Amer. Mus. Nat. Hist. Bull., 142:* 455–510.

Endler, J. A. (1973) Gene flow and population differentiation. *Science, 179:* 243–250.

Endler, J. A. (1977) *Geographic Variation, Speciation, and Clines*. Princeton, N.J., Princeton Univ. Press, 246 pp.

Ernst, C. H., and Barbour, W. (1972) *Turtles of the United States*. Lexington, Kentucky, Univ. Kentucky Press, 347 pp.

Eshelman, R. E. (1975) Geology and paleontology of the Early Pleistocene (late Blancan) White Rock Fauna from north-central Kansas. *Univ. Mich. Papers on Paleont., 13:* 1–60.

Estes, R. (1970) Origin of the Recent North American lower vertebrate fauna: An inquiry into the fossil record. *Forma et Functio, 3:* 139–163.

Fedducia, A. (1977) A model for the evolution of perching birds. *Syst. Zool., 26:* 19–31.

Fischer, A. G. (1960) Latitudinal variations in organic diversity. *Evolution, 14:* 64–81.

Fischer, A. G., and Arthur, M. A. (1977) Secular variations in the pelagic realm. *Tulsa, Okla., Society of Economic Paleontologists and Mineralogists, Spec. Publication 25:* 19–50.

Fisher, J. (1967) Fossil birds and their adaptive radiation. *In* Harland, W. B. *et al.*, eds. *The Fossil Record*, pp. 133–156.

Fisher, R. A. (1930) *The Genetical Theory of Natural Selection*. Oxford, Oxford Univ. Press, 272 pp.

Fisher, R. A. (1958) *The Genetical Theory of Natural Selection*, 2nd ed. New York, Dover, 291 pp.

Ford, E. B. (1975) *Ecological Genetics*. London, Chapman and Hall, 442 pp.

Frazer, J. F. D. (1973) *Amphibians*. London, Wykeham. 122 pp.

Frazzetta, T. H. (1970) From hopeful monster to bolyerine snakes? *Amer. Nat.*, *104:* 55–72.

Frost, S. H. (1977) Miocene to Holocene evolution of Caribbean Province reef-building corals. *Third International Coral Reef Symp., Miami, Proc.*, pp. 353–359.

Fryer, G., and Iles, T. D. (1972) *The Cichlid Fishes of the Great Lakes of Africa. Their Biology and Evolution.* Edinburgh, Oliver & Boyd, 641 pp.

Garstang, W. (1929) The origin and evolution of larval forms. *Report Brit. Assoc. Adv. Sci.* (D), 77–98.

Ghiselin, M. T. (1966) The adaptive significance of gastropod torsion. *Evolution*, *20:* 337–348.

Ghiselin, M. T. (1974) *The Economy of Nature and the Evolution of Sex.* Berkeley, Univ. of California Press, 346 pp.

Gillet, S. (1946) Lamellibranches dulcicoles, les limnocardiidés. *Rev. Sci., Paris, 84:* 343–353.

Gingerich, P. D. (1974) Stratigraphic record of Early Eocene *Hyopsodus* and the geometry of mammalian phylogeny. *Nature, 248:* 107–109.

Gingerich, P. D. (1976) Paleontology and phylogeny: Patterns of evolution at the species level in early Tertiary mammals. *Amer. Jour. Sci., 276:* 1–28.

Gingerich, P. D. (1977) Patterns of evolution in the mammalian fossil record. *In* Hallam, A., ed. *Patterns of Evolution, As Illustrated by the Fossil Record.* Amsterdam, Elsevier, pp. 469–500.

Gingerich, P. D., and Simons, E. L. (1977) Systematics, phylogeny, and evolution of early Eocene Adapidae (Mammalia, Primates) in North America. *Univ. Michigan, Mus. Paleont. Contrib., 24:* 245–279.

Glaessner, M. F. (1969) Decapoda. *In* Moore, R. C., ed. *Treatise on Invertebrate Paleontology.* Part R. Lawrence, Kansas, Univ. of Kansas and Geological Society of America, pp. 399–533.

Goldschmidt, R. (1940) *The Material Basis of Evolution.* New Haven, Conn., Yale Univ. Press, 436 pp.

Goreau, T. F., and Wells, J. W. (1967) The shallow-water Scleractinia of Jamaica: Revised list of species and their vertical distribution range. *Bull. Marine Sci., 17:* 442–453.

Gorsel, J. T. van (1975) Evolutionary trends and stratigraphic significance of the Late Cretaceous Helicorbitoides-Lepidorbitoides lineage. *Utrecht Micropaleont. Bull., 12:* 1–99.

Gould, S. J. (1966) Allometry and size in ontogeny and phylogeny. *Biol. Rev., 41:* 587–640.

Gould, S. J. (1976) Ladders, bushes, and human evolution. *Nat. Hist., 85(4):* 24–31.

Gould, S. J. (1977a) Eternal metaphors of palaeontology. *In* Hallam, A., ed. *Patterns of Evolution, As Illustrated by the Fossil Record.* Amsterdam, Elsevier, pp. 1–26.

Gould, S. J. (1977b) *Ontogeny and Phylogeny*. Cambridge, Mass., Harvard Univ. Press, 501 pp.

Gould, S. J. (1977c) The return of hopeful monsters. *Nat. Hist., 86:* 22–30.

Gould, S. J., and Eldredge, N. (1977) Punctuated equilibria: The tempo and mode of evolution reconsidered. *Paleobiology, 3:* 115–151.

Gould, S. J., Raup, D. M., Sepkoski, J. J., Schopf, T. J. M., and Simberloff, D. S. (1977) The shape of evolution: A comparison of real and random clades. *Paleobiol., 3:* 23–40.

Grant, V. (1949) Pollination systems as isolating mechanisms in angiosperms. *Evolution, 3:* 82–97.

Grant, V. (1963) *The Origin of Adaptations*. New York, Columbia Univ. Press, 606 pp.

Grant, V. (1971) *Plant Speciation*. New York, Columbia Univ. Press, 435 pp.

Greenwood, P. H. (1965) The cichlid fishes of Lake Nabugabo, Uganda. *Brit. Mus. Nat. Hist. Bull. (Zool.),* 12: 315–357.

Greenwood, P. H. (1974) The cichlid fishes of Lake Victoria, East Africa: The biology and evolution of a species flock. *Brit. Mus. (Nat. Hist.) Bull.* Suppl. 6, 134 pp.

Grüneberg, H. (1963) The Pathology of Development. New York, John Wiley, 309 pp.

Gustafsson, Å. (1946–1947) Apomixis in higher plants. *Lunds Universitets Årsskrift, 42–43:* 1–370.

Haffer, J. (1974) Avian speciation in tropical South America. *Publ. Nuttall Ornithol. Club, 14:* 1–390.

Hallam, A. (1975) Evolutionary size increase and longevity in Jurassic bivalves and ammonites. *Nature, 258:* 439–446.

Hallam, A. (1976) Stratigraphic distribution and ecology of European Jurassic bivalves. *Lethaia, 9:* 245–259.

Hansen, T. A. (1978) Larval dispersal and species longevity in lower Tertiary gastropods. *Science, 199:* 885–887.

Harper, C. W. (1975) Origin of species in geologic time: Alternatives to the Eldredge-Gould model. *Science, 190:* 47–48.

Harper, C. W. (1976) Stability of species in geologic time. *Science, 192:* 269.

Harrington, H. J. (1959) Classification. *Treatise on Invertebrate Paleontology*. Part O. *Arthropoda* 1. Lawrence, Kansas, Univ. Kansas and Geologic Society of America, pp. 145–170.

Harris, H. (1966) Enzyme polymorphisms in man. *Proc. Roy. Soc. London, 164 (B):* 298–310.

Hay, W. W., Mohler, H. P., Roth, P. H., Schmidt, R. R., and Bourdreaux, J. E. (1967) Calcareous nannoplankton zonation of the Cenozoic of the Gulf Coast and Caribbean-Antillean area, and transoceanic correlation. *Gulf Coast Assoc. Geol. Trans., 17:* 428–480.

Hazel, J. E. (1971) Paleoclimatology of the Yorktown Formation (Upper Miocene and Lower Pliocene) of Virginia and North Carolina. *Bull. Centre Rech. Pau-SNPA, 5* (suppl.): 361–375.

Hecht, M. K. (1974) Morphological transformation, the fossil record, and the mechanisms of evolution: A debate. Part II. The statement and the critique. *Evolutionary Biology, 7:* 295–303.

Hennig, W. (1966) *Fannia scalaris* Fabrisius, eine rezente Art im Baltischen Bernstein. *Stuttgarter Beiträge zur Naturkunde, 150:* 1–12.

Hertlein, L. G. (1963) Contribution to the biogeography of Cocos Island, including a bibliography. *Calif. Acad. Sci. Proc.*, *32:* 219–289.

Holman, J. A. (1976) Snakes and stratigraphy. *Michigan Academician*, *8:* 387–396.

Hooijer, D. A. (1976) Evolution of the Perissodactyla of the Omo Group deposits. *In* Coppens, Y., Howell, F. C., Isaac, G. L., and Leakey, R. E. F., eds. *Earliest Man and Environments in the Lake Rudolf Basin*, pp. 209–213.

Hooijer, D. A., and Patterson, B. (1972) Rhinoceroses from the Pliocene of northwestern Kenya. *Mus. Comp. Zool. Bull.*, *144:* 1–26.

Hooper, E. T. (1941) Mammals of the lavafields and adjoining areas in Valencia County, New Mexico. *Misc. Publ. Mus. Zool., Univ. Mich.*, *51:* 1–47.

Houbrick, R. (1978) The family Cerithiidae in the Indo-Pacific. Part I. The genera *Rhinoclavis, Pseudovertagus*, and *Clavocerithium. Monographs of Marine Mollusca*. No. 1. 130 pp.

Hubby, J. L., and Lewontin, R. C. (1966) A molecular approach to the study of genic heterozygosity in natural populations. I. The number of alleles at different loci in *Drosophila pseudoobscura. Genetics*, *54:* 577–594.

Hutchinson, G. E. (1959) Homage to Santa Rosalia, or why are there so many kinds of animals? *Amer. Nat.*, *93:* 145–159.

Hutchinson, G. E. (1965) *The Ecological Theater and the Evolutionary Play*. New Haven, Conn., Yale Univ. Press, 139 pp.

Hutchinson, G. E. (1968) When are species necessary? *In* Lewontin, R. C., ed. *Population Biology and Evolution*. Syracuse, N.Y., Syracuse Univ. Press, pp. 177–205.

Hutton, J. (1795) *Theory of the Earth with Proofs and Illustrations*, vol. I. Edinburgh, Cadell, Junior, and Davies; William Creech, 567 pp.

Huxley, J. S. (1932) *Problems of Relative Growth*. New York, Dial Press, 276 pp.

Huxley, J. S. (1939) Clines: An auxiliary method in taxonomy. *Bijdragen Tot de Dierkunde*, *27:* 491–520.

Huxley, J. S. (1942) *Evolution, The Modern Synthesis*. London, Allen & Unwin, 645 pp.

Hyman, L. H. (1951) *The Invertebrates: Acanthocephala, Aschelminthes, and Enteroprocta*. Volume III. New York, McGraw-Hill, 572 pp.

Jaanusson, V. (1973) Morphological discontinuities in the evolution of graptolite colonies. *In* Boardman, R. S., Cheetham, A. H., and Oliver, W. A., eds. *Animal Colonies, Development and Function through Time*. Stroudsburgh, Pa., Dowden, Hutchinson, & Ross, pp. 515–521.

Jackson, J. B. C. (1974) Biogeographic consequences of eurytopy and stenotopy among marine bivalves and their evolutionary significance. *Amer. Nat.*, *108:* 541–560.

Jackson, J. B. C., Goreau, T. F., and Hartman, W. B. (1971) Recent brachiopod-coralline sponge communities and their paleoecological significance. *Science*, *173:* 623–625.

Janzen, D. H. (1966) Coevolution of mutualism between ants and acacias in Central America. *Evolution*, *20:* 249–275.

Janzen, D. H. (1970) Herbivores and the number of tree species in tropical forests. *Amer. Nat.*, *104:* 501–528.

Jeannel, R. (1961) Le foissonement de certaines lignées dans le îles. *Collog. Intern. Centre Nat. Rech. Sci. Paris*, *94:* 291–294.

Johnson, M. W. (1953) The copepod *Cyclops dimorphys* Kiefer from the Salton Sea. *Amer. Midl. Nat.*, *49:* 188–192.

Jung, P. (1965) Miocene Mollusca from the Paraguaná Peninsula, Venezuela. *Bull. Amer. Paleont.*, *49*: 385–652.

Kauffman, E. G., and Sohl, N. F. (1974) Structure and evolution of Antillean Cretaceous rudist framework. *Verhandl. Naturf. Ges. Basel*, *84*: 399–467.

Keast, A. (1961) Bird speciation on the Australian continent. *Harvard Univ., Bull. Mus. Comp. Zool.*, *123*: 303–495.

Keigwin, L. E. (1978) Pliocene closing of the Isthmus of Panama, based on biostratigraphic evidence from nearby Pacific Ocean and Caribbean sea cores. *Geology*, *6*: 630–634.

Kellogg, D. E. (1975) The role of phyletic change in the evolution of *Pseudocubus vema* (Radiolaria). *Paleobiol.*, *1*: 359–370.

Kennedy, W. J. (1977) Ammonite evolution. *In* Hallam, A., ed. *Patterns of Evolution, As Illustrated by the Fossil Record*. Amsterdam, Elsevier, pp. 251–304.

Kennedy, W. J., and Cobban, W. A. (1976) Aspects of ammonite biology, biostratigraphy, and biogeography. *Palaeont. Assoc. Spec. Pap. Palaeont.*, *17*, 94 pp.

Kerfoot, W. C. (1975) The divergence of adjacent populations. *Ecology*, *56*: 1298–1313.

Kier, P. M. (1963) Tertiary echinoids from the Caloosahatchee and Tamiami Formations of Florida. *Smithsonian Miscellaneous Coll.*, *145*, 63 pp.

Kier, P. M. (1965) Evolutionary trends in Paleozoic echinoids. *Jour. Paleontology*, *39*: 436–465.

Kier, P. M. (1967) Revision of the oligopygoid echinoids. *Smithsonian Miscellaneous Coll.*, *152*: 1–147.

Kier, P. M. (1972) Upper Miocene echinoids from the Yorktown Formation of Virginia and their environmental significance. *Smithsonian Contrib. Paleobiol.*, *13*, 31 pp.

King, M. C., and Wilson, A. C. (1975) Evolution at two levels. Molecular similarities and biological differences between humans and chimpanzees. *Science*, *188*: 107–116.

Kitts, D. B. (1974) Paleontology and evolutionary theory. *Evolution*, *28*: 458–472.

Kohn, A. J. (1959) The ecology of *Conus* in Hawaii. *Ecol. Monogr.*, *29*: 47–90.

Kohn, A. J., and Orians, G. H. (1962) Ecological data in the classification of closely related species. *Syst. Zool.*, *11*: 119–126.

Krebs, C. J. (1972) *Ecology*. New York, Harper & Row, 694 pp.

Kummel, B. (1969) Ammonoids of the Late Scythian (Lower Triassic). *Harvard Mus. Compar. Zool. Bull.*, *137*: 311–701.

Kurtén, B. (1953) On the variation and population dynamics of fossil and recent mammal populations. *Acta Zool. Fenn.*, *76*: 1–122.

Kurtén, B. (1958) The bears and hyaenas of the interglacials. *Quaternaria*, *4*: 69–81.

Kurtén, B. (1964) The evolution of the polar bear, Ursus maritimus Phipps. *Acta Zool. Fenn.*, *108*: 1–30.

Kurtén, B. (1968) *Pleistocene Mammals of Europe*. Chicago, Aldine, 317 pp.

Kurtén, B. (1971) *The Age of Mammals*. New York, Columbia Univ. Press, 250 pp.

Lack, D. (1933) Habitat selection in birds. *Jour. Animal Ecol.*, *2*: 239–262.

Lack, D. (1947) *Darwin's Finches*. Cambridge, Cambridge Univ. Press, 204 pp.

Lack, D. (1954) *The Natural Regulation of Animal Numbers*. Oxford, Oxford Univ. Press, 343 pp.

Ladd, H. S. (1966) Chitons and gastropods (Haliotidae through Adeorbidae) from the Western Pacific islands. *U.S. Geol. Surv. Prof. Paper 531*, 98 pp.

Ladd, H. S. (1972) Cenozoic fossil mollusks from Western Pacific islands; Gastropods (Turritellidae through Strombidae) *U.S. Geol. Surv. Prof. Paper 532,* 79 pp.

Ladd, H. S. (1977) Cenozoic fossil mollusks from western Pacific islands; Gastropods (Eratoidae through Harpidae). *U.S. Geol. Surv. Prof. Paper 533,* 84 pp.

Lagaaj, R. (1952) The Pliocene Byozoa of the Low Countries. *Mededel. Geolog. Sticht.* Ser. C, vol. 5, no. 5, 200 pp.

Lande, R. (1976) Natural selection and random genetic drift in phenotypic evolution. *Evolution, 30:* 314–334.

Lang, J. (1971) Interspecific aggression by scleractinian corals. 1. The rediscovery of *Scolymia cubensis* (Milne Edwards & Haime. *Bull. Mar. Sci., 21:* 952–959.

Leakey, L. S. B. (1959) A new fossil skull from Olduvai. *Nature, 184:* 491–493.

Leakey, R. E. F., and Walker, A. C. (1976) *Australopithecus, Homo erectus* and the single species hypothesis. *Nature, 261:* 572–574.

Leppik, E. E. (1957) Evolutionary relationship between entomophilous plants and anthomophilous insects. *Evolution, 11:* 466–481.

Lewis, H. (1953) The mechanism of evolution in the genus Clarkia. *Evolution, 7:* 1–20.

Lewis, H. (1962) Catastrophic selection as a factor in speciation. *Evolution, 16:* 257–271.

Lewis, H. (1966) Speciation in flowering plants. *Science, 152:* 167–172.

Lewontin, R. C. (1974) *The Genetic Basis of Evolutionary Change.* New York, Columbia Univ. Press, 346 pp.

Liem, K. F. (1973) Evolutionary strategies and morphological innovations: Cichlid pharyngeal jaws. *Syst. Zool., 22:* 425–441.

Lillegraven, J. A. (1969) Latest Cretaceous mammals of upper part of Edmonton Formation of Alberta, Canada, and review of marsupial-placental dichotomy in mammalian evolution. *Univ. Kansas Paleont. Contrib., 50:* 1–122.

Lillegraven, J. A. (1972) Ordinal and familial diversity of Cenozoic mammals. *Taxon, 21:* 261–274.

Linsley, R. M. (1977) Some "laws" of gastropod shell form. *Paleobiology, 3:* 196–206.

Longacre, S. A. (1970) Trilobites of the Upper Cambrian Ptychaspid Biomere, Wilberns Formation, central Texas. *Paleont. Soc. Mem., 4,* 70 pp.

Longhurst, A. R. (1955) A review of the Notostraca. *Brit. Mus. Nat. Hist. Bull (Zool.), 3:* 1–54.

Løvtrup, S. (1972) *Epigenetics.* New York, John Wiley, 547 pp.

Løvtrup, S. (1976) On the falsifiability of neo-Darwinism. *Evol. Theory, 1:* 267–283.

Løvtrup, S., Rahemtulla, F., and Höglund, N.-G. (1974) Fisher's axiom and the body size of animals. *Zoologica Scripta, 3:* 53–58.

Lundberg, J. G. (1975) The fossil catfishes of North America. *Univ. Michigan Mus. Paleont. Papers on Paleont., 11:* 1–51.

Lyell, C. (1830) *Principles of Geology,* vol. 1. London, John Murray, 511 pp.

MacArthur, R. H. (1958) Population ecology of some warblers of northeastern coniferous forests. *Ecology, 39:* 599–619.

MacArthur, R. H. (1969) Patterns of communities in the tropics. *Biol. Jour. Linn. Soc., 1:* 19–30.

MacArthur, R. H. (1972) *Geographical Ecology.* New York, Harper & Row, 269 pp.

MacArthur, R. H., and Wilson, E. O. (1967) *The Theory of Island Biogeography.* Princeton, N.J., Princeton Univ. Press, 203 pp.

MacGillavry, H. J. (1968) Modes of evolution mainly among marine invertebrates. *Bijdragen Tot de Dierkunde*, *38:* 69–74.

Maglio, V. J. (1973) Origin and evolution of the Elephantidae. *Amer. Philos. Soc. Trans.*, *63:* 1–149.

Matthew, W. D. (1910) The paleontologic record. The continuity of development. *Popular Sci. Monthly*, *77:* 473–478.

Matthew, W. D. (1931) A review of the rhinoceroses with a description of Aphelops material from Pliocene of Texas. *Univ. Calif. Publ. Geol. Sci.*, *20:* 411–482.

Maynard Smith, J. (1966) Sympatric speciation. *Amer. Nat.*, *100:* 637–650.

Maynard Smith, J. (1969) Evolution in sexual and asexual populations. *Amer. Nat.*, *102:* 469–473.

Maynard Smith, J. (1971) What use is sex? *Jour. Theor. Biol.*, *30:* 319–335.

Maynard Smith, J., and Williams, G. C. (1976) Reply to Barash. *Amer. Nat.*, *110:* 897.

Mayr, E. (1940) Speciation phenomena in birds. *Amer. Nat.*, *74:* 249–278.

Mayr, E. (1942) *Systematics and the Origin of Species.* Reprint edition. Magnolia, Mass., Peter Smith, 334 pp.

Mayr, E. (1945) Introduction to Symposium on Age and Distribution Patterns of Gene Arrangements in *Drosophila pseudo-obscura*. *Lloydia*, (Cincinnati), *8:* 69–83.

Mayr, E. (1947) Ecological factors in speciation. *Evolution*, *1:* 263–288.

Mayr, E. (1950) Taxonomic categories in fossil hominids. *Cold Spring Harbor Symp. Quant. Biol.*, *15:* 109–118.

Mayr, E. (1954) Change of genetic environment and evolution. *In* Huxley, J., Hardy, A. C., and Ford, E. B., eds. *Evolution as a Process.* London, Allen and Unwin, pp. 157–180.

Mayr, E. (1957) Species concepts and definitions. *Amer. Assoc. Adv. Sci. Publ.*, *50:* 1–22.

Mayr, E. (1958) Behavior and systematics. *In* Roe, A., and Simpson, G. G., eds. *Behavior and Evolution.* New Haven, Yale Univ. Press, pp. 341–362.

Mayr, E. (1960) The emergence of evolutionary novelties. *In* Tax, S., ed. *The Evolution of Life.* Chicago, Univ. Chicago Press, pp. 349–380.

Mayr, E. (1963) *Animal Species and Evolution.* Cambridge, Mass., Harvard Univ. Press, 797 pp.

Mayr, E. (1967) Population size and evolutionary parameters. *In* Moorehead, P. S., and Kaplan, M. M., eds. *Mathematical Challenges to the Neo-Darwinian Interpretation of Evolution.* Philadelphia, Wistar Institute Symposium Monographs, No. 5, pp. 47–58.

Mayr, E. (1969a) Bird speciation in the tropics. *Biol. Jour. Linn. Soc.*, *1:* 1–17.

Mayr, E. (1969b) *Principles of Systematic Zoology.* New York, McGraw-Hill, 428 pp.

Mayr, E. (1970) *Populations, Species and Evolution.* Cambridge, Mass., Harvard Univ. Press, 453 pp.

Mayr, E. (1975) The unity of the genotype. *Biol. Zbl.*, *94:* 377–388.

McAlester, A. L. (1963) Pelecypods as stratigraphic guides in the Appalachian Upper Devonian. *Geol. Soc. Amer. Bull.*, *74:* 1209–1224.

McNab, B. (1963) Bioenergetics and the determination of home range size. *Amer. Nat.*, *97:* 133–140.

Mengel, R. M. (1964) The probable history of species formation in some northern wood warblers (Parulidae). *Living Bird*, *3:* 9–43.

Meyer, D. L., and Macurda, D. B. (1977) Adaptive radiation of the comatulid crinoids. *Paleobiol.*, *3:* 74–82.

Miller, R. R. (1950) Speciation in fishes of the genera *Cyprinodon* and *Empertrichthys* inhabiting the Death Valley region. *Evolution, 4:* 155–163.

Miller, R. R. (1961) Speciation rates in some fresh-water fishes of western North America. *In* Blair, W. F., ed. *Vertebrate Speciation.* Austin, Univ. Texas Press, pp. 537–560.

Miller, R. S. (1967) Pattern and process in competition. *Adv. in Ecol. Res., 4:* 1–74.

Moreau, R. E. (1966) *The Bird Faunas of Africa and its Islands.* New York, Academic Press, 424 pp.

Mortensen, T. (1928–51) *A Monograph of the Echinoidea.* Copenhagen, C. A. Reitzel, 5 vols.

Muller, H. J. (1932) Some genetic aspects of sex. *Amer. Nat., 66:* 118–138.

Murray, J. (1972) *Genetic Diversity and Natural Selection.* Edinburgh, Oliver and Boyd.

Nakazawa, K., and Runnegar, B. (1973) The Permian-Triassic boundary: a crisis for bivalves? *Alberta Soc. Petrol. Geol. Mem., 2:* 608–621.

Neff, N. A. (1975) Fishes of the Kanopolis local fauna (Pleistocene) of Ellsworth County, Kansas. *Univ. Mich. Papers on Paleont., 12:* 39–66.

Nei, M. (1975) *Molecular Population Genetics and Evolution.* Amsterdam, North-Holland, 288 pp.

Nevesskaya, L. A. (1967) Problems of species differentiation in light of paleontologic data. *Paleont. Jour., 1967:* 1–17.

Newell, N. D. (1952) Periodicity in invertebrate evolution. *Jour. Paleont., 26:* 371–385.

Newman, W. A., and Ross, A. (1976) Revision of the balanomorph barnacles; including a catalog of the species. *San Diego Soc. Nat. Hist. Mem., 9:* 1–108.

Nicol, D. (1977) Geographic relationships of benthic marine molluscs of Florida. *The Nautilus, 91:* 4–7.

Ockelmann, K. W. (1964) *Turtonia minuta* (Fabricius), a neotenous veneracean bivalve. *Ophelia, 1:* 121–146.

Oliver, W. R. B. (1955) *New Zealand Birds.* Wellington, A. H. & A. W. Reed.

Olson, E. C. (1960) Morphology, paleontology, and evolution. *In* Tax, S., ed. *Evolution after Darwin, vol. 1, The Evolution of Life.* Chicago, Univ. of Chicago Press, pp. 523–545.

Osborn, H. F. (1929) The titanotheres of ancient Wyoming, Dakota, and Nebraska. *U.S. Geol. Surv. Monogr. 55,* 953 pp.

Ovcharenko, V. N. (1969) Transitional forms and species differentiation of brachiopods. *Paleont. Jour., 1969:* 57–63.

Ozawa, T. (1975) Evolution of *Lepidolina multiseptata* (Permian Foraminifer) in East Asia. *Mem. Faculty of Science, Kyushu Univ., 23:* 117–164.

Packard, A. (1972) Cephalopods and fish: The limits of convergence. Biol. Rev., 47: 241–307.

Paine, R. T. (1966) Food web complexity and community stability. *Amer. Nat., 103:* 91–93.

Palmer, A. R. (1965a) Biomere—a new kind of stratigraphic unit. *Jour. Paleont., 39:* 149–153.

Palmer, A. R. (1965b) Trilobites of the Late Cambrian Pterocephaliid Biomere in the Great Basin, United States. *U.S. Geol. Surv. Prof. Paper 493,* 105 pp.

Patterson, B. (1975) The fossil aardvarks (Mammalia: Tubulidentata). *Harvard Univ. Mus. Comp. Zool. Bull., 147:* 185–237.

Pei, W.-C. (1974) A brief evolutionary history of the giant panda. *Acta Zool. Sinica, 20:* 188–190.

Peters, R. H. (1976) Tautology in evolution and ecology. *Amer. Nat.*, *110:* 1–12.

Petuch, E. J. (1976) An unusual molluscan assemblage from Venezuela. *Veliger, 18:* 322–325.

Pojeta, J. (1971) Review of Ordovician pelecypods. *U.S. Geol. Surv. Prof. Paper 695:* 1–46.

Porter, K. R. (1972) *Herpetology.* Philadelphia, Saunders, 524 pp.

Primoli Silva, I. (1977) The earliest Tertiary *Globigerina eugubina* zone: Paleontological significance and geographical distribution. *Segundo Congreso Latinamericano de Geologia*, pp. 1541–1555.

Pulley, T. E. (1959) *Busycon perversum* (Linné) and some related species. *Rice Inst. Pamphlet, 46:* 70–89.

Raup, D. M. (1972) Taxonomic diversity during the Phanerozoic. *Science, 117:* 1065–1071.

Raup, D. M. (1976a) Species diversity in the Phanerozoic: An interpretation. *Paleobiol.*, *2:* 289–297.

Raup, D. M. (1976b) Taxonomic survivorship curves and Van Valen's Law. *Paleobiology*, *1:* 82–96.

Raup, D. M., and Gould, S. J. (1974) Stochastic simulation and evolution of morphology—towards a nomothetic paleontology. *System. Zool., 23:* 305–322.

Raup, D. M., Gould, S. J., Schopf, T. J. M., and Simberloff, D. S. (1973) Stochastic models of phylogeny and the evolution of diversity. *Jour. Geol., 81:* 525–542.

Raup, D. M., and Stanley, S. M. (1978) *Principles of Paleontology*, 2nd ed. San Francisco, W. H. Freeman and Company, 481 pp.

Reeside, J. B., and Cobban, W. A. (1960) Studies of the Mowry Shale (Cretaceous) and contemporary formations in the United States and Canada. *U.S. Geol. Surv. Prof. Paper 355,* 126 pp.

Regal, P. J. (1977) Ecology and evolution of flowering plant dominance. *Science, 196:* 622–629.

Rensch, B. (1959) *Evolution above the Species Level.* New York, Columbia University Press, 419 pp. (transl. from the German ed. of 1954).

Rickards, R. B. (1977) Patterns of evolution in the graptolites. *In* Hallam, A., ed. *Patterns of Evolution, As Illustrated by* the Fossil Record. Amsterdam, Elsevier, pp. 333–358.

Robertson, R. (1974) The biology of the Architectonicidae, gastropods combining prosobranch and opisthobranch traits. *Malacologia, 14:* 215–220.

Romer, A. S. (1949) Time series and trends in animal evolution. *In* Jepsen, G. L., Simpson, G. G., and Mayr, E., eds. *Genetics, Paleontology and Evolution.* Princeton, N.J., Princeton Univ. Press., pp. 103–120.

Romer, A. S. (1966) *Vertebrate Paleontology.* Chicago, Univ. of Chicago Press, 468 pp.

Rudwick, M. J. S. (1972) *The Meaning of Fossils.* London, MacDonald; and New York, American Elsevier, 287 pp.

Runnegar, B., and Newell, N. D. (1971) Caspian-like relict molluscan fauna in the South American Permian. *Amer. Mus. Nat. Hist. Bull., 146:* 1–66.

Ruzhentsev, V. Ye. (1964) The problem of transition in paleontology. *Internat. Geol. Rev.*, *6:* 2204–2213.

Sage, R. D., and Selander, R. K. (1975) Trophic radiation through polymorphism in cichlid fishes. *Proc. Nat. Acad. Sci. U.S.A.* 72: 4669–4673.

Sarich, V. (1976) The panda is a bear. *Nature, 245:* 218–220.

Schindewolf, O. (1936) *Paläontologie, Entwicklungslehre und Genetik.* Berlin, Borntraeger, 506 pp.

Schindewolf, O. (1950) *Der Zeitfaktor in Geologie und Paläontologie.* Stuttgart, Schweizerbart, 114 pp.

Schopf, J. W., Haugh, Bruce N., Molnar, R. E., and Satterthwait, D. F. (1973) On the development of metaphytes and metazoans. *Jour. Paleont., 47:* 1–9.

Schopf, T. J. M. (1974) Permo-Triassic extinctions: relation to sea-floor spreading. *Jour. Geol., 82:* 129–143.

Schopf, T. J. M. (1977) Patterns and themes of evolution among the Bryozoa. *In* Hallam, A., ed. *Patterns of Evolution, As Illustrated by the Fossil Record.* Amsterdam, Elsevier, pp. 159–207.

Schopf, T. J. M., Raup, D. M., Gould, S. J., and Simberloff, D. S. (1975) Genomic versus morphological rates of evolution: Influence of morphologic complexity. *Paleobiology, 1:* 63–70.

Schuchert, C. (1914) The delimitation of the geologic periods illustrated by the paleogeography of North America. *12th Internat. Geol. Cong., Canada, Compte Rendu,* pp. 555–591.

Sepkoski, J. J. (1976) Stratigraphic biases in the analysis of taxonomic survivorship. *Paleobiology, 1:* 343–355.

Sepkoski, J. J. (1978) A kinetic model of Phanerozoic taxonomic diversity. I. Analysis of marine orders. *Paleobiology, 4:* 223–251.

Simons, E. L. (1963) Some fallacies in the study of hominid phylogeny. *Science, 141:* 879–889.

Simpson, G. G. (1944) *Tempo and Mode in Evolution.* New York, Columbia Univ. Press, 237 pp.

Simpson, G. G. (1949) *The Meaning of Evolution.* New Haven, Conn., Yale Univ. Press, 364 pp.

Simpson, G. G. (1953) *The Major Features of Evolution.* New York, Columbia Univ. Press, 434 pp.

Skevington, D. (1973) Ordovician graptolites. *In* Hallam, A., ed. *Atlas of Palaeobiogeography.* Amsterdam, Elsevier, pp. 27–35.

Slobodkin, L. G., and Sanders, H. L. (1969) On the contribution of environmental predictability to species diversity. *Brookhaven Symp. Biol., 22:* 82–95.

Smith, F. E. (1954) Quantitative aspects of population growth. *In* Boell, E., ed. *Dynamics of Growth Processes.* Princeton, N.J., Princeton Univ. Press, pp. 274–294.

Smith, G. R., and Lundberg, J. G. (1972) The Sand Draw fish fauna. *Amer. Mus. Nat. Hist. Bull., 148:* 40–54.

Smith, Maynard, J. (*See* Maynard Smith, J.)

Solbrig, O. (1971) The population biology of dandelions. *Amer. Scientist, 59:* 686–694.

Southwood, T. R. E. (1973) The insect/plant relationship—an evolutionary perspective. *Symp. Roy. Entomol. Soc. London, 6:* 3–20.

Stach, L. W. (1938) The application of the Bryozoa in Cainozoic stratigraphy. *Rep. 23rd Meet. Aust. N.Z. Assoc. Adv. Sci.,* pp. 80–83.

Stanley, S. M. (1968) Post-Paleozoic adaptive radiation of infaunal bivalve molluscs—a consequence of mantle fusion and siphon formation. *Jour. Paleont., 42:* 214–229.

Stanley, S. M. (1972) Functional morphology and evolution of byssally attached bivalve mollusks. *Jour. Paleont., 46:* 165–212.

Stanley, S. M. (1973a) An ecological theory for the sudden origin of multicellular life in the late Precambrian. *Proc. Nat. Acad. Sci. U.S.A.*, 72: 646–650.

Stanley, S. M. (1973b) An explanation for Cope's Rule. *Evolution*, 27: 1–26.

Stanley, S. M. (1973c) Effects of competition on rates of evolution, with special reference to bivalve mollusks and mammals. *Syst. Zool.*, 22: 486–506.

Stanley, S. M. (1974) What has happened to the articulate brachiopods? *Geol. Soc. Amer. Abstr. with Programs*, 6: 966–967.

Stanley, S. M. (1975a) A theory of evolution above the species level. *Proc. Nat. Acad. Sci. U.S.A.*, 72: 646–650.

Stanley, S. M. (1975b) Clades versus clones in evolution: Why we have sex. *Science, 190:* 382–383.

Stanley, S. M. (1976) Fossil data and the Precambrian-Cambrian evolutionary transition. *Amer. Jour. Sci.*, 276: 56–76.

Stanley, S. M. (1977) Trends, rates, and patterns of evolution in the Bivalvia. *In* Hallam, A., ed. *Patterns of Evolution, As Illustrated by the Fossil Record.* Amsterdam, Elsevier, pp. 209–250.

Stanley, S. M. (1978) Chronospecies' longevities, the origin of genera, and the punctuational model of evolution. *Paleobiology, 4:* 26–40.

Stanley, S. M. (1979) Predation, the opercular imperative, and torsion in snails. (Manuscript.)

Stanley, S. M. (1979) Pleistocene mass extinction of marine Mollusca in the tropical Americas. (Manuscript.)

Stanley, S. M., Addicott, W. O., and Chinzei, K. (1979) Lyellian curves in paleontology: Possibilities and limitations. (Manuscript.)

Stanley, S. M., and Newman, W. A. (1979) Competitive exclusion in evolutionary time: The case of the acorn barnacles (Crustacea: Cirripedia). (Manuscript.)

Stearns, S. C. (1977) The evolution of life-history traits. *Ann. Rev. Ecol. and System, 8:* 145–171.

Stebbins, G. L. (1950) *Variation and Evolution in Plants.* New York, Columbia Univ. Press, 643 pp.

Stebbins, G. L. (1974a) Adaptive shifts and evolutionary novelty: A compositionist approach. *In* Ayala, F. J., and Dobzhansky, T., eds. *Studies in the Philosophy of Biology.* Berkeley, Univ. of California Press, pp. 285–306.

Stebbins, G. L. (1974b) *Flowering Plants.* Cambridge, Mass., Harvard Univ. Press, 399 pp.

Steel, R. (1973) Crocodylia. *In* Kuhn, O., ed. *Encyclopedia of Paleoherpetology.* Part 16. Stuttgart, Gustav Fischer, pp. 1–116.

Steenis, C. G. J. van. (1969) Plant speciation in Milesia, with special reference to the theory of non-adaptive saltatory evolution. *Biol. Jour. Linn. Soc., 1:* 97–133.

Stehli, F. G., Douglas, R. G., and Kafescioglu, I. A. (1972) Models for the evolution of planktonic foraminifera. *In* Schopf, T. J. M., ed. *Models in Paleobiology.* San Francisco, Freeman, Cooper, pp. 116–128.

Stille, H. (1924) *Grundfragen der vergleichenden Tektonik.* Berlin, Borntraeger, 433 pp.

Stitt, J. H. (1971) Repeating evolutionary pattern in Late Cambrian trilobite biomeres. *Jour. Paleont., 45:* 178–181.

Stitt, J. H. (1977) Late Cambrian and earliest Ordovician trilobites, Wichita Mountains Area, Oklahoma. *Oklah. Geol. Surv. Bull., 124:* 1–79.

Suomalainen, E., Saura, A., and Lokki, J. (1976) Evolution of parthenogenetic insects. *Evol. Biol., 9:* 209–257.

Surlyk, F. (1972) Morphological adaptations and population structures of the Danish Chalk brachiopods (Maastrichtian, Upper Cretaceous). *Kongelige Danske Vidensk. Selsk. Biol. Skrifter, 19:* 1–57.

Sutton, R. G., Bowen, Z. P., and McAlester, A. L. (1970) Marine shelf environments of the Upper Devonian Sonyea Group of New York. *Geol. Soc. Amer. Bull., 81:* 2975–2992.

Sylvester-Bradley, P. C. (1961) Superfamily Bairdiacea. *In* Moore, R. C., ed. *Treatise on Invertebrate Paleontology.* Part Q. Lawrence, Kansas, Univ. of Kansas and Geological Society of America, p. 201–208.

Tattersall, I., and Eldredge, N. (1977) Fact, theory, and fantasy in human paleontology. *Amer. Sci., 65:* 204–211.

Taylor, D. W. (1968) Summary of North American Blancan nonmarine mollusks. *Malacologia, 4:* 1–172.

Teichert, C. (1967) Major features of cephalopod evolution. *In* Teichert, C. and Yochelson, E. L., eds. *Essays in Paleontology and Stratigraphy.* Lawrence, Kansas, Univ. of Kansas Press, pp. 162–201.

Thenius, E. (1953) Zur Analyse des Gebisses des Eisbaren, *Ursus (Thalarctos) maritimus* Phipps 1771. *Säugetierkundl. Mitteil,* 1953: 1–7.

Thoday, J. M., and Gibson, J. B. (1962) Isolation by disruptive selection. *Nature, 193:* 1164–1166.

Thompson, D'A. W. (1917) *On Growth and Form.* Cambridge, Cambridge Univ. Press, 793 pp.

Thompson, V. (1976) Does sex accelerate evolution? *Evol. Theory, 1:* 131–156.

Thorson, G. (1950) Reproductive and larval ecology of marine bottom invertebrates. *Biol. Rev., 25:* 1–45.

Tobias, P. V. (1976) African hominids: Dating and phylogeny. *In* Isaac, G. L., and McCown, E. R., eds. *Human Origins,* pp. 377–422.

Todd, N. B. (1975) Chromosomal mechanisms in the evolution of artiodactyls. *Paleobiol., 1:* 175–188.

Trewavas, E., Green, J., and Corbet, S. A. (1972) Ecological studies on crater lakes in West Cameroon Fishes of Barombi Mbo. *Jour. Zool., 167:* 41–95.

Trueman, E. R. (1966) Bivalve mollusks: Fluid dynamics of burrowing. *Science, 152:* 523–525.

Turner, B. J. (1974) Genetic divergence of Death Valley pupfish species: Biochemical versus morphological evidence. *Evolution, 28:* 281–294.

Udvardy, M. D. F. (1969) *Dynamic Zoogeography.* New York, Van Nostrand Reinhold, 445 pp.

Urbanek, A. (1960) An attempt at biological interpretation of evolutionary changes in graptolite colonies. *Acta Palaeont. Polonica, 5:* 127–224.

Uyeno, T., and Miller, R. R. (1963) Summary of late Cenozoic freshwater fish records for North America. *Univ. Michigan Mus. Zool. Occas. Pap.* 631, 34 pp.

Uzzell, T. (1970) Meiotic mechanisms of naturally occurring unisexual vertebrates. *Amer. Natur., 104:* 433–445.

Valentine, J. W. (1970) How many marine invertebrate fossil species? *Jour. Paleont., 44:* 410–415.

Valentine, J. W. (1973) *Evolutionary Paleoecology of the Marine Biosphere.* Englewood Cliffs, N.J., Prentice-Hall, 511 pp.

Valentine, J. W., and Campbell, C. A. (1975) Genetic regulation and the fossil record. *Amer. Scient., 63:* 673–680.

Van Valen, L. (1969) Evolution of communities and late Pleistocene extinctions. *North Amer. Paleont. Convention, Chicago, 1969, Proc.*, pp. 469–485.

Van Valen, L. (1970) An analysis of developmental fields. *Devel. Biol., 23:* 456–477.

Van Valen, L. (1973) A new evolutionary law. *Evol. Theory, 1:* 1–30.

Van Valen, L. (1975) Group selection, sex, and fossils. *Evolution, 29:* 87–94.

Vanzolini, P. E. (1973) Paleoclimates, relief, and species multiplication in equatorial forests. *In* Meggars, B. J., Ayensu, E. S., and Duckworth, W. D., eds. *Tropical Forest Ecosystems in Africa and South America: A Comparative Review.* Washington, D.C., Smithsonian Institution Press, pp. 255–258.

Vermeij, G. J. (1976) Interoceanic differences in vulnerability of shelled prey to crab predation. *Nature, 260:* 135–136.

Vermeij, G. J. (1977) The Mesozoic marine revolution: evidence from snails, predators and grazers. *Paleobiol., 3:* 245–258.

Voigt, E. (1930) Morphologische und stratigraphische Untersuchungen über die Bryozoanfauna der oberen Kreide. I Teile. *Leopoldina, 6:* 372–579.

Voigt, E. (1960) Zur Frage der stratigraphischen Selbstandigkeit der Danienstute. *Internat. Geol. Congr., 11:* 199–209.

Walker, B. W. (1961) The ecology of the Salton Sea, California, in relation to the sportfishery. *Calif. Dept. of Fish and Game, Fish Bull. 113:* 104–151.

Walker, E. P. (1975) *Mammals of the World.* Baltimore, Johns Hopkins Univ. Press, 1500 pp.

Waller, T. R. (1969) The Evolution of the *Argopecten gibbus* Stock (Mollusca: Bivalvia), with Emphasis on the Tertiary and Quaternary Species of Eastern North America. Menlo Park, Calif., *Paleontological Society Memoir 3*, 125 pp.

Wang, T.-K. (1974) On the taxonomic status of species, geological distribution and evolutionary history of Ailuropoda. *Acta Zool. Sinica, 20:* 191–201.

Washburn, S. L., and Moore, R. (1974) *Ape into Man. A Study of Human Evolution.* Boston, Little, Brown, 196 pp.

Wass, R. E., and Yoo, J. J. (1975) Bryozoa from site 282 west of Tasmania. *Initial Repts. Deep Sea Drilling Project, 29:* 809–813.

Waterhouse, J. B. (1977) Chronologic, ecologic and evolutionary significance of the phylum Brachiopoda. *In* Kauffman, E. G., and Hazel, J. E., eds. *Concepts and Methods of Biostratigraphy.* Stroudsburg, Pennsylvania, Dowden, Hutchinson, and Ross, pp. 497–518.

Webb, S. D. (1969) Extinction-origination equilibria in late Cenozoic land mammals of North America. *Evolution, 23:* 688–702.

Webb, S. D. (1976) Mammalian faunal dynamics of the great American interchange. *Paleobiology, 2:* 220–234.

Weidenreich, F. (1947) The trend of human evolution. *Evolution, 1:* 221–236.

Weisbord, N. E. (1967) Some late Cenozoic Bryozoa from Cabo Blanco, Venezuela. *Bull. Amer. Paleont., 53:* 1–247.

Weismann, A. (1889) *Essays Upon Heredity and Kindred Biological Problems.* Vol. II. Oxford, Oxford Univ. Press. 226 pp.

Weismann, A. (1891) The significance of sexual reproduction in the theory of natural selection (1886). *In* Poulton, E. B., Schonland, S., and Shipley, A. E., eds. *Essays upon Heredity and Kindred Biological Problems*, 2nd ed., vol. 1. Oxford, Oxford Univ. Press, pp. 257–342.

Wells, J. W. (1956) Scleractinia. *In* Moore, R. C., ed. *Treatise on Invertebrate Paleontology.*

Boulder, Colo., Geological Society of America and University Press of Kansas (joint publication), Part F, Coelenterata, pp. 328–444.

Wells, J. W. (1966) Evolutionary development of the scleractinian family Fungiidae. *Zool. Soc. London Symp. 16:* 223–246.

Westoll, T. S. (1949) On the evolution of the Dipnoi. *In* Jepsen, G. L., Simpson, G. G., and Mayr, E., eds. *Genetics, Paleontology, and Evolution.* Princeton, N.J., Princeton Univ. Press, pp. 121–184.

White, M. J. D. (1968) Models of speciation. *Science, 159:* 1065–1070.

White, M. J. D. (1973) *Animal Cytology and Evolution.* Cambridge, Cambridge Univ. Press, 971 pp.

White, M. J. D. (1978) *Modes of Speciation.* San Francisco, W. H. Freeman and Company, 455 pp.

White, M. J. D., Blackith, R. E., Blackith, R. M., and Cheney, J. (1967) Cytogenetics of the *viatica* group of morabine grasshoppers, I: The "coastal" species. *Austral. Jour. Zool., 15:* 263–302.

Wiley, E. O. (1976) The phylogeny and biogeography of fossil and Recent gars (Actinopterygii: Lepisosteidae). *Univ. Kansas Mus. Nat. Hist. Misc. Publ. 64,* pp. 1–111.

Williams, G. C. (1966) *Adaptation and Natural Selection.* Princeton, N.J., Princeton Univ. Press, 307 pp.

Williams, G. C. (1975) *Sex and Evolution.* Princeton, N.J., Princeton Univ. Press, 201 pp.

Williams, G. C., and Mitton, J. B. (1973) Why reproduce sexually? *Jour. Theor. Biol., 39:* 545–554.

Willis, J. C. (1922) *Age and Area; A Study in Geographic Distribution and Origin of Species.* Cambridge, Cambridge Univ. Press, 259 pp.

Willis, J. C. (1940) *The Course of Evolution by Divergent Mutation Rather than by Selection.* Cambridge, Cambridge Univ. Press.

Wilson, A. C., Bush, G. L., Case, S. M., and King, M.-C. (1975) Social Structuring of mammalian populations and rate of chromosomal evolution. *Proc. Nat. Acad. Sci. U.S.A., 72:* 5061–5065.

Wilson, A. C., Carlson, S. S., and White, T. J. (1977) Biochemical evolution. *Ann. Rev. Biochem., 46:* 573–639.

Wilson, A. C., Maxson, L. R., and Sarich, V. M. (1974) The importance of gene rearrangement in evolution: evidence from studies on rates of chromosomal, protein, and anatomical evolution. *Proc. Nat. Acad. Sci. U.S.A., 71:* 3028–3030.

Wilson, E. O. (1975) The Origin of Sex (Review of *Sex and Evolution,* by G. C. Williams). *Science, 188:* 139–140.

Wolpoff, M. H. (1971) Competitive exclusion among Lower Pleistocene hominids: The Single Species Hypothesis. *Man, 6:* 601–614.

Wood, A. E., and Patterson, B. (1959) The rodents of the Deseadan Oligocene of Patagonia and the beginnings of South American rodent evolution. *Harvard Univ. Mus. Comp. Zool., 120:* 279–428.

Wood, H. E. (1941) Trends in rhinoceros evolution. *New York Acad. Sci. Trans.* (II), *3:* 83–97.

Wright, S. (1931) Evolution in Mendelian populations. *Genetics, 16:* 97–159.

Wright, S. (1932) The roles of mutation, inbreeding, crossbreeding and selection in evolution. *Proc. Sixth Internat. Congr. Genetics,* pp. 356–366.

Wright, S. (1940) The statistical consequences of Mendelian heredity in relation to speciation. *In* J. Huxley, ed. *The New Systematics.* Oxford, Clarendon Press, pp. 161–183.

Wright, S. (1945) Tempo and mode in evolution: A critical review. *Ecology, 26:* 415–419.

Wright, S. (1956) Modes of selection. *Amer. Nat., 90:* 5–24.

Wright, S. (1967) Comments on the preliminary working papers of Eden and Waddington. *In* Moorehead, P. S., and Kaplan, M. M., eds. *Mathematical Challenges to the Neo-Darwinian Interpretation of Evolution.* Philadelphia, Wistar Institute Press, pp. 117–120.

Wright, S. (1977) *Evolution and the Genetics of Populations.* Vol. 3, *Experimental Results and Evolutionary Deductions.* Chicago, Univ. of Chicago Press, 613 pp.

Wynne-Edwards, V. C. (1962) *Animal Dispersion in Relation to Social Behavior.* London, Oliver and Boyd, 653, pp.

Yonge, C. M. (1950) Life on sandy shores. *Sci. Progress, 38:* 430–444.

Yonge, C. M. (1962) On the primitive significance of the byssus in the Bivalvia and its effects in evolution. *Mar. Biol. Assoc. U.K. Jour., 42:* 112–125.

Yule, G. U. (1924) A mathematical theory of evolution, based on the conclusions of Dr. J. C. Willis, F. R. S. *Roy. Soc. London Proc., 213 (B):* 21–87.

Zaret, T., and Kerfoot, W. C. (1975) Fish predation on *Bosmina longirostris:* Body-size selection versus visibility selection. *Ecology, 56:* 232–237.

Zenkevitch, L. (1963) *Biology of the Seas of the U.S.S.R.* New York, Interscience, 955 pp.

Zeuner, F. E. (1931) Die Insektenfauna des Böttinger Marmors. *Fortschr. Geol. Pal., 28:* 1–160.

Zeuner, F. E. (1958) *Dating the Past. An Introduction to Geochronology.* London, Methuen, 516 pp.

Zeuner, F. E. (1959) *The Pleistocene Period. Its Climate, Chronology and Faunal Successions,* 2nd ed. London, Hutchinson, 447 pp.

Zimmerman, E. C. (1938) Cryptorhynchinae of Rapa. *Bernice P. Bishop Mus. Bull., 151:* 1–75.

Zimmerman, E. C. (1960) Possible evidence of rapid evolution in Hawaiian moths. *Evolution, 14:* 137–138.

Index